Many gardens are recognized as being rich in wildlife but to date little quantitative study to assess their content has been attempted. This is therefore a unique documentation of the wildlife that existed in the author's Leicestershire garden over a fifteen-year period between 1972 and 1986. In total, 1782 species of animal and 422 species of plant were recorded and the diversity, abundance and yearly fluctuations of many animal groups were monitored. Results of this study lead Dr Owen to suggest that with increasing despoliation of the countryside, suburban gardens will have a considerable significance for the conservation of wildlife.

The ecology of a garden

THE ECOLOGY OF A GARDEN

GARDEN

The first fifteen years

JENNIFER OWEN

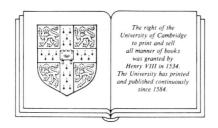

The right of the
University of Cambridge
to print and sell
all manner of books
was granted by
Henry VIII in 1534.
The University has printed
and published continuously
since 1584.

CAMBRIDGE UNIVERSITY PRESS
Cambridge
New York   Port Chester
Melbourne   Sydney

Published by the Press Syndicate of the University of Cambridge
The Pitt Building, Trumpington Street, Cambridge CB2 1RP
40 West 20th Street, New York, NY 10011–4211, USA
10 Stamford Road, Oakleigh, Melbourne 3166, Australia

First published 1991

Printed in Great Britain at the University Press, Cambridge

British Library cataloguing in publication data
Owen, Jennifer
The ecology of a garden: the first fifteen years.
1. Great Britain. Garden ecosystems
I. Title
579.5264

Library of Congress cataloguing in publication data
Owen, Jennifer.
The ecology of a garden: the first fifteen years/Jennifer Owen.
p.   cm. Includes bibliographical references and index.
1. Garden ecology—England—Leicester. 2. Garden fauna—England—Leicester.
I. Title.
QH138.L43O84 1991
574.5′268′0942542—dc20   90–44289   CIP

ISBN 0 521 34335 6 hardback

It is, I find, in zoology as it is in botany; all nature is so full that that district produces the greatest variety which is the most examined.

Gilbert White. Letter XX to Thomas Pennant
October 8th. 1768

Contents

Contents

About this book

Since 1971, when I returned to Leicester having lived abroad for 13 years, I have been investigating the fauna of my ordinary, medium-sized suburban garden, trying to find out what shares the plot with me. A non-scientific account of the garden was published by Chatto & Windus in 1983 as *Garden Life*. Continuous monitoring of selected groups in the garden since 1972 produced, up to the end of 1986, an extraordinary list of 1782 species of animals and 422 species of plants. Some are common, many are rare, and numbers fluctuate markedly both seasonally and from year to year. This book presents the first 15 years of the garden record, 1972–86, including not only species lists, but also patterns of diversity and abundance, and some assessment of what so many animals are doing in the garden. Although many are only passing through, a considerable proportion breed in the garden and hence are resident.

The book starts with two chapters that set the scene for detailed description of my garden; in them I describe urban wildlife worldwide, garden wildlife, the history of gardens, gardens in Britain, the history and geography of the county of Leicestershire, and the wildlife of the county and of the city of Leicester. In the third chapter, I describe the garden and its management, and explain how its fauna has been investigated, and the fourth chapter is a description of the garden flora. There follow nine chapters about different animal groups in the garden, their natural history, diversity and abundance, annual and seasonal fluctuations in numbers, breeding, and their status in Leicestershire. The taxonomic groups that form the subjects of chapters are of different sizes: hoverflies (the subject of Chapter 7), for instance, are a family, beetles (Chapter 10) are an order, and the non-insect invertebrates described in Chapter 12 belong to several different phyla. Chapter 14 analyses gardens as habitats and explains why they are so rich in animal life. In the final chapter, I estimate what proportion of the fauna of the British Isles occurs in a garden such as mine, and consider the significance of gardens for conservation.

Some observations or pieces of information are relevant in more than one place in the book; I have sometimes therefore repeated myself, so that each chapter might be complete on its own. This applies particularly to the suitability of the garden habitat for insects, and its dependence on the processes of gardening. In the introductory chapters, scientific names of those animals and plants that have English names are given on first mention but not repeated. In the chapters about my garden and in the concluding chapters, scientific names are used in the text wherever appropriate. The index includes the families given in taxonomic lists, but not the species.

This is not a definitive account of the ecology of gardens, no others having been investigated in detail for a prolonged period, but rather an account of what I have learnt about the one particular garden that I own and manage. Moreover, my knowledge of my garden is not exhaustive, but to a large extent selective, for identification of many animal groups depends on the readiness of specialists to examine and identify individuals caught. Consequently, there is a great deal of information about some groups of garden animals, and none at all about others. Were it not for the cooperation and help of many other people, the account of my garden's flora and fauna would not be so extensive. I gratefully acknowledge their contribution and appreciate their interest and effort.

Investigation of the garden has been a joint venture with Denis Owen, who also read the final draft of this book and made suggestions for its improvement. I am indebted to those who regularly identified large samples of insects and other invertebrates from traps: Michael Archer (wasps and solitary bees), L. R. Cole (sawflies), J. P. Dear (Calliphoridae and Sarcophagidae), Don Goddard (beetles and lacewings), Peter Kirby (Hemiptera), John Mousley (large Brachycera–Diptera), Denis Owen (moths and ladybirds), D. T. Richardson (myriapods), John Sankey (harvest-spiders), John Sellick (Psocoptera and psociform Neuroptera), Henry and Marjorie Townes (Ichneumonidae), and Phil Withers (Psychodidae). Additional identifications were made by R. Belshaw, U. Bowen, J. F. Burton, Ian Evans, Nona Finch, Don Hall-Smith, L. Jones-Walters, Richard Owen, Keith Porter, Denis Ratcliffe, Adrian Rundle, K. G. Smith, Mark Sterling, Alan Stubbs, S. L. Sutton and Derek Whiteley. Clive Stace gave invaluable help in sorting out the nomenclature of cultivated plants, and Dennis Chanter, Gilbert Mowrey and Rob Oldham patiently helped with statistics. Henry Disney advised on presentation of numerical information. Biologists at Leicestershire Museums, Arts and Records Service, particularly Jan Dawson and Derek Lott, were unstinting in answering my enquiries and providing access to county records and publications. Further help with local information came from Paul Champion, Bob Cherry, Chris Measures and Mary Swann. Allen and Edith Chinnery, Ian Evans and David Nicholls critically read earlier drafts of the introductory chapters, which were much improved by the application of their knowledge and expertise. To all these, I tender my sincere thanks. Without their endeavours, the book could not have been written, although any errors that remain are mine.

JENNIFER OWEN

1

Introducing gardens:
the broad view

This book is about a garden in the Humberstone suburb of the industrial city of Leicester in the Midlands of England. Before the garden can be described, it needs to be put into context, to draw together the many different influences on its natural history and on the development of its flora and fauna. Urban wildlife has certain features in common the world over, as does the wildlife of gardens. But it is perhaps the regional setting which has the greatest influence on a garden's flora and fauna, because it sets the limits on what is available to colonize or visit. For instance, although parallels can be drawn between the wildlife of a West African garden and of a suburban garden in Leicester, they have a very different aspect and 'feel'. Their climates differ markedly, as do their soils, and they are set in different biogeographical regions of the world. The West African garden has hibiscus, a sprawling mimosa as a ubiquitous weed, large fruit bats, fire finches, and swallowtail butterflies. That in England has roses, ivy-leafed speedwell as a rampant weed, small insect-eating bats, dunnocks, and cabbage white butterflies. The biogeographical differences between the tropics and temperate zones are so well known that no one would expect the two gardens to be the same. The same principles operate in influencing the garden faunas in different temperate countries or in different areas of England: although they share some elements, there are differences between gardens in, say, Cornwall and Yorkshire.

URBAN WILDLIFE

In common parlance, urban simply means 'town', as distinct from 'country'. The United Nations designates as urban, settlements of 20 000 or more, in an area associated with commerce and industry (Nelson 1978). It is predicted that by AD 2000, 756 million people in industrialized countries and 1411 million people in developing countries will live in cities of more than 100 000, and that more than 400 cities (mostly in developing countries) will exceed one million inhabitants (US Council on Environmental Quality and Department of State 1980). This massive urbanization, already well under way, dramatically alters the physical environment and hence the flora and fauna. Natural ecosystems are replaced by man-made ecosystems, which are organized for the well-being of man, and in which he is the dominant species. To a large extent, plants and animals are physically excluded by the way in which the land is used, but in addition the climate is modified in characteristic ways.

Urban wildlife consists firstly of those species which can survive these inhospitable conditions, and which are characterized by their tolerance of and capacity to exploit man and his activities. Some animals, such as rats, mice, cockroaches, house sparrows *Passer domesticus* and feral pigeons *Columba livia* do this remarkably well, and are found in the heart of cities the world over. These opportunists are the truly urban animals, surviving as scavengers in the depths of the concrete jungle. The second category of urban wildlife consists of those local species which find food and living space in the green areas of cities, particularly on the suburban fringe. Such animals as blackbirds *Turdus merula* and foxes *Vulpes vulpes* do particularly well in English cities, and may live at greater densities in suburbia than in the surrounding countryside.

Although a suburban garden is a very different habitat from the concrete and steel centre of a large city, the garden is as much influenced by the proximity of the city as by the surrounding countryside. The covering of large tracts of land with concrete and other hard materials, the operation of a multitude of heat-producing machines, from cars to central heating and air-conditioning, and pollution of various sorts are responsible for marked modifications of the climate of cities. Studies of the climate have largely been confined to north temperate cities, but similar trends almost certainly apply in warmer lands.

The hard surfaces of city roads and buildings over-heat at midday, and night-time cooling lags behind that of their greener surroundings, producing a 'heat island', particularly pronounced at night. Winter minima are on average 1–3°C higher than in the surrounding countryside, and there may be a temperature rise of as much as 10°C at the centre of a large city (Miess 1979). As a consequence, there is continuous water loss, and although cities receive on average 5–10 per cent more precipitation than their surroundings, the relative humidity is less, especially in summer, and soils quickly dry out. There are ten times the quantity of dust particles in the air, so that incident radiation is reduced by 15–20 per cent, and cities are on average 5–10 per cent more cloudy and may be twice as foggy (Gill & Bonnet 1973). The annual mean wind speed is 20–30 per cent less than in the surrounding countryside, but grid-iron street lay-outs cause gusts and eddies, and in summer conditions of high pressure, winds converge onto built-up areas (Miess 1979).

There is little available information as to how city climates affect wildlife, although in Leicester seasonal growth of plants seems to be as much as two weeks in advance of that in the surrounding countryside. Trees may also flower early. Birds, such as blackbirds, sing under the influence of street lighting, and starlings *Sturnus vulgaris* and pied wagtails *Motacilla alba*, which flock into the city to roost in the warmth at night, keep up a continual fidgetting and twittering under the influence of street and shop lights.

Air pollution is a major factor in cities, with on average five times as much sulphur dioxide, ten times as much carbon dioxide and 25 times as much carbon monoxide as in their rural surroundings (Gill & Bonnet 1973). Lichens are particularly sensitive to sulphur dioxide pollution and, by the mid-twentieth century, their numbers were much decreased in cities, the immediate causes of loss being acid tree bark and acid rainwater running down tree trunks (Johnsen & Søchting 1973). In Britain and the rest of Europe, and in North America, disappearance of lichens and soot pollution of tree trunks after the industrial revolution, led to population changes in the peppered moth *Biston betularia* and other moths which rest by day on tree trunks. The normal colour forms of these moths, which are pale and speckled, are at a selective advantage over rare melanic mutants as long as tree trunks are lichen-covered. On blackened, bare tree trunks, however, the normally camouflaged colour forms are conspicuous and so are exposed to predation by birds, conferring strong selective advantage on melanics. By the end of the nineteenth century, black forms of *Biston betularia* formed as much as 99 per cent of populations in, and downwind of, industrial areas in Britain. Industrial melanism evolved similarly in North America, but about 50 years later. Public outcry in Britain in the 1950s, following a report on air pollution, led to the establishment of 'smokeless zones'; by 1962, the relative frequency of the black form of the peppered moth had started to decline (Kettlewell 1973), and, by the 1980s, the area of high melanic frequency had contracted to the northeast of England (Cook, Mani & Varley 1986). This is almost certainly because lichens began to re-establish themselves in cities. The trend continues and, in the 1980s, lichen species which had not been seen in London since the eighteenth century recolonized the city after a marked fall in sulphur dioxide pollution (Hawksworth & McManus 1989).

Air pollution has also caused a decrease in the fungus *Diplocarpon rosae*, which causes 'black spot' on the leaves of roses, although sulphur and nitrogen dioxides stimulate growth of a major pest, the black bean aphid *Aphis fabae*, because of changes caused in the amino acids of host plants (Dohmen *et al.* 1984). There is some evidence that oxides of nitrogen in vehicle exhausts may increase numbers of herbivorous insects by improving the quality of their food. For instance, outbreaks of the buff-tip moth *Phalera bucephala* and of the yellow-tail moth *Euproctis similis* occur on beech *Fagus sylvatica* and hawthorn *Crataegus monogyna*, respectively, planted on the central reserves of motorways in Britain. Port & Thompson (1980) suggested that it is probably the enhanced nitrogen content of the plants that increases the insect populations. This may be why Lussenhop (1973), in a transect along 21 kilometres (km) of Chicago expressway, found no drop in invertebrate diversity towards the city centre and more *Pterostichus* (a carabid beetle) near to heavy traffic, since increased numbers of herbivores would lead to increased numbers of their predators.

Material cycles are incomplete in cities, so that wastes accumulate, causing acid soils and high levels of heavy metals and pesticide residues. Industrial spoil heaps, which tend to be low in nutrients and to contain toxic substances,

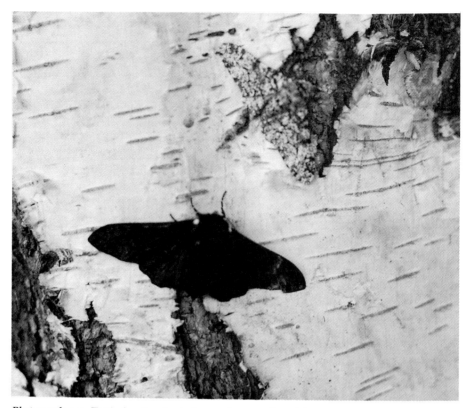

Photograph 1.1. Typical and melanic forms of peppered moth Biston betularia, *which had been captured in a mercury-vapour light trap in the Leicester garden and were placed on the trunk of a birch tree for photographing. Denis Owen.*

support characteristic plant communities, often consisting of only a few sorts of grasses and legumes. Colliery waste is acid and supports only certain grasses; alkali waste and blast furnace slag, however, often support cal-cicolous, herb-rich communities, which in northern England may include as many as nine species of orchids (Gemmell 1977). Lead emanating from exhaust gases is a major pollutant, and the tissues of urban pigeons contain two to five times the lead concentrations found in rural birds (Gill & Bonnet 1973). The many *Pterostichus* found close to a Chicago expressway, had a significantly elevated lead content, and, in general, there were fewer carabids and spiders near to heavy traffic (Lussenhop 1973).

In northern temperate cities, man's structures, activities and wastes create a range of micro-climates that accommodate organisms absent, or less com-mon, elsewhere. Refuse dumps often generate heat, either from warm ashes or by decomposition, causing the germination of exotic seeds, such as hemp or sunflowers (from bird seed), and providing ideal conditions for lizards, slow-worms, cockroaches and crickets to live, and for mice to overwinter (Owen, D. 1978e). Pigeons cluster on cold days over hot-air outlets, whereas, conversely, damp cellars may house slugs and overwintering newts, frogs and toads. The warmth of buildings shelters crickets, cockroaches, pseudoscorpions, and a host of other animals, particularly spiders, several species of which are confined to buildings in the northern parts of their ranges. A special habitat for scavengers, mostly flies and their larvae, is the faeces of urban dogs; in 1970, a minimum of 56 250 kilograms of dog droppings was deposited each day in New York, and 78 750 kg a day in London (Gill & Bonnet 1973).

Many sorts of animals are closely associated with man, either as his parasites or as scavengers on his waste, and some are present even in hygiene- and health-conscious cities in industrialized countries. Fleas, ticks, lice and bedbugs can usually be found where man lives under crowded conditions, especially if there is poverty. In less developed countries, the uncontrolled growth around cities of shanty towns, where housing is inadequate, sewage and waste disposal lacking and water supply poor or polluted, necessitating temporary water storage, has produced a so-called 'septic fringe' where mosquitoes, other vectors, and the diseases they transmit flourish. Ironically, the so-called 'affluent fringe' of cities in industrialized countries also creates health problems: encouragement of lush, natural vegetation around houses in California, for instance, has led to outbreaks of *Rickettsia* fevers, when domestic pets have picked up from opossums the fleas transmitting the pathogens, and passed them on to their owners. Flies may also be a problem in such areas, because of poor composting of garden rubbish, inadequately contained garbage, and the prevalence of dog faeces (Nelson 1978).

Relatively few animals, other than man's parasites, are truly urban dwellers found in concrete city centres, but those that do live there are often extraordinarily successful, and usually have extremely wide distributions, often having been aided in their travels by man. The house sparrow, native to Europe and parts of Asia, has been introduced worldwide, and is always found in association with man, often nesting in holes in buildings. It is suspected of displacing local sparrows from some parts of its range, scarcely surprising since one bird eats 3.6 kg of grain a year. Sparrow pie was once a regular delicacy in London, but the birds have declined somewhat in numbers there and in New York since the disappearance of horse traffic and of the scattering of grain from feed bags (Jenkins 1982).

Starlings, native to Europe and adjacent parts of Asia, have been introduced to North America, South Africa, Australia and New Zealand, and have colonized many other places. Sixty birds were released in Central Park, New York, in 1890 and a further 40 in 1891; they spread spectacularly, and are now common all over North America (Long 1981). Urban birds are commuters, roosting in city centres, and feeding largely on ground invertebrates in surrounding green areas and countryside, but the noise, mess and smell of their roosts qualify them as pests in urban sites, and vigorous, largely unavailing, attempts are made to control numbers or at least persuade them to move elsewhere.

Photograph 1.2. Feral pigeons abound in Athens, Greece, as in towns and cities around the world, despite attempts to limit their numbers. Denis Owen.

Periodic campaigns, from poisoning to contraceptive pills, are also mounted against feral pigeons, which are descended from cliff-dwelling rock doves of European, Mediterranean and Indian coasts. They were introduced to many countries as cage birds, which are often used for racing, and as table birds, which are kept free-flying in dove-cotes, and were also released in North America; many city populations probably stem from domestic stock that has gone wild. They increased greatly in numbers in the nineteenth century, and urban birds have become omnivorous (Jenkins 1982). Their droppings corrode stone and metalwork on buildings, and they are regarded as a potential health hazard, but control is hampered by the generosity with which visitors in places as far apart as New York, London and Venice feed them with specially purchased grain. The white-collared pigeon *Columba albitorques* of the high plateaux of northeastern Africa has similarly colonized Addis Ababa (Goodwin 1978). Wood pigeons *C. palumbus*, which eat buds, berries, seeds and grain and are larger than feral pigeons, are common in squares and parks of London suburbs and of other English cities.

Apart from feral cats and dogs living as scavengers, the only mammals that are truly city dwellers, on a par with feral pigeons, are the ubiquitous rats and mice. The black rat *Rattus rattus* has spread from southeast Asia over much of the world, mainly through transport in ships' cargoes. It probably reached Europe with returning Crusaders in the twelfth century, but has been driven out of many areas by the brown rat *R. norvegicus*, also from southeastern Asia. The brown rat spread over land and on ships, and reached Europe at the beginning of the eighteenth century. It is not so closely associated with man as the warmth-loving black rat, but in some areas of the world is so common that it is credited with eating half the stored food, and it also spreads various pathogens in its faeces and urine (Mourier & Winding 1975). Rats will gnaw at just about anything, including lead piping and insulation, and find shelter in vast numbers in sewers, but the record for exploiting man-made environments must go to the house mice which live, feed and breed successfully even in refrigerated stores kept permanently at −9.5°C (Jenkins 1982). House mice *Mus musculus* are native to southwestern Asia, but are now found worldwide. Their success and spread are consequences of their ability to use all man's foods, especially in storage and transit, and their adaptability and agility.

One sort of insect, the cockroach, three species of which are probably worldwide in distribution, is a truly urban dweller. Cockroaches are a tropical group, and those found indoors in temperate regions have been introduced in the last 200 years and are dependent on heated buildings (Mourier & Winding 1975). They will eat anything with an organic component, including the emulsion on photographic film, paper, paste in book bindings, and starch from laundered table linen. Furthermore, they are fast, agile, and are prolific breeders.

Within houses and other buildings, the abundance and variety of rodent and invertebrate scavengers is dependent on the tolerance, cleanliness, tidiness and, often, the affluence of the human occupants. Rats and mice, cockroaches, silverfish (Thysanura), ants, flies, mites and other invertebrates eat organic debris and stored food, and themselves support such predators as the 'rat-snakes' and geckos of warmer countries and ubiquitous feral cats, spiders and harvestmen. Stored food is also eaten by a whole range of beetles and their larvae, and by moth caterpillars, many of the species having been unwittingly disseminated worldwide by man (Mourier & Winding 1975).

Concentrations of waste material in dumps provide living space, and feeding and breeding sites for many of the scavengers found in houses, as well as Diptera such as winter gnats (Trichoceridae), hoverflies (Syrphidae), dungflies (Scatophagidae), owl midges (Psychodidae), grey flesh flies (Sarcophagidae), green-bottles and blue-bottles (Calliphoridae). Among the vertebrates that come to feed in temperate areas are gulls, particularly herring *Larus argentatus* and black-headed gulls *L. ridibundus*, which have become increasingly urban birds, feeding, roosting and often nesting away from the sea. Tropical dumps attract scavenging kites, crows and vultures, and, in Africa, marabou storks *Leptoptilos crumeniferus*.

Until the eighteenth century, kites and ravens *Corvus corax* scavenged in the streets of London, and kites, vultures and crows of various sorts are a part of the street scene in warmer parts of the world (Goodwin 1978). The urban kite of southern Europe and Africa, east to China and Australia, is the black kite *Milvus migrans*, although in natural conditions it favours fairly open country; it is something of a puzzle as to whether it was this species that once scavenged in London's streets, or the native red kite *Milvus milvus*, now found in Britain only in parts of Wales. Black vultures *Coragyps atratus* frequent dumps in the Americas, and vultures of several other species are urban scavengers in parts of southern Europe, Africa and India, and can be seen jostling for scraps or perches with crows, such as the pied crow *Corvus albus* of tropical Africa, hooded crow *C. cornix* of North Africa, and the house crow *C. splendens* of India and Sri Lanka. Jackdaws *C. monedula* and carrion crows *C. corone* may occasionally do the same in Britain. The common myna *Acridotheres tristis*, a native of India related to the starling, has been widely introduced in oriental countries, South Africa and Australia, and is almost exclusively a bird of towns and cities (Long 1981). Swifts must be included as truly urban birds because of their use of buildings for nesting; three *Apus* species in Europe, others in Africa and Asia, and the chimney swift *Chaetura pelagica* of North America nest in the hearts of towns and cities and, since they hawk for aerial plankton high in the air, find at least a proportion of their food over cities.

Scavenging mammals are usually nocturnal and so less obvious than birds, but urban foxes thrive in Europe, North America and in Australia, the latter being introductions from Britain. Coyotes are common on the outskirts of a number of North American cities, and jackals and hyaenas in India and other hot countries. Badgers *Meles meles* and beech martens *Martes foina* also occur in European suburbs, the latter often sheltering in the roof spaces of buildings. Raccoons *Procyon lotor* exploit garbage in North American cities, and opossums of different species do likewise in North America and in Australia. Hyaenas are tolerated, and even encouraged, as street cleaners in some African and Indian towns, but the most spectacular urban scavengers are undoubtedly the polar bears *Ursus maritimus*, which raid dustbins and pick over the city dump in Anchorage, Alaska, and other Arctic settlements.

Buildings provide a variety of living spaces and roosting and breeding sites for animals. Ledges on urban 'cliffs' are used as roosting or nesting sites by house sparrows, starlings and feral pigeons, occasionally by kestrels *Falco tinnunculus*, and in New York and London, by peregrine falcons *F. peregrinus* (which prey on pigeons). Black redstarts *Phoenicurus ochruros* exploited similar sites in the undisturbed wasteland of the bombed sites of London in the 1940s and 1950s, and have nested in industrial wasteland in a number of cities (Fitter 1945). White storks *Ciconia ciconia* return every year, although in diminishing numbers, to roost on towers and chimneys in continental European towns, and cartwheels may be specially provided as nest bases. In cities in many parts of the world, buildings that allow access to roof spaces or crevices house roosting or breeding bats and such birds as swifts, often without the knowledge of the human occupants. Drains and sewers harbour rats and mice, and occasionally colonies of bats, or, in Florida, stray alligators *Alligator mississipiensis*.

The range and availability of spatial habitats and breeding sites for animals varies with the age of a city and the degree of urbanization; the more modern and urbanized a city, the fewer animals find a home there. In general, animal diversity, whether of birds and mammals or of invertebrates, tends to decrease towards the built-up city centre, where there is less space and there are fewer feeding opportunities. Since the late 1960s, however, an increasing number of European and North American cities have had an element of wildlife conservation in their planning, which means that habitats are created, nature reserves are set aside, and there is increasing tolerance and, often, active encouragement of urban wildlife. This is particularly true of Britain, where wildlife and conservation groups and local authorities are making progress in greening urban areas, estimated by the Civic Trust in 1976 to include more than 100 000 hectares of wasteland. The fast-growing fund of knowledge and expertise on creation and management of urban nature areas has been made widely available by such publications as that produced by the Ecological Parks Trust (Emery 1986).

Green spaces are found in all cities, and are important refuges for wildlife. Wasteland, railway embankments and roadside verges are soon colonized by a variety of plants (Gilbert 1983), whose foliage, flowers and seeds provide food for insects and birds, particularly Lepidoptera and their larvae, and finches, such as linnets *Carduelis cannabina*, greenfinches *Chloris chloris* and goldfinches *Carduelis carduelis*. The typical wild plants of urban areas, that grow wherever there is cleared land, or even from crannies in walls or cracks in pavements, are species that produce large quantities of small seed, are fast-growing, and are tolerant of a wide range of physical conditions and pollution. Many of these are introductions from other countries: the composite, *Tridax procumbens*, found throughout the tropics, originated in America; goldenrod *Solidago canadensis*, rampant in Birmingham and Berlin, comes from North America; and the ubiquitous *Buddleia davidii* of European cities and tree-of-heaven *Ailanthus altissima* of New York were introduced from China. The devastated sites left by the London 'blitz' of 1940–41 were colonized by buddleia, rowan *Sorbus aucuparia*, elder *Sambucus nigra* and 123 other species of flowering plants (Fitter 1945). Rosebay willowherb *Epilobium angustifolium* was particularly abundant, leading to great numbers of the spectacular elephant hawk moth *Deilephila elpenor*, whose large caterpillars feed on the leaves. The green corridor along a railway or stream may be an important route to and from the centre of a city for foxes and other large animals, and railways aid the dispersal of plants. Oxford ragwort *Senecio squalidus* was introduced to Oxford Botanic Garden from the slopes of Mt Etna in the late eighteenth century, and has spread throughout Britain, finding the clinker of railways a hospitable substrate.

Parks, gardens and ornamental plantings incorporate a range of spatial habitats and feeding opportunities for animals, particularly for birds and insects, but the predilection of planners for monocultures has often been the cause of herbivorous insects getting out of hand. Lime trees lining English streets drip with honeydew discharged by millions of aphids, the hairy caterpillars of vapourer moths *Orygia antiqua* occasionally reach plague densities on plane trees in London squares (Owen, D. F. 1975c), as do gypsy moth *Lymantria dispar* and other caterpillars on North American street trees (Olkowski *et al.* 1978), and the caterpillars of an *Acraea* butterfly regularly defoliate *Musanga cecropioides* trees surrounding a university quadrangle in Kampala, Uganda. *Jacaranda, Hibiscus, Calliandra, Allamanda, Nerium oleander,*

9

Photograph 1.3. Buddleia davidii *growing in Cripplegate, London, on a site devastated in the London 'blitz' of 1940–41.*

Bougainvillea and many other trees and shrubs planted in gardens and squares in tropical cities are exploited by nectar-feeding (and pollinating) insects and such birds as humming birds and sunbirds. Tall trees in European cities provide nesting sites for birds such as crows and magpies (which are town birds in Scandinavia), and those in tropical cities provide roosting sites for fruit bats. Open, grassy areas are feeding grounds for gulls, starlings and other birds, exploiting leatherjackets (Tipulidae) and other insects in temperate cities, and for ibis, such as the hadada *Hagedashia hagedash* of East Africa, in tropical countries.

Rivers, canals, reservoirs and ornamental lakes accommodate a range of fish and other aquatic life, together with many birds, even in the heart of a city. The lake in St James's Park in central London, for instance, not only supports introduced, and pinioned, pelicans, cormorants and ducks, but large numbers of pochard *Aythya ferina* and tufted duck *A. fuligula* visit to feed and both have bred in the park; coot *Fulica atra* and moorhen *Gallinula chloropus* are resident, and herons *Ardea cinerea* call in to feed (Burton 1974). Fish, ducks and wading birds are again abundant in and on the River Thames in London since an efficient anti-pollution programme was instituted in 1959 to restore the river from the filthy state it was then in (Harrison & Grant 1976). The only ducks to breed regularly within New York city limits are black duck *Anas rubripes* and wood duck *Aix sponsa*, but a whole variety of other water birds frequent lakes in Central Park, reservoirs, and marsh and swamp reserves, where rails, night herons and bitterns breed (Kieran 1959).

A variety of other sorts of animals are typical of suburban and greener areas of towns and cities, and may occasionally stray into the densely built-up centre. These include opportunists, such as foxes and collared doves *Streptopelia decaocto* in Britain, and also species abundant in the environs of a city, such as hedgehogs *Erinaceus europaeus* and robins *Erithacus rubecula*, which are either enclosed within city boundaries by development around their habitat, or are able to extend their range into the peripheral greener areas of a city. Large vertebrates are generally excluded by human settlement, but snowy owls *Nyctea scandiaca* hunt during winter on Toronto airport, elk *Alces alces* (the moose of North America) not infrequently stray into Moscow, and alligators wander into suburban Miami, sometimes taking over swimming-pools (Jenkins 1982). Deer are maintained in semi-wild condition in large parks in some European cities, such as the fallow *Dama dama* and red deer *Cervus elaphus* of Richmond Park, London. Exotic species sometimes become naturalized from introductions or escapes. Monk parakeets *Myiopsitta monachus*, native to temperate parts of South America, have established themselves in suburban areas of eastern North America and in California, and rose-ringed parakeets *Psittacula krameri*, native to India and parts of Africa, are feral in several parts of the United States, and in southeastern England (Long 1981).

GARDEN WILDLIFE

Consideration of urban wildlife leads naturally on to consideration of gardens because gardens are particularly important for the animals of city and suburb. Many of the animals found in parks and green spaces in cities will also be found in gardens, especially the larger ones. In gardens the world over, birds and insects exploit the feeding and breeding opportunities offered by dense growth, plant diversity and an abundance of flowers, fruit and seeds. Even in the vast urban sprawl of London, bee-keepers are able to maintain hives on top of office blocks because of the abundance of flowers from which bees can collect nectar and pollen. Local bird species, particularly opportunists, may be abundant in town gardens, whether the bulbuls (Pycnonotidae), wax finches (Estrildidae), robin chats (Muscicapidae), barbets (Capitonidae) and sunbirds (Nectariniidae) of Africa, the robins *Turdus migratorius*, blue jays *Cyanocitta cristata*, leaf warblers (Parulidae), cardinals *Richmondena cardinalis* and chickadees (Paridae) of North America, or the thrushes (Turdidae), tits (Paridae), robins, greenfinches and warblers (Sylviidae) of an English garden.

Small, ground-feeding doves find a niche around human settlements and are found in gardens and parks in cities all over the world: laughing doves *Streptopelia senegalensis* in Africa, mourning doves *Zenaidura macroura* in North America, introduced spotted *Streptopelia chinensis* and laughing doves in Australia, and collared doves in Europe. By the sixteenth century, the collared dove had extended its range westwards from northern India to southeastern Europe. In 1930 it began to spread rapidly across Europe, and reached Britain in 1955 (Owen, D. 1978e), since when it has become abundant as a ground-gleaner, particularly in towns and suburbs. It often feeds in enormous flocks around grain silos, so that it sometimes has to be controlled (Coombs *et al.* 1981).

The population of English blackbirds may be more than 20 times higher in gardens than in nearby woodland (Perrins 1974), and their breeding success is greater (Snow 1958). The American robin (also a *Turdus* sp.) is equally successful in towns. Many gardens in European and North American cities have bird feeding tables, which are visited by a variety of birds, including blue jays, cardinals and chickadees in North America, and tits, blackbirds, robins and greenfinches in Britain. The provision of food undoubtedly attracts birds to gardens and may improve their chances of survival in bleak winters.

Exactly what frequents a garden depends on its locality and size and on the country, but birds and insects in particular, because they fly, are present in variety and abundance even in inner-city gardens. There is much published on gardens and their wildlife, but for the most part it is generalized or anecdotal, and there have been few detailed studies. To most people, garden wildlife seems to mean vertebrates (particularly mammals and birds), butterflies, and only a few of the more spectacular among other insects and invertebrates. Parasitic wasps (Ichneumonidae) or owl midges (Psychodidae), for instance, are rarely noticed, and aphids are seen merely as plant pests or as bird food. There is even a tendency to exclude plants from consideration, although as many as 95 species of 'wild' plants came up in one Leicester garden over a 25-year period (Tutin 1973). Most professional ecologists, moreover, dismiss gardens as man-made and artificial and of no relevance to wildlife. Elton (1966) considered them 'in the direction of biological deserts' as he assumed that 'the multitude of exotic plant species ... have so few animals attached to them'. How wrong he was, as investigation of my garden demonstrates.

There is a fashion dating from about 1980 for 'wildlife gardening', although this is usually interpreted as emulating the countryside. It is far from being a case of *laissez faire*, for the wildlife gardener puts a lot of time and labour into creating wildflower meadows, marshy pools, or even woodland walks. Few people attempt to improve, as wildlife habitats, ordinary gardens that are productive and attractive in the conventional sense; instead, an effort is made to create outside the backdoor a most ungarden-like habitat. This is admirable in its own way, but overlooks the possibilities of having both the productive garden that most people want, and a wildlife habitat at the same time.

I first realized the wildlife potential of a conventional garden while living at

Mount Aureol on the outskirts of Freetown, Sierra Leone, in West Africa. Around the house we made a garden which abutted onto areas of mown grass with scattered flowering trees and shrubs and was close to rainforest maintained as a Botanic Garden. Among the many different sorts of animals that visited or lived in the garden was a wealth of butterflies. Astonishingly, nearly 300 different species were recorded, more than in nearby tropical rainforest. The reason was that savanna butterflies, as well as forest butterflies, visited the garden. This immediately suggested one feature of gardens that makes them particularly rich as habitats. Incorporating as they do a mosaic of open and shady places, they can attract or harbour animals characteristic of both types of habitat. Moreover, the continual necessity to cut back shrubs, in order to maintain a garden in the hot, humid climate of Freetown, had the effect of promoting new growth and prolonging the breeding season of those butterflies that used the shrubs as larval foodplants. This suggested that a gardener's management techniques may actually improve the habitat for certain animals. With these experiences of a West African garden, I returned to England with an inkling that gardens *per se* may be of great significance for wildlife, that they can be enhanced as habitats by appropriate management practices, and hence may be of considerable significance for conservation (Owen, J. 1983a).

Although many books describe in a generalized way the rich fauna of gardens and advise on enhancing them as wildlife habitats (see e.g. Baines 1984, 1985, Chinery 1977, 1986, Gibbons & Gibbons 1988, Jones *et al.* 1981, Wilson 1979), there is remarkably little reliable and documented information available on what actually occurs. Journals contain many short notes each on one particular record of, say, an unusual beetle or a scarce moth taken in a garden, but there is no main body of knowledge to put such records of garden insects into context. Local natural history societies often have a great deal of information on record, but it is not generally available. In the mid-1940s, Fitter (1945) published a fascinating account of London's natural history in which he recognized the richness of gardens as habitats, particularly for birds. He listed 'the true birds of the garden association, that can be found in most gardens of any size in the London area' as starling, greenfinch, chaffinch *Fringilla coelobs*, house sparrow, great tit *Parus major*, blue tit *P. caeruleus*, mistle thrush *Turdus viscivorus*, song thrush *T. ericetorum*, blackbird, robin, dunnock *Prunella modularis*, wren *Troglodytes troglodytes*, house martin *Delichon urbica* and swift *Apus apus*. He continued: 'A great many other species, such as the linnet, bullfinch (*Pyrrhula pyrrhula*), pied wagtail, spotted flycatcher (*Muscicapa striata*), willow warbler (*Phylloscopus trochilus*), great spotted woodpecker (*Dendrocopus major*), tawny owl (*Strix aluco*) and wood pigeon are found in the larger gardens from time to time.' A list of the birds of the London area (186 species) was given as an appendix. He then considered briefly insects as food for birds, particularly those insects that the gardener views as pests, and went on to mention a few butterflies, and the moths found in a South Hackney garden in the 1880s. More complete, and hence interesting, is his list of the weed flora of London gardens and of those garden escapes, such as Michaelmas daisy *Aster novi-belgii*, which have been added to the flora of London. Seventeen of the 27 weeds he listed have occurred in my Leicester garden.

Davis (1978) reviewed some of the changes in the insect fauna of London with urbanization. He mentioned that 72 species of aculeate Hymenoptera were caught in six gardens around Hampstead by Guichard and Yarrow, and referred to agromyzid flies utilizing garden plants and to moths that feed on garden conifers. He also summarized the detailed investigations of A. A. Allen, who collected beetles in his southeast London garden from 1926, when it was still meadow, to 1973; more than 700 species were recorded, including 64 species of Carabidae (Coleoptera), two of them new to Britain. Davis found that the number of species of carabid beetles, myriapods, Isopoda and Opiliones caught in pitfall traps increased with the age of a garden, with its distance from the centre of London and, particularly, with the amount of open space within a 1 km radius. Carabidae have been collected in pitfall traps in other London gardens and in other parts of the country e.g. Morrison (1974) operated traps for two years in the cultivated and bracken-covered parts of a garden in Perthshire, Scotland, catching 25 species of Carabidae, some in considerable numbers.

A survey, necessarily rather limited, of the natural history of Buckingham Palace garden was undertaken in the 1960s by the South London Entomological and Natural History Society (Bradley & Mere 1966, McClintock *et al.* 1964). This cannot be rated as a typical garden because of its size (about 39 acres) and its seclusion, but it is in a densely built-up area of London. Over 260 species of flowering plants were found growing wild in the garden, 59 of them unknown elsewhere in central London, and 26 mosses, five liverworts and two lichens. There was a maximum of 21 species a year of breeding birds (excluding feral pigeon and the red-crested pochard *Netta rufina*, which were descendants of the pinioned birds kept in the Royal Parks), and an additional 37 species visiting the garden, 60 in total. Frogs and toads appeared to be scarce and there were no newts; the only mammals positively identified were the house mouse and brown rat. A total of 788 species of insects, including 367 Lepidoptera, and 107 species of other invertebrates were recorded.

There are no comparable lists for gardens elsewhere in Britain and only selective reports from continental Europe. A 25 square kilometre area in the suburbs of Liège, Belgium, yielded 256 species of solitary bees and wasps between 1890 and 1972 (Leclercq & Remacle 1974). Pappa (1976) listed three species of Heteroptera, 69 Homoptera, 14 Hymenoptera, 12 Coleoptera, 48 Lepidoptera and 29 Diptera from 100 species of ornamental plants in Hamburg, Germany.

Frankie & Ehler (1978), in a review of the ecology of insects in urban environments, comment that 'Our ignorance in this area can be attributed, in large part, to the general dearth of relevant ecological studies'. The purpose of their review was 'to evaluate and synthesize information important to an understanding of the ecological bases for the reproduction and survival of insects in urban environments'. They refer to some ecological studies of urban insects, including my investigation of the Leicester garden, e.g. Owen & Owen (1975), but evidently found little published information on gardens apart from analyses of turfgrass ecosystems. One detailed study of a garden insect fauna is Frank Lutz's celebrated description of the insect fauna of his 'yard' in suburban New York (Lutz 1941).

When pleading with the Director of the American Museum of Natural History for more entomological staff, Frank Lutz, the Curator of Insects, based his argument on the vast number of different sorts of insect, even in his 75 × 200 foot (roughly 23 × 61 metre) yard (or garden). In the course of the discussion, he suggested that the museum raise his salary by ten dollars a year for every species above 500 that he could find on his 'lot', and volunteered to take a ten dollar reduction for every species short of 500. In the event, the bargain was not made, probably because the Director was convinced there were many fewer than 500 species, but in the ensuing search, over several years, Lutz found no less than 1402 species on his 'lot'. He did not trap systematically, but was very familiar with his piece of land and was constantly vigilant. It is, for instance, remarkable that he caught a long-tongued euglossid bee, the first to be seen north of the West Indies and Mexico. He was not, however, collecting insects in such a way that he could give any indication of numbers, other than a subjective assessment.

The distribution between insect orders of Lutz's 1402 species necessarily reflects the size of the different orders and their conspicuousness, and thus his list includes 467 Lepidoptera, 259 Coleoptera, 258 Diptera, 167 Hymenoptera, 75 Homoptera and 62 Heteroptera. Beetles, far and away the largest order, were certainly under-represented in his collection. Much of the information on my Leicester garden comes from trapping using various methods, and is therefore selective in a different way, especially as it depends on access to a specialist willing to identify a particular group. It is, however, interesting to compare the numbers of species in the same orders from among the 1602 species of insects identified in the Leicester garden: 364 Lepidoptera, 251 Coleoptera, 140 Diptera, 709 Hymenoptera, 42 Homoptera and 52 Heteroptera. As the figures to some extent indicate, I have had collaboration with a world expert on Ichneumonidae (Hymenoptera).

My garden list is probably more biased towards or away from particular groups than is Lutz's list, because I have concentrated on some groups to the exclusion of others. For instance, I have collected 91 species of syrphid (Diptera), a group in which I am particularly interested, whereas Lutz collected only 39. Other differences, however, reflect genuine zoogeographical differences, such as 35 butterflies on Lutz's lot but only 21 recorded from the Leicester garden. There is also a greater preponderance of aquatic insects, from dragonflies (Odonata) to caddisflies (Trichoptera) on Lutz's list, suggesting proximity to a lake, river or stream. It is, however, noteworthy that Lutz's garden was managed in the way that most people look after their suburban lot: in the four consecutive years in which he entered for a garden contest conducted by one of the New York newspapers, he won a silver medal for first place, two bronze medals for second place and a 'certificate of achievement'. Evidently, in the United States, as in England, a fairly conventional suburban garden can be rich in insects, given sympathetic management.

HISTORY OF GARDENS

The first cultivated plots that can fairly be called gardens, each consisting of an enclosed area with a shade tree, a trained vine, vegetables grown in rows and provision of water for irrigation, were probably established by the Sumerians in the rich land of Mesopotamia between the Rivers Tigris and Euphrates during the third millennium BC. By 1500 BC, decorative as opposed to purely utilitarian gardens were being created in Egypt and, a little later, in Persia. Homer's *Odyssey* includes a detailed description of what seems to have been an attractive and productive garden, but the earliest archaeological record of a Greek garden is from the Temple of Hephaistos, dating from 300–200 BC. Although the Romans developed both market and decorative gardening, and were the first culture to use gardens for sheer pleasure, such as eating outside, for many centuries the only gardens in Britain were those of monasteries, where herbs, vines, flowers, fruit and vegetables were grown. The *Domesday Book* (1086) mentions gardens, but there was at first little room for gardens in fortified and walled medieval towns, although by the reign of Henry II (1154–89) the senior citizens of London had large gardens. Generally, interest in gardening revived in the twelfth and thirteenth centuries, perhaps encouraged by contact with Arab countries during the Crusades, and by this time gardens were both productive and decorative, with a strong emphasis on herbs, which could be used for cooking, medicinally, or for strewing on floors. The English love of flowers for their beauty's sake can be traced back to 1300, although it was about 1600 before the richly varied cottage garden became traditional (Fisher 1982).

Commercial plant nurseries were first established in the sixteenth century, when many towns included gardens, and a 'garden of delight', as opposed to one for herbs and vegetables, was a well-established luxury. In the seventeenth century, vast sums were spent by private garden-owners on plants and seeds, and by 1690 London had at least 15 nurseries. Gardening up to this time had been mostly done for gentlemen by employees, and head gardeners commanded high salaries, but by the late seventeenth century gardening extended far beyond the world of the rich and fashionable. Small-scale domestic flower gardening gradually established itself as one of the most characteristic attributes of English life, and in the eighteenth century it became one of the most popular recreations among people from all walks of life. About 600 new gardening books were produced in the eighteenth century, and by the beginning of the nineteenth century there was no other country in which flower-gardening held as socially wide an appeal as in England (Thomas 1983). Gardens purely for pleasure have, however, been developed only when a culture has both wealth and time, and until relatively recently they tended to be the prerogative of the rich.

Many aspects of modern gardening have a long history. The setting aside of parts of the garden for vegetables and herbs, planted in lines or rectangles to allow ease of cultivating and harvesting, dates from ancient Egypt. The earliest recorded plant-collecting expedition was instigated by Queen Hatshepsut (1495 BC), who caused incense trees to be brought from Somalia to Luxor, and Tilglash-Pileser I of the Assyrian civilization in Mesopotamia boasted in about 1100 BC of planting 'trees that none of the kings my forefathers have possessed . . . carried from other countries I conquered . . . in the parks of Assyria have I planted them'. Scientific irrigation was a Sumerian invention, independently invented in China, and the use of manure is recorded in Egypt in the third millenium BC. From earliest times, gardeners have struggled to combat pests of one sort or another, such as the locusts described in Chapter 10 of the Book of Exodus, and bees have been kept since Egyptian times, although the realization that they pollinate flowers is relatively recent.

Grass has formed an important part of European gardens since the Middle Ages, although the medieval ideal was a meadow spangled with flowers. Turf was originally cut from natural grassland, managed by scything twice a year, and re-turfed, every three or four years, when it got out of hand. The popularity of bowls in the sixteenth century led to the development of better turf, although it was not until about 1900 that satisfactory grass seed mixtures were available. The establishment in the early eighteenth century of large landscaped gardens in which expanses of smooth grass were a feature of the design, called for regular precision-mowing, but this was still done with scythes, with horse-drawn rollers used to flatten the turf. The major innovation in lawn management was the invention in 1830 of the cylinder or reel mower by Edward Budding, an engineer in the textile trade, who saw the possibilities in a device used to trim the nap of cloth. Nowadays, the lawn may be the most intensively looked after part of a garden, but it would not form such a typical part of even the smallest garden, had it not been for Budding's invention.

Specialized cultivation of particular types of flowers goes back to at least the fourth century, when the Chinese concentrated on chrysanthemums and later on peonies, hollyhocks and other plants. One of the first European specialist crazes was for tulips, leading to the Dutch tulipomania of 1634–37, when the bulbs of prize blooms commanded incredible prices. There are now more than 40 specialist societies in Britain, concerned mostly with perennials, but over the years, fashions in flowers and vegetables, and their social cachet have changed. In the 1820s, for instance, it was taken for granted that pinks, roses and polyanthus belonged by cottages, whereas villas would aspire to geraniums, dahlias and clematis (Thomas 1983). For the ordinary gardener with a small suburban plot, the milestone of flower-growing was the influx in the second part of the nineteenth century of half-hardy or tender annual plants, such as *Tagetes* species from tropical America, and the subsequent establishment of nurseries for growing bedding plants. At least 3000 plant species, excluding hybrids, cultural varieties and vegetables, are in general cultivation in Britain today, although at least five times as many have been introduced at one time or another, with no more than marginal success (Fisher 1982).

This brief outline, derived largely from Huxley (1978), plots some of the developments that have led up to the suburban gardens of the 1980s. Gardening practice is based on methods devised and experience gained in many countries, by different cultures, and in the large gardens of monasteries and of the wealthy. What is the situation in Britain today?

GARDENS IN BRITAIN

Largely because the average householder in Britain wants to own and work his or her plot of land around or behind the house, the total area of garden is astonishingly large, and a greater proportion of the land area than in any other country. A few gardens are very large, some are very small, but the average size is said to be 186 square metres (Hessayon & Hessayon 1973). The number, and hence extent, of gardens has undoubtedly increased since Hessayon & Hessayon stated that there were 14.5 million gardens covering 710 000 acres, or nearly 287 500 hectares. According to Arvill (1973), in 1965 there were 4.3 million acres of urban land in England and Wales, of which nearly a quarter was devoted to gardens, amounting to perhaps a million acres (or 405 000 ha). On the basis of an unpublished garden survey in 1986 by the City Wildlife Project, 27.6 per cent of the area of Leicester is under gardens; if this percentage is extrapolated to Arvill's figure for the area of urban land in England and Wales, these two countries have nearly 1.2 million acres, or more than 485 000 ha, of garden in urban areas alone. Even if this is an exaggerated figure for urban areas, there are, in addition, gardens in rural areas, and in the rest of Britain and, furthermore, urban areas have expanded since 1965. It seems safe to conclude that 485 000 ha is a conservative estimate for the total area of gardens in Britain in 1987; it is noteworthy that this is more than three times the area of National Nature Reserves (Nature Conservancy Council figures) and nearly 20 times the holdings of the 46 County Naturalists' Trusts in 1987 (Perring 1987). Gardens may indeed occupy more than 3 per cent of the total area of England and Wales. This is a very significant proportion, particularly now that many people, concerned about modern farming methods, are beginning to manage their private plots in a way that is sympathetic to wildlife.

Over 70 per cent of British gardens have a vegetable plot, 34 per cent have fruit trees or bushes (Huxley 1978), 82 per cent have a lawn (Hessayon & Hessayon 1973) and, in addition, most people have flower beds or borders. A basic pattern is repeated again and again in the small- to medium-sized plots of extensive suburbia; a well-manicured lawn – however small, flower beds, a vegetable patch, ornamental shrubs, usually a tree or two, and often a privet hedge. Although some gardeners put most of their land under grass, a few specialize in certain plants, such as roses, some grow a range of vegetables, and others abandon their plot to weeds, the gardens of England collectively

Photograph 1.4. Gardens, such as this well-kept suburban plot in Leicester, cover more than 485 000 hectares in built-up areas in England and Wales.

are a highly diversified and productive area of land. For many, gardening is a rewarding pastime and major preoccupation, if not an obsession; in 1973, £100 million was spent annually on plants and tools in Britain, a figure said to be increasing by 10 per cent per annum (Thorns 1973), so in 1987 the outlay on maintenance and care of gardens was probably £400 million. Some of the care lavished on gardens undoubtedly has social significance – a desire to have the best-kept garden in the street, to grow the earliest beans or the largest marrows, or a compulsion to exhibit an ordered, attractive aspect to the world at large, rather than to appear unkempt and uncaring. As early as 1838, John Claudius Loudon was emphasizing that the art of gardening should be displayed in the front garden (Fisher 1982). For few, particularly in suburbia, is gardening an economic enterprise in terms of produce, and it may indeed, in some instances, be a way of demonstrating affluence as much as a source of enjoyment. But for whatever reason a garden is maintained, the vast amount of care lavished on the large area of gardens in Britain, results in a significant area of land with considerable value, or at least potential, for wildlife.

2

Geographical, historical and biological background to the Leicester garden

The regional setting of Leicester

The limits of the pre-1974 county of Leicestershire, before amalgamation with Rutland, were defined by the Anglo-Saxons, and in the Domesday Survey (1086) it was estimated as having a value of £827.4s.7½d. per annum (Leicestershire Regional Planning Report 1932). Leicestershire was below the average in shire wealth for the England of 1086, in terms both of manorial values (a manor was a territorial unit of lordship) and also of valuations for each recorded man and plough-team. It is impossible now to tell whether the 'poverty' was real or a result of the way the records were presented (Phythian-Adams 1986). The post-1974 county, including Rutland, is a compact area of 2548 square kilometres with the conurbation of greater Leicester, with a population of about 500 000, in the centre (Hickling 1978).

Leicestershire was traditionally thought of as a land of rolling pastures and wooded valleys. Bryan (1933) identified 'the main cultural forms' of the county's grasslands as 'the grass fields with their herds of cattle and sheep, the farmsteads and scattered houses of the herdsmen, the well-cared for whitethorn hedges, the broad road spaces with their wide grass margins, the scattered villages usually smaller than those of tilled country, and the prevalence of brick as a building material.' This was written more than 50 years ago, and refers to the pre-1974 county of Leicestershire and not to Rutland, where yellow ironstone is the traditional building material. The aspect of the countryside has changed considerably since then, as hedges have been removed to enlarge fields given over to arable crops.

Pre-1974 Leicestershire, 'this modest English county' (Hoskins 1957), is roughly divided by the River Soar, running from south to north through the centre, into two distinct regions differing in both their physical geography and in their historical development and modern character (Pye 1972a). The city of Leicester, a pre-Roman settlement and Roman military centre at an important river crossing, is situated in the wide, flat river valley. The town was originally established on river gravel terraces, but in the last hundred years or so has extended out onto the alluvial soils of the flood plain and up onto the higher land beyond. My garden lies in the eastern part of the city at an altitude of about 72 metres on the slopes above the older parts of the city, whose average altitude is 45.7 m. The underlying rock of the city and much of the county west of the Soar is Triassic Keuper Marl, but from about 15 kilometres east of the river the base rock is Jurassic. Rounded ironstone hills form the higher land (maximum elevation 230 m) in the east of the county, and Precambrian rocks of the Charnian series, thrusting through the Keuper Marl, form the craggy outcrops of Charnwood Forest (maximum elevation 278 m) in the west. Most of the south of the county, the land to the east of the Soar, and that between southeastern Charnwood and the river, is overlain with Pleistocene boulder clay, in places 21 m deep, within which are patches of sand and of gravel. In most of eastern Leicestershire, the boulder clay forms deep soils that support rich grassland. The soil immediately to the east of the city, and including my garden, is rather poorly drained clay soil, which is graded as Class 3 agricultural land, and has a tendency to waterlog.

The climate tends towards the continental, being relatively dry with well-marked seasonal temperature differences, as the county is the part of Britain furthest away from the sea and hence from maritime influences (Pye 1972b). Weather records for 1972–86 for Newtown Linford, just to the northwest of the city of Leicester, show mean maximum and mean minimum July temperatures of 21.0 and 10.7 °C and mean maximum and mean minimum January temperatures of 5.8 and 0.2 °C. (Table 2.1); the extreme maximum and minimum temperatures were 32.3 and −14.6 °C for July 1976 and January 1982, respectively. There is a growing season of 230 days from late March to early November, during which mean temperatures exceed 6.0 °C, and evapotranspiration reaches a maximum in July. Leicester city temperatures show a marked heat island effect: at 23.15 on 19 August 1966, when the temperature at the edge of the city was 17.8 °C, that in the centre was 21.1 °C (Pye 1972b). My garden is situated on the heat island map for 19 August 1966, between the 18.9 and 19.4 isotherms, and so is evidently influenced climatically by proximity to the city.

The rainfall varies considerably from year to year (Table 2.2), although distribution over the year is remarkably uniform (Table 2.1). Annual average rainfall for Newtown Linford, 11.5 km from my garden, was 665.5 millimetres for the period 1972–86; monthly averages varied from 42.4 mm in July to 68.5 mm in December. According to Pye (1972b), the period July to January is the wettest, the total duration of rain in a month and also the number of days with rain being greater in November, December, January and February than in the rest of the year. In an analysis of 40 years of rainfall figures, Pye recognized many dry spells and droughts (15 or more consecutive days with less than 0.2 mm of rain) but few rain spells and no wet spells (15 or more consecutive days with 1.0 mm or more of rain).

The monthly average rainfall for July (1972–86) is lower than might be expected largely because of low rainfall in July 1975, 1976 and 1977. The summers of 1975 and 1976 are remembered as long and hot, and 1976 was 'the drought year' when water use was restricted and supply limited in many parts of the country. Mean maximum monthly temperatures for June, July

Table 2.1 *Monthly minimum and maximum temperatures, rainfall and sunshine for the 15-year period 1972–86 at Newtown Linford, Leicestershire, 11.5 km from the garden*

	Mean minimum temp. (°C)	Mean maximum temp. (°C)	Mean rainfall (mm)	Mean sunshine (hours)
Jan.	0.2	5.8	59.8	49.3
Feb.	−0.1	5.6	50.8	50.2
Mar.	1.6	8.9	55.1	87.2
Apr.	3.1	11.7	42.8	121.6
May	5.8	15.2	59.1	159.3
June	8.7	18.6	62.8	169.1
July	10.7	21.0	42.4	169.4
Aug.	10.6	20.6	59.2	163.3
Sept.	8.6	17.5	56.2	125.7
Oct.	6.0	13.4	51.6	93.2
Nov.	3.0	9.2	57.4	63.4
Dec.	1.4	7.0	68.5	46.4

Table 2.2 *Annual rainfall 1972–86 and 15-year mean annual rainfall at Newtown Linford*

	Rainfall (mm)
1972	600.4
1973	670.8
1974	696.9
1975	493.1
1976	510.4
1977	756.4
1978	695.6
1979	729.5
1980	786.1
1981	679.5
1982	630.3
1983	689.9
1984	659.5
1985	693.5
1986	690.6
15-year mean	665.5

and August in 1975 and 1976 were markedly higher than the 15-year mean maxima for these months, the mean maximum for August 1975 being 24.7 °C (cf. Table 2.1). The years 1975 and 1976 were also particularly sunny, with total hours of sunshine of 1393.2 and 1454.3, respectively, compared with 15-year mean annual sunshine of 1298.1 hours. For the period 1972–86, June was the sunniest month with an average of 5.6 hours per day, and December the dullest with a daily average of 1.5 hours (Table 2.1).

There is a marked dominance of southwesterly winds in all months, more

Table 2.3 *Seasonal weather in different years, 1972–86, at Newtown Linford.*
Each description refers firstly to temperatures (cold, cool, ave, mild, warm, hot),
secondly to rainfall (wet, damp, ave, dry), and thirdly to sunshine
(dull, ave, bright). ave = average

	Winter	Spring	Summer	Autumn
1972	mild, ave, dull	mild, dry, dull	cool, dry, dull	ave, ave, dull
1973	warm, dry, ave	ave, wet, bright	warm, damp, bright	cool, dry, ave
1974	warm, ave, ave	ave, dry, ave	cool, damp, ave	cool, wet, dull
1975	warm, dry, dull	ave, wet, dull	hot, dry, bright	cool, dry, bright
1976	warm, dry, ave	warm, dry, bright	hot, dry, bright	ave, wet, dull
1977	cold, wet, ave	cool, dry, bright	cool, damp, ave	ave, dry, ave
1978	cool, damp, ave	mild, dry, bright	cool, damp, dull	warm, dry, ave
1979	cold, wet, bright	cool, wet, ave	ave, dry, dull	warm, ave, bright
1980	cool, wet, bright	mild, dry, bright	cool, wet, dull	cool, damp, ave
1981	ave, dry, ave	ave, wet, dull	ave, dry, dull	ave, damp, ave
1982	cold, dry, dull	warm, dry, bright	mild, damp, dull	mild, ave, ave
1983	mild, ave, ave	ave, wet, dull	hot, dry, bright	warm, ave, ave
1984	ave, damp, bright	ave, dry, ave	hot, dry, bright	warm, wet, dull
1985	cold, dry, bright	ave, ave, ave	cool, damp, ave	warm, dry, ave
1986	cold, damp, bright	cool, damp, ave	ave, damp, ave	ave, damp, bright

than half of the annual total having a westerly component; annual mean wind speed is 7.8 knots, being slightly greater from November to April than in the rest of the year. On average, snow falls on 22 days a year, with the maximum number of snowy days in February; snow falls in April in three years out of four, but the probability of early autumn snowfall is small. Most thunderstorms occur in summer, reaching a maximum in August, and most hail falls in January, February and March (Pye 1972b).

In later chapters of this book, when discussing annual fluctuations in numbers of insects and other invertebrates, 1972–86, reference is frequently made to the weather in different years. An attempt has therefore been made to summarize the seasonal weather year by year, using records from the Newtown Linford meteorological station. Winter is taken as December, January and February, spring as March, April and May, summer as June, July and August, and autumn as September, October and November. Mean maximum and mean minimum temperatures, total rainfall and total hours of sunshine for each month were compared with 15-year means, 1972–86, to arrive at an overall description of the seasonal weather for each year. These descriptions are given in Table 2.3. Inevitably, these are generalizations, as each season includes three months, one of which may have been cooler or warmer, wetter or drier, and duller or brighter than the others.

HISTORY OF LEICESTERSHIRE

What is now Leicestershire was probably once densely covered with mixed deciduous woodland, which it is assumed was dominated by small-leaved lime *Tilia cordata*, and ash *Fraxinus excelsior*, with possibly some pedunculate oak *Quercus robur*. In the early Neolithic, between 4000 and 5000 years ago, agriculture was introduced to Britain, substantial forest clearance began, and indeed most of the suitable land in lowland Britain had been brought under the plough before the Roman conquest. Certainly, by the end of the Roman era, the wildwood was only a dim memory in England (Colebourn & Gibbons 1987). Finds of rotary querns and modern techniques of aerial archaeology indicate widespread Iron Age and early Romano-British occupation of Leicestershire and considerable agricultural activity (Liddle 1982a). The distribution of Roman sites around Leicester is still uncertain, but it was a major market town in Roman Britain, centre of an important and densely settled agricultural area. Anglo-Saxon settlement, from about AD 400, seems to have been under the control of the basically urban Roman or sub-Roman authorities, and there was no sudden change in the countryside. Fields and complete estates were probably taken over by the new Saxon masters, giving continuity of agriculture. However, the Anglo-Saxons were largely subsistence farmers, with a culture in many ways similar to that of the early Iron Age British, and social and commercial organization of the Roman type did not survive the Saxon take-over, not least because towns had little place in their culture (Liddle 1982b). Saxon settlers found England as a whole only about 30 per cent wooded but with an uneven distribution of woodland; intensive Late Saxon farming reduced woodland cover to perhaps 15 per cent, about twice what it is now (Colebourn & Gibbons 1987).

An invasion from Denmark in the ninth century led, in AD 877, to Leicestershire becoming part of the Danelaw, which has left a linguistic legacy in the names of many villages and of some streets in Leicester. By the time of Domesday (1086), most of the county's villages existed and Leicestershire was one of the most densely peopled and extensively cultivated counties in medieval England (Hoskins 1957). Recent reconstructions of population figures from Domesday suggest an average density of as many as 47 people per square mile (18 per square kilometre), with four out of five living east of the Soar. Only a quarter of the county was wooded, mostly in the west, and another quarter, mostly in the east was pasture woodland where stock grazed among the trees, thus inhibiting regeneration (Phythian-Adams 1986). In the twelfth century, much of the county was declared 'forest' (i.e. a hunting reserve, which might be pasture woodland or waste, but rarely woodland), although it was exempt from this in 1235, and from then on there was further clearance of woodland and extension of farming. The fourteenth century saw the beginning of enclosures and an extension of sheep and cattle grazing, the open fields were reorganized, arable land was converted to pasture and there was a further reduction of 'waste', i.e. woodland, scrub and other common land.

The most striking feature of the English landscape around 1500 was that there were three sheep to every person: 8 million sheep and 2½ to 3 million people (Hoskins 1973). By the period 1535–43, much of the south and east of pre-1974 Leicestershire was open country, although more woodland persisted in the north and west and in Rutland. More than a thousand years of cultivation, clearing, draining and establishment of open field systems meant that by the late sixteenth century, despite some enclosures, the land had reached its ultimate development as wide open countryside (Hickling 1978), although the situation must be as extreme now, with vast arable 'prairies'. There was further enclosure with the appearance of permanent grassland in the seventeenth century, and a contemporary account in 1622 talks of the 'want of wood and fuel for fire' in the southeast (Ratcliffe & Conolly 1972). At the same time, towns other than Leicester increased in size and importance, whereas hitherto the county had been entirely rural apart from Leicester. In the eighteenth century, ploughed land was left to grass over and become pasture, there was some re-seeding and 'improvement', and by 1870 the area was once again noted for its grassland (Ratcliffe & Conolly 1972). In summary, apart from the loss of Leicester Forest and the Forest of Rutland, and the contraction of some woods, there was little change in the general aspect of the county from Domesday to the time of the parliamentary enclosures.

A considerable proportion of the county had been enclosed by landowners before the onset of parliamentary enclosures in the second half of the eighteenth and first quarter of the nineteenth centuries, when the rest was enclosed. The parliamentary enclosure awards usually stipulated that the land should be hedged within 12 months. Large landowners, however, only complied as far as the outer perimeter hedge was concerned, and hedges internal to the award came much later. One of the first effects of enclosure was reduction in trees, as they were cut down and used as posts for the new fences, along which quickset (hawthorn) *Crataegus monogyna* was planted; it also meant a reduction in heathland (Hoskins 1973). However, a number of spinneys planted on arable land around Leicester first appear on maps produced after 1800, and in the county as a whole there are more than a thousand woods exceeding 0.4 hectare which were probably planted in the eighteenth or nineteenth centuries. The parliamentary enclosures produced a landscape with fewer roads and more field paths than counties such as Devon which were enclosed in medieval times, as Georgian county-planners only made roads out of the most heavily used tracks; the main roads, field paths and bridle roads of Leicestershire date from before the Norman conquest (Hoskins 1957). Parliamentary enclosures, however, led to construction of the new straight roads with wide verges that are so characteristic of Leicestershire; these were the main thoroughfares, with side roads to each of the villages, for, even in the twentieth century, Leicestershire is a landscape of villages. The great width of the roads increased the chances that a firm course could be found by carriages and horsemen at even the muddiest season. The old thoroughfares have in many instances been preserved as wide verges flanking the relatively narrow tarmac strip of twentieth century roads. Despite the stipulations of the Enclosure Awards, the characteristic small fields and dense hedges of many parts of Leicestershire owe as much to the influential eighteenth century agriculturalist, Robert Bakewell, who farmed in the county, and advocated the partitioning of large tracts of grazing land into fields of 10 acres.

Organized fox-hunting developed in Leicestershire during the 1770s, just in time to enjoy the exhilaration of galloping over miles of unfenced country, although care was necessary as the undrained pasture was often very boggy. Subsequent enclosure changed the gallops to what became known as 'pewy' country, which the hunts gradually accepted, however begrudgingly at first, for the excitement of the jumps. Few trees were left in old hedges in fox-hunting areas, as they got in the riders' way, but over most of the Midlands, ash and elm *Ulmus procera* were dominant hedgerow trees. An effect of fox-hunting on the countryside, subsequently beneficial for wildlife, was the establishment of fox coverts – spinneys of roughly 1–8 hectares (2–20 acres) with dense understoreys.

In 1900, four out of every five acres of cultivated land were under permanent grass, but there was a temporary change from pasture to arable in World War I and then much more dramatically in World War II. Since World War II, the county's landscape has changed yet again, with grubbing up of hedges, felling of woods, drainage of marshes, the disappearance of the farm horse, and increasing use of pesticides and fertilizers on the large open fields, which create a landscape curiously reminiscent of the mid-eighteenth century (Hickling 1978). In 1957, ploughed fields were twice as common as in the 1930s, and much land was under the plough that had not been since 1600 (Hoskins 1957). There are, however, still many areas of grassland, although now it is mainly uniformly green 'improved' pasture rather than the flower-spangled meadows and pastures that once flourished; the prevalence of ridge-and-furrow contours shows that much of the remaining grass was once under the plough, long before the enclosures. However, despite modern developments, the landscape is still a landscape of villages with comparatively small fields (particularly in the west), large hedges and rough corners, and farming is the major land use in the county, occupying at least 80 per cent of the land.

THE WILDLIFE OF LEICESTERSHIRE

Leicestershire has never been renowned for its wildlife; there are richer and more spectacular woodlands, grasslands, moorlands, wetlands and aquatic habitats elsewhere in the British Isles. It is, however, a county of remarkably varied terrain and soils, despite the uniformity imposed by modern agricultural techniques, and the county's apparent lack of interest for the naturalist may be largely a consequence of lack of knowledge. Increased investigation recently is leading to more and more discoveries both of new species, particularly of animals, for the county, and of species that were thought to have been lost.

In terms of distribution of plants in particular, but also of animals, centrally placed Leicestershire is at a crucial position in Britain with regard to the decline in rainfall from northwest to southeast, the fall in mean summer

Photograph 2.1. Pastures and arable fields surrounded by dense hedges with tall trees are typical of countryside east of the River Soar in Leicestershire.

temperature from south to north, and the fall from west to east in mean winter temperature. This is superimposed on the geological contrast between the Precambrian Charnwood rocks with their acid soils in the northwest, and the ironstones and oolitic limestones in the east. Consequently, western and northern species are restricted to Charnwood, and 'continental' species to the limestone. Many species are at the limits of their distribution in or near Leicestershire: northern and western species such as cowberry *Vaccinium vitis-idaea* (present until High Sharpley bog was drained in the mid-1950s), western and southwestern species such as navelwort *Umbilicus rupestris*, eastern and northeastern species such as purple milk-vetch *Astragalus danicus*, southern and eastern species such as nettle-leaved bellflower *Campanula trachelium* and southern species such as wayfaring tree *Viburnum lantana* (Ratcliffe & Conolly 1972). The raven, with a northern and western distribution, bred in Charnwood Forest until the 1820s (Hickling 1978), northern birds such as goosander *Mergus merganser* and whooper swan *Cygnus cygnus* regularly winter on Leicestershire reservoirs, and moths of northern moorland, such as the glaucous shears *Hadena glauca*, occur at Charnwood Lodge Nature Reserve. The dormouse *Muscardinus avellanarius*, with a southern and western distribution, although less common than at one time, still occurs in one Leicestershire wood, and there are species with southeastern distributions, such as the nightingale *Luscinia megarhynchos* and a woodlouse, *Trachelipus rathkei*.

The *Flora of Leicestershire* (Primavesi & Evans 1988) provides an up-to-date assessment of habitat in the pre-1974 county, i.e. excluding Rutland, and some of this section on wildlife is based on its findings. Leicestershire, excluding Rutland, is one of the most poorly wooded counties in lowland Britain, with only 2.3 per cent of its area covered by woodlands of over 5 acres (2.02

hectares), only four counties having less woodland and one the same. There are, however, 1200 woods exceeding 0.4 ha, their total area being just over 4850 ha. At least 50 of these are thought to be primary woodlands directly continuous with the wildwood of prehistoric times, although they have been much utilized and modified by man. Some of the others, while secondary, are classed as ancient, i.e. dating from before about 1700, often having been planted on medieval ridge and furrow, but the majority were planted in the eighteenth or nineteenth centuries as game or fox coverts, and many are small. None the less, they are important for wildlife, two-thirds of the land birds of the British Isles, more than half the butterflies, and one-sixth of the flowering plants being exclusively or chiefly dependent on woodland (Mabey 1981).

Most of the ancient woods are ash/maple *Acer campestre* on the Peterken stand-type system (Colebourn & Gibbons 1987), although in the eighteenth and nineteenth centuries pedunculate oak was planted in many. They have a rich ground flora dominated by dog's mercury *Mercurialis perennis* and blue-bell *Hyacinthoides nonscripta*, and including giant bellflower *Campanula latifolia*, herb Paris *Paris quadrifolia*, woodruff *Galium odoratum*, sanicle *Sanicula europea*, ramsons *Allium ursinum* and toothwort *Lathraea squamaria*. Some woodland flowers, including wood anemone *Anemone nemorosa* and such orchids as common twayblade *Listera ovata*, broad-leaved helleborine *Epipactis helleborine* and early-purple orchid *Orchis mascula*, are good colonizers and occur in even the recent woodlands established in or since the eighteenth century. The bird fauna of the ash/maple woods, of the sessile oak *Quercus petraea*/birch *Betula pendula* woods on the acid soils of Charnwood Forest and of many recent woods is rich and varied, 40 species having been recorded from Swithland Wood northwest of the city.

Fifty years ago Leicestershire was famous for its extensive grasslands, but the landscape is now drastically changed as a consequence of conversion of pasture to arable, and a general tidying up of the countryside. Prior to 1939 only about a ninth of the county's farmland was ploughed, but during World War II the proportion of arable land rose to half, and by 1982 only 36 per cent of post-1974 Leicestershire (including Rutland) was permanent pasture. In the interest of farming efficiency, the conversion to arable has been accompanied by an increase in field size, with the removal of many miles of hedgerow. Furthermore, the remaining pasture lands have been 'improved' by appli-cation of fertilizers and herbicides, so that species-rich grassland is now rare in the county. Although neutral grassland predominates, its characteristic species, such as adder's-tongue *Ophioglossum vulgatum*, yellow rattle *Rhinan-thus minor*, and pepper-saxifrage *Silaum silaus*, are now unusual finds, and the once abundant cowslip *Primula veris* is much reduced in numbers. The county naturalists' trust regards the preservation of the few remaining unimproved grassland areas as a high priority and has been able to acquire a number as nature reserves, including an area of limestone grassland in the northeast and fairly extensive areas of acid grassland and heath on Charnwood Forest. Another encouraging occurrence is that species typical of grassland but no longer common there, such as cowslip and many calcicoles, colonized and have survived in railway verges and abandoned railway cuttings. The latter quickly become overgrown with scrub, but one rich cutting is leased and managed by the county naturalists' trust.

The tidying-up of rough corners and headlands on agricultural land has led to a decline in such species as gorse *Ulex europaeus* and the whinchat *Saxicola rubetra*, although gorse has often been planted in game coverts. Rough grass-land at the margins of reservoirs harbours large populations of field voles *Microtus agrestis*, which attract such predators as kestrels, barn owls *Tyto alba* and, particularly at the newly created Rutland Water, short-eared owls *Asio flammeus*. As a consequence of pasture 'improvement' there are now few localities where pasture is covered with nest-mounds of the yellow ant *Lasius flavus*, but it occurs in small numbers on railway embankments (Evans & Block 1972) and is widespread in gardens.

Wetlands are unusual habitats in the county and are declining as more areas of farmland are drained, although the county naturalists' trust owns and manages a number of marshes. Moth species more typical of East Anglia have been recorded from one of these near Leicester (Evans & Block 1972). The area of fresh water in the county is greater than ever before in historic times, not only because of the creation of reservoirs, notably Rutland Water, but also because of the number of water-filled gravel pits. These, the rivers, streams and canals, to some extent compensate for the filling-in or abandoning of field ponds, once a major habitat, although there are problems with pollution, dredging and bank straightening works on some waterways. By and large, the many fresh-water habitats have a rich and varied fauna.

Charnwood Forest, with its sandy soil, heath and moorland vegetation, and sessile oak/birch woodlands, has always attracted particular interest among the county's naturalists. It was the last part of the county to be enclosed, and still has an extensive deer park, used as a favourite recreational area by the people of Leicester. This deer park has over 500 species of beetles, including the tiger beetle *Cicindela campestris*, the minotaur *Typhaeus typhoeus* and 16 dung beetles of the genus *Aphodius*; wood- and stem-nesting aculeate Hymenoptera are as well represented as anywhere in England, but there is surprisingly a dearth of earth-nesting species with a preference for light soils. The Forest is rich in amphibians and reptiles, including the palmate newt *Triturus helveticus*, the adder *Viperus berus* and the slow-worm *Anguis fragilis*, and there is a distinctive invertebrate fauna in the small acid ponds of the highest areas and on relict heath and moorland, including 141 species of spiders on Bardon Hill, the highest point in the county (Evans & Block 1972).

Leicestershire has been farmed since the Iron Age and intensively farmed since Saxon times or earlier; the basic pattern of woodland in the county was established by Roman times. It is therefore hardly surprising that all the landscape shows clearly the impact of man and his activities. Since World War II, the pace of change has quickened with use of modern tools and equipment, and liberal application of chemicals, particularly to grasslands, and consequently the rate of species loss has increased. Thirty-four plant

Photograph 2.2. Fallow deer Dama dama *are abundant in Bradgate Park to the northwest of the city of Leicester on Charnwood Forest.*

species were lost in the eighteenth and nineteenth centuries, but a further 61 have gone since 1900. The greatest number (16) have disappeared this century from grassland, 11 from aquatic and marginal habitats and 10 from marsh and bog; the other 24 are woodland, heath, ruderal (growing on wasteland or land disturbed by man) and arable species. The effects of man on the landscape are in part explained by the demands made by an increasing population. In 1801, the county had a population of 130 000, 17 000 of whom lived in Leicester; in 1901, 269 000 people lived in the county, 95 000 of them in the city; and in mid-1986, 875 000 lived in the county, including 281 000 in the city of Leicester (greatly expanded in area since 1801). Man has dominated the landscape throughout historic times, but does so particularly now.

With the loss of traditional habitats has come the creation of new, man-made habitats, including gardens. Some traditionally minded nature conservationists dismiss these man-made habitats as artificial and therefore of no interest or significance. But many of them are extremely rich habitats and, as has also been shown for railway verges and cuttings, they may serve as refuges for animal and plant species displaced from agricultural land or other parts of the countryside. The common spotted orchid *Dactylorhiza fuchsii* and the bee orchid *Ophrys apifera* have colonized spoil heaps left by mineral extraction so successfully that their populations in the county have increased, and common spotted and, in one case, southern marsh orchid *Dactylorhiza praetermissa* have rapidly colonized areas of coal-mining dereliction. Small toadflax *Chaenorhinum minus*, pale toadflax *Linaria repens* and common toad-flax *L. vulgaris* have increased in the last 50 years on railway ballast. Oxford ragwort is now abundant along railways and, together with such species as rosebay willowherb, American willowherb *Epilobium adenocaulon* and pine-appleweed *Chamomilla suavolens*, on waste ground. The 'rough' of the 20 golf courses in the county provides relatively undisturbed sites for plants and animals, and the 800 ha Longcliffe course has large areas of heather *Calluna vulgaris*.

Sometimes, man-made habitats reach a peak of interest and richness and then decline unless managed for wildlife. In the couple of decades after the widespread Beeching closures of railway lines in the 1960s, the disused tracks, particularly east of Leicester, supported varied butterfly populations, including ringlet *Aphantopus hyperantus*, green hairstreak *Callophrys rubi*, Duke of Burgundy *Hamearis lucina*, dingy skipper *Erynnis tages*, grizzled skip-per *Pyrgus malvae* and chequered skipper *Carterocephalus palaemon* (Evans & Block 1972). Now they are less interesting; the Duke of Burgundy has gone, and the chequered skipper is extinct in England. Grasshopper warblers *Locustella naevia*, redpolls *Carduelis flammea* and turtle doves *Streptopelia turtur* appear to have benefited from felling and re-planting in woodland. Nightjars *Caprimulgus europaeus* persisted in one such site on Charnwood Forest until 1965 then disappeared, although they bred in one wood in the far east of the county (i.e. in Rutland) until 1976. There has been a trend with felling and re-planting for butterflies to disappear, although the comma *Polygonia c-album* and white admiral *Ladoga camilla* reappeared in the 1950s, the latter disappear-ing again then reappearing in 1986.

Hedgerow removal must have had disastrous effects on breeding birds, as hedgerow length has been shown to be an important factor in determining population densities. A Common Birds Census carried out annually since 1964 near to Leicester on a 87-hectare (215-acre) farm, with dense stock-proof hedges and many mature hedgerow trees, recorded a total of 56 breeding species up to the end of 1986. The peak year was 1975, with 445 territorial males of 43 species. The top ten birds of Leicestershire farmland were reported by Hickling (1978) to be, in descending order, blackbird, wren, yellowhammer *Emberiza citrinella*, dunnock, blue tit, skylark *Alauda arvensis*, song thrush, robin, chaffinch and great tit, although the wood pigeon, thought to be the commonest, was not included in the census. This list demonstrates how critical for birds is hedge removal, affecting the numbers of all these except the skylark. Luckily, Leicestershire is still largely a county of comparatively small fields, large hedges and rough corners, like vastly extended woodland edge, and hence is ideal for birds (Hickling 1978).

Badgers are common particularly in the east, with more than 500 setts, and foxes are plentiful, not least because of protection by the hunts. The muntjac deer *Muntiacus reevesi* is spreading and the harvest mouse *Micromys minutus* is widespread, if local. Lowe *et al.* (1933) described the red squirrel *Sciurus vulgaris* as not so frequent as formerly and stated that 'several examples of the grey squirrel (*S. caroliensis*) have been seen or shot, but so far their numbers have not given cause for alarm.' Now the red squirrel has disappeared, although interestingly its flea, *Monopsyllus sciurorum*, persisted on grey squirrels, which are common, until 1964 (Evans & Block 1972). Bell (1970) found the frog *Rana temporaria* and smooth newt *Triturus vulgaris* to be generally distributed in the county and in most places common, the toad *Bufo bufo* to be not so universally distributed or as common, the palmate newt to be restricted to Charnwood Forest, and the crested newt *Triturus cristatus* to be sparsely distributed and abundant in very few localities. The palmate newt is now known also to occur in east Leicestershire, and the crested newt appears to be widely though patchily distributed.

Existing knowledge about the wildlife, particularly the animals, of the county is inconclusive rather than demonstrating it to be as impoverished as was once thought. A growing body of amateur and professional naturalists of high calibre is making new discoveries all the time. Many good habitats and wildlife sites are gone for good, but the work of the Leicestershire and Rutland Trust for Nature Conservation, both directly as a landowner and also through the local authorities and by influencing the public at large, is helping to safeguard what is left.

THE CITY OF LEICESTER: HISTORY AND DEVELOPMENT OF HUMBERSTONE SUBURB

Leicester is an ancient city with the remains of a Roman Baths complex, and a fourteenth century Guildhall. To put my garden into the context of the city, however, we can start with the early nineteenth century expansion of Leicester. Land was freely available for building, because the fields on the east had been enclosed before the expansion of population and industry, so Leicester never had the terrible crowded slums that blighted Nottingham. In 1845 there were many large gardens in the town, and the average working class house of the nineteenth century had a little garden (Hoskins 1973). Between 1860 and 1890 the town doubled in size, and rows of brick terrace houses swamped the surrounding fields. A prosperous middle class suburb developed on higher ground in the southeast in the 1870s and 1880s (Hoskins 1957). The town grew particularly to the east and northeast (my garden in Humberstone is on the eastern side of the city), although a map of 1875 shows that by that time it extended less than half the distance from the centre to the site of the garden (Ellis 1948). Beyond this were the villa-residences of the employers of labour, and roads leading to surrounding villages, such as Humberstone, which was originally built on an island of sand and gravel amid the boulder clay. In 1892, the boundaries of the County Borough of Leicester were extended to include substantial parts of six surrounding ancient parishes. Four of these, including Humberstone, had been entirely rural, and showed no significant increase in population before 1871. Only a quarter of Humberstone parish, not including the site of my garden, was incorporated into the borough because of a wrangle about rateable values (Simmons 1974).

Railways and other public transport were not as important in the development of suburbs in Leicester as in other towns. The expansion to the east was
undoubtedly facilitated, however, by the opening of stations on the Great
Northern Line at what was then the eastern edge of the town in 1875 and
further out in villages beyond Humberstone in 1883. However, although
connected to Leicester by ribbon-building along the main Humberstone Road,
Humberstone, like the centres of the other five absorbed parishes, remained a
village community until World War I or later. It was a thriving community,
and as early as 1830 had three public houses and tradesmen such as a baker
and a blacksmith. Not that it was a backwater, as early in the twentieth
century it was the location of a planned social development. Anchor Tenants
Ltd, an offshoot of the Anchor Boot and Shoe Productive Society (founded in
1893), acquired 17 acres of land in 1907 for the development of Humberstone
Garden City. By 1915, 350 people lived there in 95 houses, with three shops, a
meeting room, tennis courts and a bowling green (Simmons 1974). Unwin,
the chief architect of the first 'garden city' at Letchworth, laid down a
standard pattern of house and garden density of 12 to the acre, which later
became the norm of inter-war suburban building (Thorns 1973), and the
Humberstone development seems to have adhered to this. In 1935, all of
Humberstone parish was taken over by the city, but until that time, my
garden was outside the city boundary.

Thus, the suburb of which my garden is a part, began to develop in the
early twentieth century as a mixture of out-of-town residences for the middle
class, and planned and enlightened working class housing, both grafted onto
an ancient rural village. Since then, the surrounding and intervening land has
been built on, mainly in the inter-war years, to form a typical suburb as
described by Thorns (1973), whose growth was determined by improvements
in public transport, land availability, easy credit through building societies,
the development of local authority housing estates and the social aspirations
of those who settled there. It is very varied in housing type and age, but by
and large has remained a green area, almost all the houses having gardens,
often quite large ones. An aerial view would show a much greater green area
than that occupied by buildings, with parks, bowling greens and playing
fields in addition to gardens. And, because most of the suburb dwellers are
owner-occupiers, there is pride of ownership, demonstrated by their
individual abilities and investment, which extends from the house to the
garden.

THE WILDLIFE OF LEICESTER

A great deal is known about the city's wildlife habitats as a consequence of the
Leicester Habitat Survey (Lomax 1987) conducted by the City Wildlife Project,
until recently part of the Leicestershire and Rutland Trust for Nature Conservation. The following account is derived largely from the Habitat Survey
and associated documents. Of the city's 7337 ha, 64.9 per cent is built-up,
including gardens, 18.8 per cent is managed as formal open space, such as

public parks, sports pitches and golf courses, 11.3 per cent is agricultural land, including allotments, and 5 per cent is land left to nature, either intentionally or by neglect. The Habitat Survey enabled the ranking of sites within the city in terms of their ecological value, ranging from A* (sites of city-wide ecological importance, the best examples of habitat type(s) in the city, with a great diversity of flora and fauna and locally rare or uncommon species), through A (sites of local importance and significant ecological value) and B (sites of some ecological value) to C (sites of limited ecological value). On this ranking, the 35.1 per cent of open space comprises 1.9 per cent of A* land, 7.5 per cent of A, 14.4 per cent of B and 11.3 per cent of C. Of the 1379 ha of formal open space, 26.8 per cent is classed as A* or A, as is 13.1 per cent of the 829 ha of agricultural land and 56.2 per cent of the 367 ha of land left to nature. In addition, 41.6 per cent of the 133 230 m of hedgerow is ranked as A* or A, as is 4.3 per cent of the 72 410 m of road verges, 65.8 per cent of the 47 297 m of river, stream and canal, and 46.2 per cent of the 15 721 m of used and disused railways.

Less than a third of the formal open space has high ecological value, as it consists largely of mown perennial rye grass *Lolium perenne* with isolated trees or groups of trees, and exotic shrubs. However, the roughs on the 250 ha of golf courses harbour voles which support stoats, weasels, kestrels and tawny owls, and cowslip, primrose *Primula vulgaris*, dropwort *Filipendula vulgaris*, cuckooflower *Cardamine pratensis* and lady's-mantle *Alchemilla vulgaris* grow on former rich pasture land. The trees include oak, beech *Fagus sylvatica*, ash, birch, Scots pine *Pinus sylvestris*, holly *Ilex aquifolium* and yew *Taxus baccata*, as well as exotics, and there is an increasing trend towards planting groups of native trees and dense shrubbery, all of which improves the formal open space as bird habitat.

The city's birds are well known thanks to the active and thriving Leicestershire and Rutland Ornithological Society, and the city appears to be rich in birds, mainly because it is well supplied with parks and other green areas. Between 1935 and 1952, 144 species were recorded within the city boundaries, which at that time included an old-fashioned sewage farm and extensive flood-meadows (Hickling 1978). On a 'Leicester Birdwatch' on 11 May 1986, about 150 observers amassed a total of 73 species within the city boundary, including such excitements as a hobby. Abbey Park and Abbey grounds, 89 acres of good bird habitat with large trees and dense shrubberies as well as mown lawns and flowerbeds, are further enhanced by a stretch of the River Soar with two weirs, which maintain a large area of ice-free water where an astonishing variety of water birds gather, particularly in the coldest winters. A smaller park of some 34 acres, about 2 km from my garden in a densely built-up area of late nineteenth century terrace housing and small factories, had 17 breeding species of birds during the period 1960–68, and a further 35 species were recorded flying over (Hickling 1978). Apart from parks, Leicester has more than 24 000 street trees, and there are tree-lined walks and squares in the heart of the city. Blackbirds, song and mistle thrushes and other birds breed, pied wagtails roost in the city centre, and tawny owls can often be heard.

Five green wedges of agricultural land extend into the city forming, together with allotments, 32.2 per cent of the 2575 ha of open space, comprising 13.7 per cent of grassland, 13.1 per cent of arable and 5.4 per cent of allotment. The grassland is used for grazing cattle, horses, ponies and goats, and for hay production, and has been much 'improved' by drainage, re-seeding and use of fertilizers; widespread change from grazing by cattle to horses has further reduced the interest of the grassland because of physical damage. The few old unimproved meadows on drier calcareous boulder clay have agrimony *Agrimonia eupatoria*, betony *Stachys officinalis*, burnet-saxifrage *Pimpinella saxifraga*, greater burnet-saxifrage *P. major*, pepper-saxifrage *Silaum silaus*, spiny restharrow *Ononis spinosa* and pignut *Conopodium majus*, while those on damper, neutral soils have ragged-robin *Lychnis flos-cuculi*, cuckoo-flower, common meadow rue *Thalictrum flavum*, great burnet *Sanguisorba officinalis* and bugle *Ajuga reptans*. There is a variety of animal life on these unimproved meadows, such as the large flocks of snipe that feed and yellow wagtails that breed alongside the River Soar in the south of the city. As pasture land tends to be ploughed up, the area of arable land is increasing, and none of this ranks as A* or A. Little of the allotment land reaches the A grade.

Land left to nature includes abandoned and disused land – the unofficial wilderness, and also land which is maintained and managed as nature reserves – the official wilderness. It comprises old woodland and plantations, disused allotments and agricultural land, and ponds and wetlands. Woodland is regarded as one of the most important habitat types in the city, mainly because there is little of it, and consequently 77.7 per cent of the area of woodland is ranked as A* or A. There are 117 woodland areas in the city, ranging from 0.5 ha to 4.8 ha in size, and totalling 78.42 ha. Apart from two ancient remnants of the thirteenth century Leicester Forest, consisting of oak and ash, with a rich shrub layer and a herb layer including wood anemone, bluebell and dog's mercury, most of the woodland is derived from plantations. Furthermore, one of the ancient woodlands is now dominated by sycamore *Acer pseudoplatanus* and Norway maple *Acer platanoides*. Nevertheless, some woods are rich in birds, including woodpeckers *Picus viridis, Dendrocopus major* and *D. minor*, nuthatches *Sitta europaea*, treecreepers *Certhia familiaris* and warblers, such as willow warblers and chiffchaffs *Phylloscopus collybita*, and these woods are also home to small mammals and foxes. Most, however, have been so neglected that the trees are spindly, shrubs are lacking, and they are dominated by bramble *Rubus fruticosus*, cow parsley *Anthriscus sylvestris* and nettles *Urtica dioica*. More than 50 per cent of the 286.8 ha of disused and unmanaged land is A* or A quality. It includes rich and varied areas on which natural succession is proceeding towards woodland of birch, hawthorn, willow *Salix* spp., ash and sycamore. Flocks of mixed finches fed on the abundant weed seeds on areas of wasteland created by demolition for re-building and road-widening in the 1960s and 1970s. There is also unmanaged agricultural grassland, alongside the river, where such plants as cuckooflower and meadowsweet *Filipendula ulmaria* grow, stoats hunt for small mammals and reed buntings nest.

In the same area in the south of the city as the rich, unmanaged grasslands alongside the River Soar and the canal, are some areas of wetland, which overall is a scarce habitat in the city. Despite modifications to reduce flooding and to aid navigation, many stretches of canal and river are diverse and rich. The same cannot be said of streams, many of which have been straightened and concrete-lined, and decrease in water quality towards the city centre. A few, however, retain a rich enough flora and fauna to be visited by herons and kingfishers *Alcedo atthis*. Apart from garden ponds, the number of ponds in the city has declined dramatically during the last century. A map dated 1884 shows 438, but in 1985 there were only 93, of which only 42 were former field ponds. Most of these have declined in quality from various forms of pollution, but a third of the former field ponds have a diverse flora and are rich in aquatic insects, fish and amphibians, including the crested newt. This reduction in ponds emphasizes and increases the ecological value and importance for conservation, particularly of amphibians, of garden ponds.

Old field boundaries persist as hedgerows in some areas of the city, but most have deteriorated, are no longer stock-proof, and have reduced wildlife interest. They are primarily hawthorn, elder and blackthorn *Prunus spinosa*, but the better ones include such shrubs as guelder-rose *Viburnum opulus*, and such trees as ash and field maple. The ground flora of a few suggests that they are remnants of ancient woodland. Hedgerows, even in a city, are important to mammals and birds as breeding sites, and as part of the green network linking good habitats. Most road verges, however, with a few notable exceptions, are poor habitats, as they are managed like formal open space.

Densely built-up areas and the city centre also support a surprisingly varied fauna, although there have been some changes with modernization. In the late nineteenth century, there was a large marshy area and reed bed on the river right in the centre of the city, where there were several nesting reed warblers *Acrocephalus scirpaceus* and sedge warblers *A. schoenobaenus* (Hickling 1978). The late Victorian terrace houses and factories supplied nesting sites for even more house martins and swifts than there are now. Substantial areas have been cleared of this sort of building, but swifts still scream over the city in summer and house martins flicker through the air, as many old buildings remain and these birds have colonized many areas of 1930s housing. The buildings of the city also furnish nesting sites for swallows *Hirundo rustica*, spotted flycatchers *Muscicapa striata* and kestrels, as well as the ubiquitous feral pigeons, house sparrows and starlings. Bats of three species, pipistrelle *Pipistrellus pipistrellus*, long-eared *Plecotus auritus* and noctule *Nyctalus noctula*, hawk for insects over the city, and Daubenton's bats *Myotis daubentoni* hunt over the almost rural expanse of meadows by the river. There are grey squirrels in all the parks and squares. Apart from pest species, such as cockroaches, and the exotic fauna associated with imported fruit, most of the

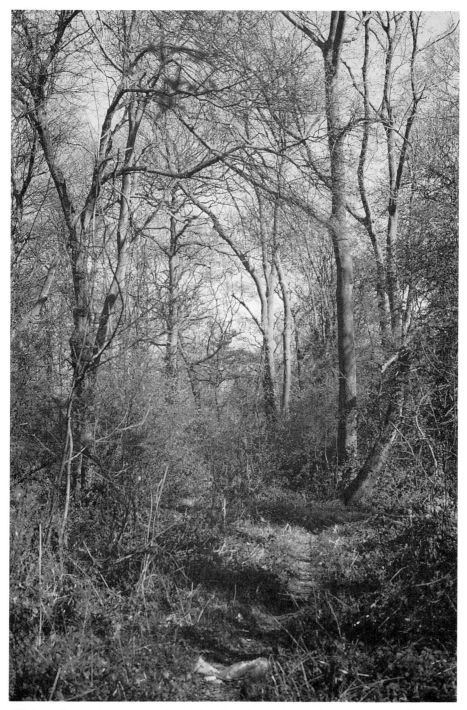

Photograph 2.3. A remnant of the thirteenth century Leicester Forest survives within the city of Leicester.

invertebrate records for the built-up area are of striking species that attract attention, e.g. silver-striped hawkmoths *Hippotion celerio* and wood wasps *Sirex gigas* and *S. noctilio* (Evans & Block 1972).

In the city as a whole, 2025 ha, or 27.6 per cent of the total area, are under garden of one sort or another. A partial survey by the City Wildlife Project indicates that 230 ha, more than 10 per cent, of the gardens are good wildlife habitat because of diversity, management, the attitudes of the owners, or all three. 'Growth and maturing of suburban gardens since the 1920s' is one of the factors that Hickling (1978) considered had led to increases in bird numbers in the county as a whole. The common city birds, according to a City Wildlife Project survey, are dunnock, blackbird, song thrush, robin, starling, house sparrow, blue tit, great tit, greenfinch and wood pigeon, but there are also woodpeckers, nuthatches, treecreepers, tawny owls and sparrowhawks. One large garden, about 3 km from the centre of Leicester on a busy main road, was investigated from 1953 to 1962, and found to have 22 species of breeding birds.

The wildlife value of inner-city gardens is limited, but those of suburbia are rich in insects, birds and amphibians. A survey of amphibians in the city, where they are using garden ponds for breeding, showed that frogs and smooth newts are particularly common (Mathias 1975). Toads seem to be colonizing garden ponds more slowly, and although crested newts have been recorded in city ponds, it seems less likely, in view of their preference for large ponds and lakes for breeding, that they will become common. Gardens are also a regular haunt of the one mammal for which some information is available – the fox. In early 1984, Leicester Wildlife Group (affiliated to the Leicestershire and Rutland Trust for Nature Conservation) conducted a survey of Leicester's urban foxes (Leicester Foxwatch) over the same 6-week period that Stephen Harris of Bristol University was systematically collecting information on foxes through the city's schools. Most of the Leicester Foxwatch records were from the eastern and southeastern sectors of the city, areas of larger gardens, but foxes have been seen at night in the centre of the city. Leicester Foxwatch estimated that there were 19 fox family units (FFU) in the city as a whole, the average family being one dog, one vixen and 4.8 cubs, often with a barren female in attendance. The Bristol University group conducted a door-to-door survey in 5.1 square kilometres of the eastern part of the city, where they found 7 FFU, and they estimated that there were 38 FFU in the city as a whole, a density of 0.43 FFU/sq km (Owen, J. 1985). The true value is probably a figure intermediate to their figure and the 0.22 FFU/sq km estimated by Leicester Foxwatch. Some towns and cities in southern and western England, such as Cheltenham and Bristol, have considerably higher densities of foxes, and there are estimated to be 1.16 FFU/sq km in the West Midlands (Harris & Rayner 1986a). Perhaps Leicester's relatively low density is a reflection of the encouragement of foxes in the countryside in a fox-hunting county. It is probably significant that many more respondents to Leicester Foxwatch expressed positive approval of urban foxes than were opposed to them.

Photograph 2.4. Twenty species of butterflies have been recorded in this Leicester inner-city nature garden created by the City Wildlife Project.

The Leicester Habitat Survey has not only established the wildlife and conservation value of the city, it has also been used by the City Wildlife Project, in collaboration with local authority officers, to formulate a Leicester Ecology Strategy. This incorporates recommendations for nature conservation and habitat improvement and, importantly and significantly, in 1986 became official policy of Leicester City Council. More and more about Leicester's wildlife is being discovered and recorded, the public is becoming better informed, and habitats are being improved or created through the work of the City Wildlife Project. On an inner-city nature garden that it created, 20 butterfly species have been recorded, and 27 species of birds, including tawny owl, sparrowhawk, rook *Corvus frugilegus*, nuthatch and treecreeper, bred in the 1980s in a small nineteenth century spinney that it manages. With increased awareness of and interest in the environment, the outlook is good for Leicester's wildlife, despite its inauspicious geographical position.

3

The Leicester garden

Visitors who have heard or read of my garden's richness in wildlife, frequently express surprise when they first see it. They expect an ungarden-like wilderness, the popular conception of a wildlife garden as a microcosm of the countryside. Instead, they see an apparently conventional suburban garden, with a well-trimmed lawn, neat and well-swept paths, a profusion of colourful flowers and abundant vegetables, herbs and salad stuffs. True, the vegetation tends to be denser and more luxuriant than in many gardens, but it is none the less neat, attractive and productive, and does not differ markedly from neighbouring gardens. Although the garden surrounds the detached house, it is not particularly large, and, moreover, incorporates all the elements of the typical suburban garden as described in Owen, J. 1983b. It has a lawn, flowerbeds, herbaceous borders, rockeries, vegetable patches, fruit bushes and an apple tree, flowering and evergreen shrubs, deciduous trees, conifers, a pond, a compost heap, a greenhouse, paths and paved areas.

HISTORY OF THE GARDEN AND ITS ENVIRONS

So well-established is my garden and so settled is the pattern of the surrounding suburb, that it takes a considerable imaginative leap to picture what the area was like a hundred years ago, let alone two centuries ago or even further back. Tracing the history of the site before the land was acquired for building has not been easy, and there are considerable gaps in the story. It is worth telling, to illustrate the difficulties there are in attempting to establish former land use as an adjunct to understanding the present status of an area.

The deeds to the house show that the land was once part of the Humberstone Hall estate, belonging to the Paget family. In October 1919, the Paget family sold the Hall and estate of just over 77 acres to Sir Arthur Wheeler, Baronet, of Woodhouse Eaves, Leicestershire. Sir Arthur Wheeler sold the site in November 1921 to a chemical manure manufacturer, and in August 1923 he sold it to a builder and, as sub-purchaser, a coal merchant, who had the house built. The plot is not quite rectangular, being 50 feet (15.24 metres) on the south side, where there was a fence and ditch between it and Scraptoft Lane, 46 ft 9 in (14.25 m) on the north, 202 ft (61.57 m) on the west, and 198 ft 8 in (60.55 m) on the east side along which runs another road, constructed some time after 1923. The house and land changed hands again in 1925 and in 1927, and it was almost certainly only then that the garden was designed and laid out. It changed hands twice more before I moved in, but there were no alterations to the basic garden plan before I took it over in 1971.

Working backwards from the sale of the Humberstone Hall estate by the Paget family in 1919, the latest previous record is an Ordnance Survey map dated 1915. That shows the Hall and grounds, with a mixed plantation on the site of my garden. In 1915, town development extended no closer to the site than 550 m. Furthermore, the map shows the Hall drive, which, together with its double row of Wellingtonias *Sequoiadendron giganteum*, has been absorbed into subsequent housing development as Pine Tree Avenue. Ordnance Survey maps of 1904 and 1887 also show a mixed plantation on the site of the garden, as part of the Humberstone Hall estate. Directories for 1883, 1878, 1876 and 1861 listed Thomas Paget as living at the Hall, and the entry on Humberstone in an 1846 directory of Leicestershire (White 1846) stated that a 'great part of the soil belongs' *inter alia* to Thomas Paget, who had a 'neat residence'. It therefore seems likely that the Hall was in existence by 1846, and that the plantation dates from that time, although there is no firm evidence of it before 1887. According to the Victoria County History (Martin & Bird 1958), Thomas Paget, a Leicester banker, began to acquire land in the parish of Humberstone in 1816. Certainly, the first edition Ordnance Survey map, dated 1814, shows nothing but fields on the site of Humberstone Hall and its estate. The parish was enclosed in 1788, and, interestingly, the map accompanying the Enclosure Award shows that the land where my garden is now situated belonged to T. Paget Esq.

In summary, it would appear that prior to 1923, the site of the garden was a mixed plantation, certainly from 1887 and possibly from 1846. The plantation was surrounded by the well-wooded parkland of the estate to the west and north, and by open fields to the east and south. After 1923, by which time the plantation must have been felled, the surrounding area gradually became built up, although, at the time of building, my house and its immediate neighbours represented the most easterly extension of the town. In 1989, the garden is nearly 2 km within the city boundary, surrounded by suburbia; it is a corner plot on a busy road, only 3.8 km from the city centre. However, there are open fields about 800 m away to the north (although development began on these in 1989), a small brook flanked by tall trees passes within 125 m, and there are two mature parks with old trees within a radius of 450 m. The surrounding gardens are well-established and many are densely planted; some of the larger gardens have extensive areas of orchard with long grass and old hedgerows.

Photograph 3.1. The Leicester garden in spring 1980, looking south from the garage, showing the general lay-out and Malaise trap. Since 1980, the birch tree Betula pendula *close to the house has grown much larger and the willow* Salix fragilis *in the foreground has been cut back to a coppice stool. Trevor Grewcock.*

GARDEN LAY-OUT AND CHANGES IN LAND USE

As indicated earlier, the garden (Grid reference SK 624054) was laid out about 1927, and the basic design has been unchanged since then. As shown in Fig. 3.1, the plot is a long rectangle, with the house set well back from the road, almost in the middle. The design of the garden and its dense and varied planting make it look far bigger than it is, even though the house is in the centre of the 930 square metre plot and part is taken up by the garage and its forecourt. The area of the garden (plot area − area of house and garage) is 741 square metres.

The site slopes slightly down from north to south necessitating some terrac-

Garden lay-out and changes in land use

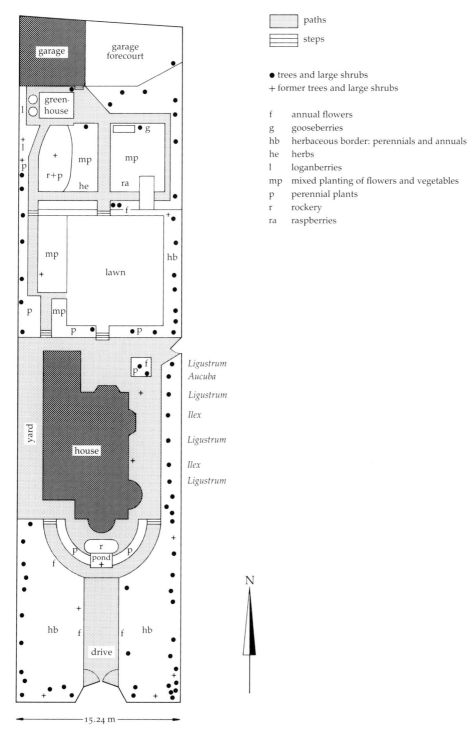

▭	paths
▤	steps

● trees and large shrubs
+ former trees and large shrubs

f	annual flowers
g	gooseberries
hb	herbaceous border: perennials and annuals
he	herbs
l	loganberries
mp	mixed planting of flowers and vegetables
p	perennial plants
r	rockery
ra	raspberries

Fig. 3.1 Plan of entire garden.

ing. The house and its immediate paved surrounds are four steps up from the front garden and four steps down from the back garden. Half-way up the back garden are a further three steps, and three more up to the garage at the northern end. These are shallow steps, and the slope is little more than 1.5 m in the entire length. The retaining walls of the terraces in front of the house, at its back, half-way up the back garden, and by the garage are mainly made of yellowish York stone. The front drive and the house surrounds consist of mundane, granite-like paving slabs. The main path up the centre of the north end of the back garden, that to the west of the lawn and the paved area around the greenhouse are yellowish York stone slabs, and various smaller paths, including the semi-circular ones at the front of the house, are a crazy-paving of off-cuts. The back garden is separated from the neighbouring garden to the west by a 1.8 m fence, and the front garden by an insubstantial low fence. On the north and east of the back garden is a 1.7 m fence which, in 1971, extended the full length of the front garden as well. In the winter of 1975/76, the section of fencing to the east of the house and front garden blew over in high winds, and, as the shrubbery was so dense, it was replaced by a 1 m fence. A 1 m-high wall of York stone backed with brick runs along the southern boundary on either side of the front gates.

Tall Lawson's cypress *Chamaecyparis lawsoniana*, ornamental *Prunus*, and dense shrubbery to a large extent screen the front of the house from the main road (Fig. 3.2). A short, paved drive-in for cars leads from the front gates towards the house, and at the end of this, up a step and below the front bay of the house, is a square pond surrounded by a low wall and backed by a rockery in the centre of a semicircular terrace overgrown with wild thyme *Thymus praecox*. On either side of the drive-in are dense stands of perennial plants, such as dame's-violet *Hesperis matrionalis*, goldenrod *Solidago canadensis*, *Phlox paniculata* and *Delphinium* sp., and of self-seeding biennials, such as forget-me-nots *Myosotis alpestris* and honesty *Lunaria annua*, behind a border of annuals, such as French marigolds *Tagetes patula* and *Nemesia strumosa*. A semicircle of paths curves round the pond and up steps to either side of the house, that to the left leading to the front door, near which is a massive bush of mock orange *Philadelphus coronarius*. To the left (west) of the house is a paved yard, to the right (east) a path flanked by shrubbery (Fig. 3.1), consisting mainly of privet *Ligustrum ovalifolium* and holly *Ilex aquifolium*, which screens the house from the side road. At the back of the house is a paved area in which is set a small, square flowerbed containing a bush of *Buddleia davidii* (Fig. 3.1). Central steps lead up to the lawn and another flight, on the left, lead to a path running along the west side of the garden (Fig. 3.3.).

A large birch tree *Betula pendula* beside the steps leading up to the lawn, shelters and shades it at certain times of day. The lawn is further screened from the house by a rowan *Sorbus aucuparia*, and low-growing shrubs, such as shrubby cinquefoil *Potentilla fruticosa* and sage *Salvia officinalis*, below which grow *Tulipa* sp., daffodils *Narcissus* sp. and snowdrops *Galanthus nivalis*. To the right (east) of the lawn is a herbaceous border backed by hawthorn *Crataegus monogyna*, holly, *Pyracantha coccinea*, buddleia and climbing roses on

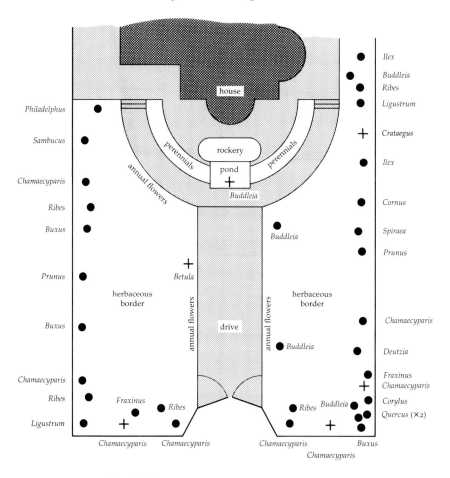

Fig. 3.2 Plan of front garden.

the fence. To the left (west) of the lawn are a small and a larger bed given over to mixed planting of annual flowers and vegetables and backed by runner beans *Phaseolus coccineus* on a trellis. A path runs between the bean trellis and a narrow border containing honeysuckle *Lonicera periclymenum*, a rowan, and *Forsythia intermedia*, beneath which grow meadow cranesbill *Geranium pratense* and lilies-of-the-valley *Convallaria majalis*. The house end of this border is wider, with a large beauty bush *Weigela florida*, a climbing rose and various perennials and bulbs. Narrow beds, largely given over to annuals, border the north side of the lawn and, beyond this, a broad path leads towards the garage.

Proceeding up this path, on the right there is first an area of perennials and self-seeded annuals surrounding the raspberry canes *Rubus idaeus*, then an area of mixed planting of flowers and vegetables, and finally a compost heap,

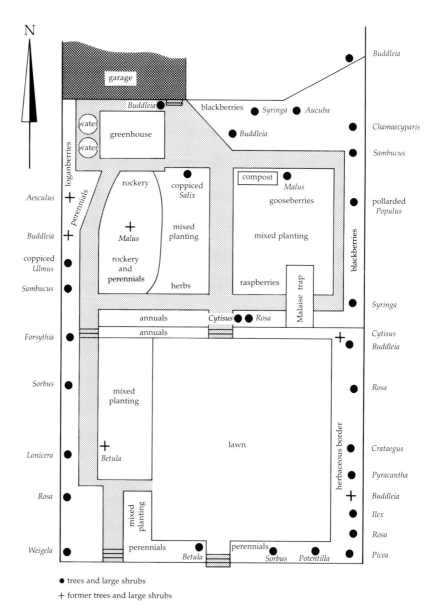

N

• trees and large shrubs
+ former trees and large shrubs

Fig. 3.3 Plan of back garden.

below a large apple tree and in front of a gooseberry patch. Further to the right, beyond a narrow path, brambles *Rubus fruticosus* are trained along the fence, behind a lilac *Syringa vulgaris* and a poplar *Populus nigra*. To the left of the central path, there is first a collection of herbs dominated by tall lovage *Levisticum officinale*, then an area of mixed planting, and a coppiced willow *Salix fragilis*. Further to the left is a rockery largely overgrown with marguerite

Photograph 3.2. The Leicester garden in June 1981, looking northwest from the northeast corner of the house. The trellis (left centre) was erected every year on the west side of the lawn bed to support runner beans Phaseolus coccineus.

daisies *Chrysanthemum frutescens*, and, beyond a narrow path, a bed of perennials and self-seeded annuals and biennials, behind which brambles and loganberries *Rubus loganobaccus* are trained on the fence. At the top (north) of the garden, a greenhouse stands on a paved area to the left, in front of the garage, and the central and right-hand (eastern) part is a mass of buddleia, lilac and laurel *Aucuba japonica*, with a tangle of brambles beyond the rubbish heap which lies beneath the Lawson's cypress in the northeastern corner.

The apple tree, the cypresses, the holly, the *Prunus serratula* on the east side of the front garden, the *Prunus cerasifera* var. *atropurpurea* on the west side, and the majority of the shrubs were in existence and mature when I acquired the garden in 1971. At that time there were elegant lawns flanked by flower borders on either side of the front drive, and the back lawn extended almost to the path on the west side, from which it was divided by a border containing roses and dahlias. The beds in the upper, northern half of the back garden were divided by paths, one curving round the rockery, and the other running from west to east across the middle of the eastern bed. There was no greenhouse, so that most of the area immediately south of the garage was an open York stone terrace, with a small flowerbed on its western side. The compost heap just west of the old apple tree was already in use in 1971, and an area in the far northeastern corner, below the Lawson's cypress, which is used now for temporary storage of woody and thorny garden rubbish prior to burning, appears to have been used as a rubbish dump for many years to judge from old pottery and bottle fragments unearthed there. The garden is basically unchanged since I took it over in 1971, although there have been changes in planting and luxuriance of vegetation. In 1971, it was a conventional, rather open garden, although there were many mature trees and shrubs. Now it is much more densely planted, and very lush.

The main losses of permanent woody plants that were present in 1971 are a young apple tree which was removed from the back garden rockery in the mid-1970s (Fig. 3.3), three *Chamaecyparis* in the front garden, which fell and had to be cut down in 1982 (Fig. 3.2), and a large broom *Cytisus* × *kewensis* at the northeast corner of the back lawn, which died in 1982. Other woody plants have been or have become established since 1971 but have then died or been intentionally removed. A horse chestnut *Aesculus hippocastanum* planted to the northwest of the back garden in the early 1970s, was cut down in 1981 as it created too much shade, self-seeded birches that became established by the back lawn, by the front drive and beside the northeastern corner of the house were removed after four or five years because of their thirsty roots, and buddleia bushes, established in various places from seedlings found in the garden, subsequently died. One fine buddleia grew for some years in the drained well of the pond. The concreted pond bottom apparently cracked in the winter of 1975/76. As fast as the pond was topped up, the water level fell again, and concern as to where the water was going made it prudent to drain the pond. In early 1984, the pond was re-established by fixing a pond liner into the original well.

The most marked changes in land use since 1971 have been the digging up of part of the back lawn in early 1975 to create beds for mixed planting, and the digging up of the front lawns to create herbaceous borders. That on the east was dug over in the winter of 1977/78, that on the west was left uncut as 'meadow' in 1978 and then dug up in winter 1978/79. The erection of the greenhouse in winter 1981/82 also represents a major change, because it allows the cultivation of seedlings of many annual flowers and vegetables for planting out in the garden. Another marked, although short-term, change was the cutting back of the shrubbery to the east side of the house in the winter of 1985/86. Since 1971, the only pruning had been that required to keep clear the path beside the house, and to prevent the shrubs encroaching on the pavement outside. The shrubbery had become a luxuriant tangle of vegetation where many birds nested. However, the privet bushes were becoming thin and spindly lower down, and it was judged necessary to take steps to promote more bushy growth. Consequently, in a major clearance, the privet bushes were cut back to about 1.5 m of main stem, the lower branches of the holly pruned, and the entangling brambles cleared to their main stems. This let light on to the ground for the first time for many years, and a thick cover of ground vegetation sprang up, including feverfew *Tanacetum parthenium*, spear thistles *Cirsium vulgare* and goat's-beard *Tragopogon pratensis*.

A number of the trees and shrubs that are shown in Figs. 3.2 and 3.3 have been planted, or have established themselves, since 1971. The elders and hawthorns are self-seeded plants that have been left to grow and, in winter 1986/87, several hawthorn seedlings from various places in the garden were planted along the east side of the house to supplement the screen of shrubbery. Hawthorn, elder, birch, holly, Lawson's cypress, flowering currant *Ribes sanguineum* and buddleia are continually appearing as self-set seedlings. A poplar, several wych elm *Ulmus glabra* cuttings, a willow cutting, and two seedlings of oak *Quercus petraea* were planted in the back garden in the early 1970s. The oaks did not survive, and all but one elm succumbed to Dutch elm disease, but the poplar thrives as a pollard, and the one elm and the willow, both as coppice stools. The screens of vegetation along both east and west sides of the back garden have largely developed since 1971. On the west, apart from the chestnut, buddleia, elm and elder already mentioned, the *Forsythia*, rowan and honeysuckle were planted in 1972. On the east side, the hawthorn and holly are self-set, the buddleia was planted in 1983, a dense tangle of *Rosa canina* grew from the root stock of a cultivar, the Norway spruce *Picea abies* was the family Christmas tree of 1973, and the *Pyracantha* was planted in 1981. The birch by the steps to the back lawn, now a magnificent tree, was self-set in 1974, the rowan to the south of the lawn was planted in 1975, and the extant broom *Cytisus scoparius*, beside the steps on the north of the lawn, was self-set in 1975. One or two of the flowering currants were present in the front garden in 1971, the others are self-seeded since then. The ash trees *Fraxinus excelsior* were self-set in the mid-1970s, the buddleia grew from seedlings in 1978, and the hazel *Corylus avellana* and oaks *Quercus robur* were planted in 1983.

Describing the changes in woody vegetation and in land use over-emphasizes differences between the garden of 1971 and as it is now. The overall aspect of the garden has not been dramatically altered, because the basic lay-out and the large trees, the skeleton of the garden, remain the same, as is particularly evident in winter. The most evident change, which almost certainly affects the fauna, is in the variety and density of the vegetation. The vegetation is more diverse than in most natural habitats, and compares with the richest natural area in luxuriance.

GARDEN MANAGEMENT

The density and variety of the garden vegetation are direct consequences of the way in which the garden has been managed since 1971. To some extent this represents a departure from conventional gardening practice, although, as has been emphasized, the garden is neat, attractive and productive, and is by no means a wilderness.

Every effort is made to maintain, for as much of the year as possible, an abundance of flowers at which insects can feed. Additional shrubs, such as buddleia and pyracantha, have been planted, perennials, such as delphiniums and goldenrod, have been established, such self-seeded plants as elder and borage *Borago officinalis* have been encouraged, and a wide diversity of annuals, including viper's bugloss *Echium plantagineum*, black-eyed Susan

Rudbeckia hirta, and mallow *Lavatera trimestris* are grown. Furthermore, unharvested brassicas are left to bolt, producing a wealth of yellow flowers attractive to butterflies, and such flowering 'weeds' as feverfew, hogweed *Heracleum sphondylium*, rosebay willowherb *Epilobium angustifolium* and great willowherb *E. hirsutum* are encouraged. Insects find it difficult to feed at the double flowers prized by horticulturalists, because the froth of petals blocks access to nectar and pollen, and so the chrysanthemums and dahlias present in 1971 have been phased out, and as ornamental rose bushes have died they have not been replaced. The shorter-tongued insects, such as hoverflies and the smaller bees, can only use flat, open flowers, and the clusters of florets that form the 'flowers' of composites are ideal, as insects can walk all over them. The larger bees are well catered for with sage, delphiniums, mallow, hollyhocks *Althaea rosea*, snapdragons *Antirrhinum majus* and many others, and buddleia, in particular, attracts butterflies.

The flowering period of many plants is extended by removing wilted flowers, and 'dead-heading' seems particularly effective with buddleia, perennial cornflower *Centaurea montana* and pot marigold *Calendula officinalis*. The profusion of garden plants, such as Michaelmas daisy *Aster novi-belgii* and iceplant *Sedum spectabile*, that flower in late summer and autumn, makes gardens particularly attractive to flower-feeding insects at a time of year when there are few flowers in the countryside; such flowers in the garden are constantly alive with bees, butterflies and hoverflies.

The increased diversity of plants in the garden, including the cultivation of such native trees as poplar, birch and willow, increases the opportunities for leaf-feeding caterpillars and other insects, and the varied garden flora, including such 'weeds' as nettle *Urtica dioica* and rosebay willowherb, provides the bases for many food chains. An effort has been made to increase the structural diversity of the garden by planting trees and encouraging a many-layered vegetation, a network of open and shady places. As well as providing food, this increases the living spaces and available ecological niches, thus adding to the richness of the habitat. Pollarding of the poplar, and coppicing of the elm and willow have been used as a means of accommodating the trees in a restricted space, and ensuring regular new leaf growth.

A conscious effort has also been made to provide plant food for birds. Members of the thrush family, in particular, feed on the berries of flowering currant, holly, pyracantha, *Cotoneaster* sp., elder and *Mahonia aquifolium*. Winter-visiting fieldfares *Turdus pilaris* and redwings *T. musicus* eat windfalls left beneath the apple tree, and, as the fruits ferment and soften in the autumn, they also provide a source of juices for wasps and such butterflies as red admirals *Vanessa atalanta*. Herbaceous perennials are not cut down as soon as they have finished flowering but left to go to seed, and seed-eating linnets feed at the Michaelmas daisies, greenfinches at sage and goldfinches at teasel *Dipsacus fullonum*. Leaving the dead standing vegetation until after Christmas also ensures that insects have the opportunity to complete their annual life cycles without disruption.

Pruning and clearing of woody vegetation and weeding are always kept to

Photograph 3.3. Overgrown shrubbery in the Leicester garden east of the house in May 1983, with blossom of Prunus serratula *top left.*

the minimum necessary for healthy growth and minimal competition. This means that the shrubby vegetation, the soft fruits and, indeed, much of the herbaceous vegetation, develop into wellnigh impenetrable tangles where birds nest and many insects and spiders find living spaces. The disposal of garden rubbish is never a problem; all soft garden rubbish is simply piled on the compost heap, together with soil clinging to the roots of weeds, and no special care is taken over the heap's composition. Woody or thorny cuttings are piled up in the northeastern corner to dry, until a bonfire is made on the vacant lawn bed after Christmas. Any further prunings that are necessary can be taken then, and added straight to the fire, and, although some nutrients go up in smoke, there is still abundant rich ash to boost the nutrients in the soil of the bed in preparation for planting annual flowers and vegetables. In spring, the compost heap is broken open and spread over one of the larger areas used for mixed planting. Areas that receive no compost are given a liberal application of a 'Growmore' (NPK 7:7:7) fertilizer in granular form, the only commercial chemical used in the garden.

Every effort is made to maintain good ground cover, so that there is little bare soil other than during winter. Ground cover is achieved by dense planting, by mixed planting of flowers and vegetables, by encouragement of such plants as wild strawberry *Fragaria vesca* in the herbaceous border and greater periwinkle *Vinca major*, spotted deadnettle *Lamium maculatum* and creeping-

Jenny *Lysimachia nummularia* among other perennials, and by tolerating limited growth of non-vigorous weeds among vegetables. Apart from avoiding baking of the soil surface in hot weather, provision of ground cover increases the numbers of predators, such as carabid beetles, centipedes and spiders, which shelter beneath the vegetation by day, and at night hunt soft-bodied insects, including pest species on crop plants.

In experimental set-ups, mortality of cabbage white *Pieris rapae* caterpillars on Brussels sprouts *Brassica oleracea* was found to be greater in weedy plots because of increased numbers of carabid beetles in particular (Dempster 1969). Similarly, numbers of *Pieris* caterpillars, cabbage aphids *Brevicoryne brassicae* and cabbage rootfly *Erioischia brassicae* were reduced on Brussels sprouts interplanted with clover *Trifolium* spp. (Dempster & Coaker 1974), and, in the Philippines, predatory spiders caused greater mortality to corn-borers *Ostrimia furnacalis* when the corn *Zea mays* was interplanted with peanuts *Arachis hypogaea* (Litsinger & Moody 1976). A further advantage of good soil cover arises because some herbivorous insects find the appropriate plants for themselves or their larval stages by using sight. Their ability to locate their foodplants depends on the background to a crop, so that more cabbage aphids colonize Brussels sprouts in weed-free plots (Smith 1976). This may be an additional reason why Dempster & Coaker (1974) found fewer pests on Brussels sprouts interplanted with clover, and may, together with the increase in predators, explain why Theunissen & den Ouden (1980) found that intercropping sprouts with corn spurrey *Spergula arvensis* at three different densities gave a graded reduction in total populations of four species of caterpillars. As well as limiting problems with pests, good ground cover adds a further layer of vegetation, thus diversifying and enriching the habitat, and augmenting the supply of plant food and living spaces.

All the management techniques listed above may be regarded as creating a somewhat unorthodox garden, but perhaps the most radical departure from usual gardening practice is rigorous exclusion of insecticides, herbicides and other pesticides. Such poisons have no place in a garden whose wildlife is being documented, monitored and investigated. This means that other methods have to be employed to maintain the appearance and productivity of vegetables and other garden plants. As shown, maintaining good soil cover is one way of achieving this, but there are additional advantages to be gained from mixed planting, or intercropping. This is a traditional practice in many cultures, from the apparent jumble of an African peasant farmer's plot to the variety of an English cottage garden, and it is advocated by adherents of the companion plant thesis (see Owen, J. 1984). That one sort of plant benefits from the presence of another is questionable, but mixed planting has some justification in terms of biological control of pests, and so I use it extensively (Owen, J. 1983b).

While some herbivorous insects, including pest species, locate their food-plants by sight, others respond to characteristic plant odours. The glucosino-late, sinigrin, and its mustard oil products, found in Cruciferae, act as attractants and feeding stimulants for crucifer-feeding Lepidoptera, Homop-

Photograph 3.4. Mixed planting in the Leicester garden, with pot marigolds Calendula officinalis *growing among vegetables in the lawn bed. To the right are marguerites* Chrysanthemum frutescens.

tera and Coleoptera (Root 1973). It would seem reasonable to suppose, therefore, that presentation of a battery of different aromas might confuse, disorient and deter herbivores. Admittedly, experimental evidence for this is conflicting, but Tahvanainen & Root (1972) found that monocultures of brassicas were colonized more rapidly by the chrysomelid beetle, *Phyllotreta cruciferae*, and were damaged more by feeding than were stands interplanted with tomatoes *Lycopersicon esculentum* and tobacco *Nicotiana tabacum*. Risch (1980), too, found that the presence of non-host plants reduced numbers of chrysomelid beetles on crop plants, and seed catalogues and gardening journals advocate planting of basil *Ocimum basilicum* and French marigolds to deter pests. I find that interplanting of vegetables with marigolds and other flowers, and intercropping brassicas with tomatoes, onions *Allium cepa* and leeks *A. Porrum* keeps pest depredations down to tolerable proportions.

A further benefit of mixed planting of flowers and vegetables derives from the habits of hoverflies. The larvae of the most abundant garden species are voracious predators of aphids, and egg-laying females are attracted either to the host plant or to aggregations of aphids (Smith 1976). Most hoverflies eat some pollen, rich in protein, when visiting flowers, and it is known that females of many species require a pollen meal for maturation of their eggs (Pollard 1971). Provision of flat composite flowers, such as asters *Callistephus chinensis*, *Rudbeckia hirta* and marigolds, close to beans, cabbages and other plants susceptible to aphid infestations, increases the likelihood that egg-laying female hoverflies will visit and lay their eggs. This ensures that predatory hoverfly larvae will exercise a degree of biological control of aphids on the crops. As a consequence of such mixed planting, I find that by high summer my brassicas are clean of cabbage aphids, there being instead a high density of hoverfly pupae, particularly of the common species *Episyrphus balteatus*. Extensive use of mixed planting, as an alternative to insecticides, seems to work, and it certainly contributes to the richness and small-scale variety of the garden as a habitat.

METHODS OF INVESTIGATION

The fauna, particularly the insects, of my garden has been investigated since 1971 in a number of ways. As every ecologist knows, a great deal can be gained from regular observation and surveillance of a study area, and it is in this respect that a private domestic garden is a particularly satisfactory site for investigation, being close-at-hand and accessible. Information is gleaned on every walk around the garden, while gardening, and sometimes simply by glancing out of the window. Whenever insect feeding damage to leaves is noticed, or its tell-tale signs, such as frass on a path beneath a plant, is seen, a search is made for the perpetrator. Adults, such as beetles, are collected for identification, and larvae are reared in jars on the appropriate foodplant, until adult, when they, too, can be identified. Predatory hoverfly larvae are also collected, and reared to maturity. Beating vegetation out onto a tray (or upturned umbrella) yields further larvae for rearing. Sometimes the end result is not the adult expected, say a moth, but instead an ichneumonid, which can also be identified. Rearing of larvae found in the garden has produced a wealth of information about feeding relationships, and facilitated recognition of food chains.

Detailed investigation of the fauna, of its abundance, diversity and changes from season to season or year to year, necessitates use of various trapping methods. The simplest of these is hand-netting of butterflies, which were marked and released during the insect flight seasons of 1972 to 1979 inclusive. Butterflies were marked on the wings with waterproof ink, using an ordinary felt-tip marker, so that individuals already caught and scored could be recognized. Although prone to some bias, because of variations in time spent catching butterflies and in effort put into catching individuals, and because netting was unsystematic, this method none the less gives a reasonable estimate of changes, with season and year, in numbers of individuals and species visiting the garden.

One of the most useful and profitable trapping techniques has proved to be operation of a Malaise trap (Fig. 3.4). This catches flying insects non-selectively. It uses no bait or other attractant, simply intercepting insects in the course of their normal activities. Moreover, it is kept continuously in position, by day and by night, in all weathers. The type of Malaise trap used is that designed and described by Townes (1972a). It is an open-sided, tent-like structure of meshed fabric, about 2 m long and 1.3 m wide. The pitched roof rises obliquely to a peak almost 2 m high at one end, where there is an opening leading into a collecting jar containing 70 per cent ethyl alcohol as killing agent and preservative. An internal baffle of netting runs the length of

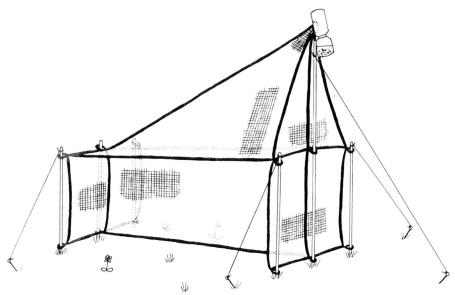

Fig. 3.4 Malaise trap.

the long axis, and also divides the roof space almost up to the opening into the collecting jar. The trap is supported on poles held upright by guy ropes. The side openings are about 1.1 m high and the length of the trap. Flying insects that wander into the open sides and come up against the centre baffle, tend to clamber or fly upwards towards the light, and eventually fall into the collecting jar. This tendency is encouraged by using black netting for the lower parts of the trap, and white for the roof. The collecting jar is emptied and the catch sorted at least once a week.

Not all insects that enter the open sides of the trap end up in the collecting jar. It is generally believed that larger beetles drop to the ground on striking the baffle, and then walk away. An unknown proportion of other groups also find their way out again; the trap is estimated, for instance, to capture only 20 per cent of the Ichneumonidae that enter its 2.9 cubic metres of air space (Owen, J., Townes & Townes 1981). Trapping is, moreover, generally believed to have an effect on the community sampled, because of the consequences of removing individuals. Removal of each leaves a vacancy which is promptly filled by immigration. Consequently, trapping in itself boosts the size of the trap sample. It none the less provides a large, unselected sample of flying insects in a standardized way, and can be used to compare abundances, diversities and species composition at different times.

The Malaise trap in the garden has been operated every year since 1972 from 1 April to 31 October in the same site in the back garden (Fig. 3.3). One open side faces a lilac bush and other vegetation close to the eastern fence; the other side is open to the main part of the garden. Below the trap is low vegetation, including spotted dead-nettle, creeping-Jenny, perennial corn-flower and various annual flowers. Observation suggests that it is situated

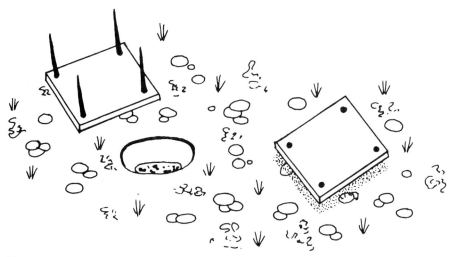

Fig. 3.5 Pitfall trap.

across a flight path of insects, particularly butterflies. In 1978, a second
Malaise trap was operated on the overgrown lawn (meadow) on the western
side of the front garden to compare the effectiveness of trapping in the two
sites. The comparison was made using hoverflies (Syrphidae). The total catch
from the second trap was smaller, the relative frequency of species was
different, and a new hoverfly to the garden was caught. This suggested firstly
that the fauna differs slightly in different parts of the garden, and secondly,
and more importantly, that continued use of the original site is justified.

Another trapping method used continuously from 1 April to 31 October
every year since 1979, is operation of pitfall traps (Fig. 3.5). These consist of
plastic drinking beakers set into permanently positioned lengths of sunken
piping and one-third to one-half filled with a 4 per cent formalin solution as
killing agent and preservative. Over each trap is positioned a square of hard-
board mounted on nails, which can be driven into the soil to give a clearance
of about one centimetre, sufficient to exclude frogs and larger animals and to
keep out rainwater and dead leaves, but to admit carabid beetles, centipedes,
spiders and other ground-dwelling invertebrates. A total of 11 pitfalls is
operated, two in the back garden rockery, two in the back garden herbaceous
border, two in the bed of perennials by the westernmost steps up to the back
garden, two below the shrubbery to the east of the house, and three beneath
the Malaise trap. The traps are emptied, cleaned and re-filled with formalin
when necessary, usually every two weeks.

Most of the information about moths in the garden comes from operation of
a standard mercury-vapour light trap (Fig. 3.6), comprising a circular, black
box, 60 centimetres in diameter and 25 cm high, topped by a truncated cone of
transparent plastic, in the centre of which is the light bulb mounted in a metal
funnel. Moths attracted to the light, which may be effective only over dis-

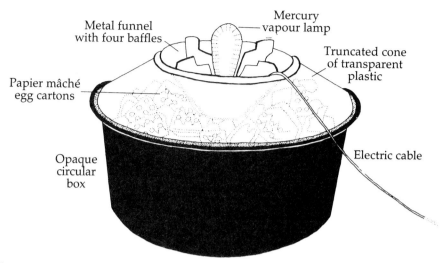

Fig. 3.6 Mercury-vapour light trap.

tances of up to 3 m (Baker & Sadovy 1978), strike the baffles in the metal funnel as they circle, and drop into the body of the trap, which is packed with papier mâché egg cartons on which they come to rest. When the trap is unpacked in the morning, most moths can easily be identified, and only 'difficult' specimens are collected. The bulk are kept in the covered trap in a cool place, to be released the following evening. The trap is usually operated on the back lawn, occasionally at the north end of the garden, near the garage. Sometimes the mercury vapour light is suspended against a white sheet for inspection of moths as they are attracted during the night.

A baited trap is used occasionally to capture particular sorts of insects. The design usually used consists of a cylinder of white netting fabric, about 30 cm in diameter and 60 cm in height, closed with netting at the top, and supported top and bottom by circles of stiff wire, below which is suspended a board of equal diameter (Fig. 3.7). The trap is hung from a branch by a handle attached to the top, and bait is placed on the board. The gap of about 3 cm between board and trap is sufficient to admit insects as large as butterflies, but not so large as to be an obvious exit. When insects leave the bait, they tend to fly up into the cylinder of netting, and can be removed by lifting down the trap so that it collapses. A different design of baited trap is held rigid and upright by longitudinal wire supports, and the top narrows down to an oblique funnel which leads into a collecting jar. Baited traps have been used principally for collecting moths, butterflies and hoverflies at fermented fruit, and blow-flies (Calliphoridae) and grey flesh flies (Sarcophagidae) at fish heads and skins.

Continuous use of a Malaise trap since 1972 and of pitfall traps since 1979, regular use of a mercury-vapour light trap and of hand-netting, and occasional use of baited traps has resulted in the collection of a great deal of information about invertebrates, particularly insects, in the garden.

Fig. 3.7 Baited trap.

Abundance, diversity, fluctuations in numbers, seasonality, and appearances and disappearances have been monitored. Continual observations and rearing of larvae provide the information that 'fleshes out' the trap data, and gives an idea of what insects are doing in the garden.

4

Flora

The occurrence in the garden of different species of flowering plants, whether intentionally grown or growing of their own accord, has been recorded annually since 1975. In many cases, particularly with perennials, it is also known whether or not they were present in 1971. A total of 78 plant families, excluding grasses, was recorded up to the end of 1986; some are well represented, whereas 36 are represented by a single species. The largest families are Compositae with 56 species, Labiatae (24), Rosaceae (21), Cruciferae (20), Liliaceae (16), Leguminosae (15), Solanaceae (14), and Caryophyllaceae, Ranunculaceae and Scrophulariaceae, each with 11 species. The plant list is cumulative, and the species did not all occur at the same time. E.J. Salisbury's remarks about gardens (Salisbury 1935) are particularly apposite here. Gardens 'comprise an artificial assemblage of many kinds of plants, the nature and numbers of which are mainly dependent on our arbitrary whims and fancies. Some that we grow are only too patently unfitted for the conditions in which they live, and many more fail to produce offspring and so when they perish must perforce be replaced by artificial propagation or by importation. The plants of a garden grow not where they will but where they must.' He goes on to describe the need to combat competition from weeds, plants that come in of their own accord, pointing out that 'The garden is, in fact, an unstable community which only continual labour maintains in seeming stability.'

Table 4.1 shows for each year, 1975–86, the numbers of alien and of native species, the numbers that were intentionally planted and that came in of their own accord, and the total number of species. Although many aliens are cultivated and most weeds are native species, it is important to remember that not all aliens are intentionally grown, neither do all native plants come in of their own accord as 'weeds'. For instance, weeds include such aliens as buttonweed *Cotula coronopifolia* from South Africa, Canadian fleabane *Conyza canadensis* from North America, spotted dead-nettle *Lamium maculatum* from continental Europe and large-flowered pink-sorrel *Oxalis corymbosa* from South America, whereas cultivated plants include such natives as rowan *Sorbus aucuparia*, Jacob's-ladder *Polemonium caeruleum*, shrubby cinquefoil *Potentilla fruticosa* and cabbage *Brassica oleracea*. Some aliens are ancient

Table 4.1 *Annual numbers of alien and native species of plants other than grasses, of those intentionally grown (cultivated) and of those that came in of their own accord (spontaneous introductions) 1975–86*

	Alien	Native	Cultivated	Spontaneous introductions	Total
1975	132	90	163	59	222
1976	130	95	158	67	225
1977	131	99	160	70	230
1978	140	86	169	57	226
1979	133	94	161	66	227
1980	134	96	160	70	230
1981	139	105	168	76	244
1982	149	94	180	63	243
1983	144	99	180	63	243
1984	157	107	197	67	264
1985	158	110	196	72	268
1986	150	106	177	79	256
1975–86	214	146	266	94	360

Photograph 4.1. Dense growth of Buddleia davidii *and marguerites* Chrysanthemum frutescens, *with blackberry* Rubus fruticosus, *teasel* Dipsacus fullonum, *hollyhock* Althaea rosea *and cranesbill* Geranium cinereum.

introductions, such as sycamore *Acer pseudoplatanus*, probably brought from central and southern Europe by the Romans, but such naturalized species are considered as aliens.

A species was scored for every year that it was present, but was scored in terms of the manner of its first appearance, i.e. either cultivation or spontaneous introduction. Many cultivated plants are perennial and so persist for many years, and others, such as borage *Borago officinalis* and pot marigold *Calendula officinalis*, continue because they seed themselves in the garden. Similarly, some plants that have come in of their own accord, such as ash *Fraxinus excelsior* and elder *Sambucus nigra*, persist indefinitely, and others, such as goat's-beard *Tragopogon pratensis* and wood avens *Geum urbanum*, seed themselves year after year. All such plants have been scored in terms of the manner of their first arrival, for every year when they were present. For instance, borage was first cultivated from seed before 1975, and has since self-seeded every year: it has been scored as a cultivated plant every year. On the other hand, some plants that came in of their own accord, such as creeping-Jenny *Lysimachia nummularia*, spotted dead-nettle, feverfew *Tanacetum parthenium* and hogweed *Heracleum sphondylium*, have been encouraged for their value as ground cover or their attractiveness, and persist: they have been scored every year as spontaneous introductions.

The annual total number of plant species tended to rise over the 12-year period, with 222 in 1975 (the lowest) and 256 in 1986 (the highest being 268 in 1985). This was not simply a rise in the number of cultivated species, although this did reach a maximum of 197 in 1984, but also reflects an increase in the number of species that came in of their own accord, which reached a maximum of 79 in 1986. There was little variation in the proportion of alien to native species, and both reached a maximum in 1985 with 158 aliens and 110 natives: the greatest excess of aliens over natives occurred in 1982, when there were 149 aliens and 94 natives. Although the garden undoubtedly contained more plant species in the period 1984–86 than in the 1970s, the trend was neither steady nor particularly marked. The figures given in Table 4.1 are necessarily generalizations that hide appearances and disappearances of particular 'weeds' and decisions to start or stop cultivating particular plants. In other words, the composition of the flora differed considerably in different years, and by no means all the cultivated plants and 'weeds' that were present in 1972 were also present and included in the higher figures for 1986.

The majority of cultivated plants were identified using Gault (1976), Hay & Synge (1971), Nicholson *et al.* (1972), and Mitchell (1974) for trees; scientific names were updated using the extensive library of the University of Leicester Herbarium. Native and naturalized plants were identified using Fitter, Fitter & Blamey (1974) and Keble Martin (1967), but nomenclature, both scientific and vernacular, mostly follows Dony, Jury & Perring (1986), except where scientific names differ from those used in the *Flora of Leicestershire* (Primavesi & Evans 1988), when the latter has been followed.

Cultivated plants

Species of flowering plants, excluding grasses, that have been cultivated or intentionally grown in the garden are listed by family below. There is great variation between species in growth form, in duration of cultivation, and in their contribution to the total quantity of plant material, i.e. plant biomass, in the garden. T = trees, S = shrubs and woody climbers, P = herbaceous perennials, L = low-growing perennials, A = annuals and biennials, B = bulbs, corms and other monocotyledons, and W = plants growing in pond. 1 = present for 12 or more years, 2 = present for 6–11 years, 3 = present for 3–5 years, 4 = present for 1–2 years, * = those that make a significant contribution to the biomass.

Amaryllidaceae

    1. Peruvian lily *Alstroemeria aurea* Graham – B 3
    2. Snowdrop *Galanthus nivalis* L. – B 1
    *3. Daffodil *Narcissus* sp. – B 1

Apocynaceae

    *4. Greater periwinkle *Vinca major* L. – L 1

Aquifoliaceae

    *5. Holly *Ilex aquifolium* L. – T 1

Araliaceae

    *6. Ivy *Hedera helix* L. – L 1

Asclepiadaceae

    7. *Araujia sericifera* Brot. – A 4
    8. Milkweed *Asclepias curassavica* L. – A 4
    9. Milkweed *Asclepias syriaca* L. – A 4

Balsaminaceae

    10. Busy Lizzie *Impatiens balsamina* L. – A 4

Begoniaceae

    11. *Begonia semperflorens* Link & Otto – A 4

Berberidaceae

    12. *Berberis darwinii* Hook. – S 1
    *13. Oregon grape *Mahonia aquifolium* (Pursh) Nutt. – S 1

Betulaceae

  14. Hazel *Coryllus avellana* L. – T 3

Boraginaceae

  15. Alkanet *Anchusa azurea* Mill. – P 3
  16. *Anchusa capensis* Thunb. – A 4
 *17. Borage *Borago officinalis* L. – A 1
 *18. Viper's bugloss *Echium plantagineum* L. – A 2
 *19. Forget-me-not *Myosotis alpestris* F. W. Schmidt – A 1
  20. Lungwort *Pulmonaria officinalis* L. – P 3
 *21. Comfrey *Symphytum* × *uplandicum* Nyman – P 2

Buxaceae

 *22. Box *Buxus sempervirens* L. – S 1

Callitrichaceae

  23. Common water-starwort *Callitriche stagnalis* Scop. – W 4

Campanulaceae

  24. Clustered bellflower *Campanula glomerata* L. – P 3
 *25. Creeping bellflower *Campanula rapunculoides* L. – P 1
  26. *Lobelia erinus* L. – A 4

Caprifoliaceae

  27. Himalayan honeysuckle *Leycesteria formosa* Wall. – S 2
 *28. Honeysuckle *Lonicera periclymenum* L. – S 1
 *29. *Weigela florida* (Bge.) DC. – S 1

Caryophyllaceae

 *30. Border pink *Dianthus* × *allwoodii* Hort. – L 1
 *31. Sweet William *Dianthus barbatus*, L. – A 2
  32. Maiden pink *Dianthus deltoides* L. – L 4
  33. Cheddar pink *Dianthus gratianopolitanus* Vill. – L 2
  34. Pink *Dianthus plumarius* L. – L 1
  35. Baby's breath *Gypsophila elegans* Bieb. – A 4
  36. Campion *Silene coeli-rosa* Godron – A 4

Celastraceae

  37. Spindle *Euonymus europaeus*, L. – S 3

Chenopodiaceae

 *38. White beet *Beta vulgaris* L. – A 2
 *39. Spinach *Spinacia oleracea* L. – A 4

Compositae

*40. Fernleaf yarrow *Achillea filipendulina* Lam. – P 2
41. Tarragon *Artemisia dracunculus,* L. – P 3
42. Michaelmas daisy *Aster amellus* L. – P 1
*43. Michaelmas daisy *Aster novi-belgii* L. – P 1
44. Michaelmas daisy *Aster spectabilis* Hook. & Arn. – P 1
*45. Pot marigold *Calendula officinalis* L. – A 1
46. Aster *Callistephus chinensis* (L.) Nees – A 2
47. Cornflower *Centaurea cyanus* L. – A 3
*48. Perennial cornflower *Centaurea montana* L. – P 1
49. *Chrysanthemum carinatum* Schousboe – A 4
*50. Pyrethrum *Chrysanthemum coccineum* Willd. – P 1
*51. Marguerite *Chrysanthemum frutescens* L. – P 1
52. *Chrysanthemum* sp. – P 2
53. *Cosmos bipinnatus* Cav. – A 4
*54. *Dahlia* sp. – P 2
55. Plantain-leaved leopard's-bane *Doronicum plantagineum* L. – P 1
56. *Echinacea purpurea* (L.) Moench – A 4
57. Globe thistle *Echinops ritro* L. – P 2
*58. Michaelmas daisy *Erigeron* sp. – P 1
59. *Gazania rigens* (L.) Gaertner – A 4
60. Sneezeweed *Helenium autumnale* L. – P 2
61. Sunflower *Helianthus annuus* L. – A 4
62. *Heliopsis helianthoides* (L.) Sweet – P 2
*63. Lettuce *Lactuca sativa* L. – A 1
64. Gayfeather *Liatris* sp. – P 3
*65. Black-eyed Susan *Rudbeckia hirta* L. – A 2
*66. Lavender cotton *Santolina chamaecyparissus* L. – P 1
67. Sea ragwort *Senecio bicolor* (Willd.) Tod. – P 4
68. Oxford ragwort *Senecio squalidus* L. – A 2
*69. Canadian goldenrod *Solidago canadensis* L. – P 1
70. African marigold *Tagetes erecta* L. – A 2
*71. French marigold *Tagetes patula* L. – A 1
72. *Tagetes tenuifolia* Cav. – A 4
*73. Tansy *Tanacetum vulgare* L. – P 2
74. *Zinnia elegans* Jacqu. – A 4

Convolvulaceae

75. Dwarf convolvulus *Convolvulus tricolor* L. – A 2

Cornaceae

*76. *Aucuba japonica* Thunb. – S 1
77. Dogwood *Cornus alba* L. – S 1

Crassulaceae

78. English stonecrop *Sedum anglicum* Huds. – L 3

79. Reflexed stonecrop *Sedum reflexum* L. – L 1
80. Iceplant *Sedum spectabile* Boreau – P 1
81. Caucasian stonecrop *Sedum spurium* Bieb. – L 1

Cruciferae

82. *Alyssum saxatile* L. – L 2
*83. *Arabis caucasica* Willd. – L 1
*84. *Aubrieta deltoidea* DC. – L 1
*85. Cabbage etc. *Brassica oleracea* L. – A 1
86. Wallflower *Cheiranthus cheiri* L. – P 1
87. Sea-kale *Crambe maritima* L. – P 3
*88. Dame's-violet *Hesperis matronalis* L. – P 2
*89. Candytuft *Iberis sempervirens* L. – L 1
90. Candytuft *Iberis umbellata* L. – A 3
*91. Honesty *Lunaria annua* L. – A 2
92. Virginian stock *Malcolmia maritima* (L.) R. Br. – A 4
*93. Brompton stock *Matthiola incana* (L.) R. Br. – A 3
94. Water-cress *Nasturtium officinale* R. Br. – W 4
*95. Radish *Raphanus sativus* L. – A 2

Cucurbitaceae

96. Melon *Cucumis melo* L. – A 4
97. Cucumber *Cucumis sativus* L. – A 4
*98. Marrow *Cucurbita pepo* L. – A 2

Cupressaceae

*99. Lawson's cypress *Chamaecyparis lawsoniana* (A. Murr.) Parl. – T 1

Ericaceae

*100. Heather *Erica tetralix* L. – L 1

Euphorbiaceae

101. Castor oil plant *Ricinus communis* L. – A 4

Fagaceae

102. Sessile oak *Quercus petraea* (Mattuschka) Liebl. – T 4
103. Pedunculate oak *Quercus robur* L. – T 3

Fumariaceae

104. Yellow corydalis *Corydalis lutea* (L.) DC. – P 3
105. *Dicentra formosa* Walp. – P 1

Geraniaceae

*106. *Geranium cinereum* Cav. – L 1
*107. Meadow cranesbill *Geranium pratense* L. – L 3

Grossulariaceae

108. Black currant *Ribes nigrum* L. – S 1
109. Red currant *Ribes rubrum* L. – S 2
*110. Flowering currant *Ribes sanguineum* Pursh. – S 1
*111. Gooseberry *Ribes uva-crispa* L. – S 1

Guttiferae

112. Rose of Sharon *Hypericum calycinum* L. – P 3
113. *Hypericum inodorum* Mill. – S 2

Haloragaceae

114. Spiked water-milfoil *Myriophyllum spicatum* L. – W 3

Hippocastanaceae

*115. Horse-chestnut *Aesculus hippocastanum* L. – T 2

Hydrangeaceae

116. Common hydrangea *Hydrangea macrophylla* (Thunb.) Ser. – S 1

Hydrophyllaceae

117. Baby blue eyes *Nemophila menziesii* Hook. & Arn. – A 4

Iridaceae

*118. Montbretia *Crocosmia crocosmiiflora* N. E. Br. – B 1
*119. *Crocus chrysanthus* (Herbert) Herbert – B 1
120. *Crocus flavus* Weston – B 1
121. *Crocus vernus* (L.) Hill – B 1
*122. Yellow iris *Iris pseudacorus* L. – B 1
123. Bearded iris *Iris variegata* L. – B 1
124. Dutch iris *Iris xiphium* L. × *tingitana* Boiss. & Reuter – B 2

Juglandaceae

125. Walnut *Juglans regia* L. – T 4

Labiatae

126. Large-flowered hemp-nettle *Galeopsis speciosa* Mill. – A 4

*127. Hyssop *Hyssopus officinalis* L. – P 3
 128. Lavender *Lavandula angustifolia* Mill. – P 1
*129. Balm *Melissa officinalis* L. – P 2
*130. Peppermint *Mentha × piperita* L. – P 1
*131. Spear mint *Mentha spicata* L. – P 1
 132. Bergamot *Monarda didyma* L. – P 2
 133. Cat-mint *Nepeta mussinnii* Spreng. – P 1
 134. Sweet basil *Ocimum basilicum* L. – A 3
 135. Bush basil *Ocimum minimum* Hort. = *basilicum* – A 4
*136. Marjoram *Origanum majorana* L. – P 2
 137. Rosemary *Rosmarinus officinalis* L. – P 2
*138. Sage *Salvia officinalis* L. – P 1
 139. Clary *Salvia sclarea* L. – P 4
 140. Scarlet sage *Salvia splendens* Sello – A 4
 141. Lemon thyme *Thymus × citriodorus* (Pers.) Schreber – P 3
*142. Wild thyme *Thymus praecox* Opiz – L 1
 143. Thyme *Thymus vulgaris* L. – P 2

Leguminosae

*144. Broom *Cytisus × kewensis* Bean – S 2
 145. Broom *Cytisus scoparius* (L.) Link – S 1
*146. Sweet pea *Lathyrus odoratus* L. – A 1
 147. Asparagus pea *Lotus tetragonolobus* L. – A 2
 148. Lupin *Lupinus polyphyllus* Lindl. – P 1
*149. Runner bean *Phaseolus coccineus* L. – A 1
 150. French bean *Phaseolus vulgaris* L. – A 4
*151. Pea *Pisum sativum* L. – A 3
 152. Broad bean *Vicia faba* L. – A 4

Lemnaceae

 153. Common duckweed *Lemna minor* L. – W 3

Liliaceae

 154. African lily *Agapanthus africanus* (L.) Hoffmansegg – B 2
 155. Onion *Allium cepa* L. – B 1
 156. Egyptian onion *Allium cepa* var. *viviparum* L. – B 1
 157. Few-flowered leek *Allium paradoxum* (Bieb.) G. Don – B 2
 158. Leek *Allium porrum* L. – B 3
 159. Chives *Allium schoenoprasum* L. – B 2
*160. Lily-of-the-valley *Convallaria majalis* L. – B 1
 161. Bluebell *Hyacinthoides non-scripta* (L.) Rothm. – B 1
 162. Hyacinth *Hyacinthus orientalis* L. – B 1
 163. Red-hot poker *Kniphofia uvaria* (L.) Hook. – B 3
 164. Madonna lily *Lilium candidum* L. – B 1
*165. Grape hyacinth *Muscari armeniacum* Leichtlin ex Baker – B 1
 166. Solomon's-seal *Polygonatum multiflorum* (L.) All. – B 1
 167. Butcher's-broom *Ruscus aculeatus* L. – S 3
 168. *Tulipa* sp. – B 1

Limnanthaceae

  169. Poached-egg plant *Limnanthes douglasii* R. Br. – A 3

Loganiaceae

*170. *Buddleia davidii* Franchet – S 1

Malvaceae

*171. Hollyhock *Althaea rosea* Cav. – P 1
*172. *Lavatera trimestris* L. – A 2
*173. Common mallow *Malva sylvestris* L. – P 2

Nyctaginaceae

  174. Marvel of Peru *Mirabilis jalapa* L. – A 4

Nymphaceae

  175. Water-lily *Nymphaea* sp. – W 3

Oleaceae

*176. *Forsythia intermedia* Zabel – S 1
*177. Garden privet *Ligustrum ovalifolium* Hassk. – S 1
*178. Lilac *Syringa vulgaris* L. – S 1

Onagraceae

  179. *Clarkia amoena* (Lehm.) Nels. & Macbr. – A 4
  180. *Clarkia unguiculata* Lindl. – A 3
  181. *Fuchsia magellanica* Lam. – S 1
*182. Evening primrose *Oenothera biennis* L. – A 1

Papaveraceae

  183. *Hunnemania fumariifolia* Sweet – A 4
  184. Alpine poppy *Papaver alpinum* L. – A 3
*185. Opium poppy *Papaver somniferum* L. – A 2

Passifloraceae

  186. Passion flower *Passiflora caerulea* L. – S 2

Pinaceae

  187. Norway spruce *Picea abies* (L.) Karst. – T 1

Polemoniaceae

188. *Phlox douglasii* Hook. – L 1
*189. *Phlox paniculata* L. – P 1
190. Jacob's-ladder *Polemonium caeruleum* L. – P 3

Polygonaceae

*191. Rhubarb *Rheum rhaponticum* L. – P 1
*192. Sorrel *Rumex acetosa* L. – P 2

Primulaceae

193. Primrose *Primula vulgaris* Hudson – L 1
194. Primula *Primula vulgaris* × *veris* L. – L 1

Ranunculaceae

195. Japanese anemone *Anemone hupehensis* Lemoine – P 2
*196. Columbine *Aquilegia hybrida* Hort. – P 1
*197. Columbine *Aquilegia vulgaris* L. – P 1
198. *Clematis* sp. – S 2
199. Larkspur *Consolida ambigua* (L.) P. W. Ball & Heywood – A 3
*200. *Delphinium* sp. – P 2
201. Winter aconite *Eranthis hyemalis* (L.) Salisb. – P 2
*202. Peony *Paeonia officinalis* L. – P 1
203. Meadow-rue *Thalictrum dipterocarpum* Franchet – P 1

Rosaceae

204. Lady's-mantle *Alchemilla mollis* (Buser) Rothm. – P 3
*205. Goat's beard *Aruncus dioicus* (Walter) Fernald – P 2
206. Japonica *Chaenomeles japonica* (Thinb.) Spach – S 2
*207. Hawthorn *Crataegus monogyna* Jacq. – T 1
*208. Strawberry *Fragaria virginiana* × *chiloensis* (L.) Duchesne – P 1
*209. Apple *Malus sylvestris* Mill. – T 1
*210. Shrubby cinquefoil *Potentilla fruticosa* L. – S 1
*211. Myrobalan plum *Prunus cerasifera* Ehrh. – T 1
212. Almond *Prunus dulcis* (Miller) D. A. Webb – T 4
*213. Japanese cherry *Prunus serratula* Lindl. – T 1
*214. Firethorn *Pyracantha coccinea* Roem. – S 2
*215. *Rosa* spp. – S 1
*216. Blackberry *Rubus fruticosus* L. – S 1
*217. Raspberry *Rubus idaeus* L. – S 1
*218. Loganberry *Rubus loganobaccus* L. H. Bailey – S 1
*219. Rowan *Sorbus aucuparia* L. – T 1
*220. *Spiraea* × *vanhouttei* (Briot) Zabel – S 1

Rubiaceae

221. *Phuopsis stylosa* Benth. & Hook. – P 4

Rutaceae

222. Mexican orange-blossom *Choisya ternata* H.B.K. – S 2
223. Rue *Ruta graveolens* L. – P 3

Salicaceae

*224. Lombardy poplar *Populus nigra* L. var. *italica* – T 1
*225. Crack willow *Salix fragilis* L. – T 1

Saxifragaceae

226. Pigsqueak *Bergenia cordifolia* (Haw.) Sternb. – P 3
*227. *Deutzia scabra* Thunb. – S 1
*228. Mock orange-blossom *Philadelphus coronarius* L. – S 1
229. Pink saxifrage *Saxifraga* sp. – L 1
*230. White saxifrage *Saxifraga* sp. – L 1

Scrophulariaceae

231. Snapdragon *Antirrhinum majus* L. – A 1
*232. Foxglove *Digitalis purpurea* L. – A 1
233. Blood-drop-emlets *Mimulus luteus* L. – P 4
234. *Nemesia strumosa* Benth. – A 3
235. Brooklime *Veronica beccabunga* L. – W 4
236. Spiked speedwell *Veronica spicata* L. – P 1
237. *Veronica* sp. – P 3

Solanaceae

238. Deadly nightshade *Atropa bella-donna* L. – P 3
239. Sweet pepper *Capsicum annuum* L. – A 4
*240. Thorn apple *Datura stramonium* L. – A 2
241. Henbane *Hyoscyamus niger* L. – A 4
*242. Tomato *Lycopersicon esculentum* Mill. – A 1
*243. Flowering tobacco *Nicotiana alata* Link & Otto – A 1
244. *Nicotiana glauca* R.C. Graham – A 4
245. *Petunia hybrida* Vilm. – A 3
246. *Salpiglossis sinuata* Ruiz & Pavón – A 3
247. Chilean potato tree *Solanum crispum* Ruiz & Pavón – S 4
248. Eggplant *Solanum melongena* L. – A 4
249. Potato *Solanum tuberosum* L. – P 1

Tamaricaceae

250. Tamarisk *Tamarix ramosissima* Ledeb. – S 2

Taxaceae

251. Yew *Taxus baccata* L. – T 2

Thymelaeaceae

  252. Mezereon *Daphne mezereum* L. – S 1

Tropaeolaceae

  253. Nasturtium *Tropaeolum majus* L. – A 2

Ulmaceae

  254. Wych elm *Ulmus glabra* Huds. – T 1

Umbelliferae

  255. Chervil *Anthriscus cerefolium* (L.) Hoffm. – A 3
  256. Celery *Apium graveolens* L. – A 3
  257. Carrot *Daucus carota* L. – A 3
  258. Sea holly *Eryngium maritimum* L. – P 3
*259. Fennel *Foeniculum vulgare* Mill. – P 1
*260. Lovage *Levisticum officinale* Koch – P 2
  261. Parsnip *Pastinaca sativa* L. – A 4
  262. Parsley *Petroselinum crispum* (Mill.) A. W. Hill – A 1

Valerianaceae

  263. Valerian *Centranthus ruber* (L.) DC. – P 1

Violaceae

  264. Pansy *Viola hybrida* Hort. – P 1
  265. Pansy *Viola tricolor* L. – P 1

Vitaceae

  266. Grape vine *Vitis vinifera* L. – S 3

The list includes 18 trees, of which 11 were either present in 1971, or
planted soon after, and are still growing: holly, Lawson's cypress, Norway
spruce, hawthorn, apple, Myrobalan plum, Japanese cherry, rowan, willow,
Lombardy poplar and elm. In addition, a horse chestnut was present for 11
years. All these, except Norway spruce and elm, have made a significant
contribution in terms of quantity of plant material, or biomass, to the availability of food for herbivores, and also to the appearance of the garden.
Seedlings of holly, Lawson's cypress and hawthorn are abundant in most
years and are usually removed, although some have been left when they fill a
gap.

  At one time or another, there have been 40 different species of shrubs or
woody climbers of which 26 were present in 1971, or planted soon after, and

are still growing. Nineteen of these make a significant contribution to the plant biomass. Some, such as box, are foliage shrubs; others, such as mock orange, also have conspicuous flowers; and a few, such as gooseberry, fruit prolifically. The climbers include species such as honeysuckle and blackberry, important for foliage, flowers and fruit. Seedlings of Oregon grape, flowering currant, rose and bramble are frequent, as, to a lesser extent, are the seedlings of buddleia; the latter are often nurtured and removed to more suitable positions but the others have to be controlled.

Seventy-six herbaceous perennials have been cultivated at some time, 34 of them for 12 years or more, and 19 of these are significant in terms of biomass. Important species include plants grown for their flowers, such as hollyhocks, and plants with culinary uses, such as mint, rhubarb and strawberries. Fourteen species, such as goat's beard (*Aruncus*) and lovage, grown for less than 12 years, make a significant contribution in terms of biomass. A number of perennials need rigorous control, among them comfrey (which grows from leaf and root fragments), marguerite daisies (which seed freely), tansy and creeping bellflower (which put out suckers) and the common Michaelmas daisy *Aster novi-belgii* (which spreads by both seed and suckers).

Twenty-two low-growing and rockery perennials have been grown, 17 of them for 12 or more years, and ten of these are significant in terms of biomass. They include ground-hugging and non-flowering ivy, which has to be rigorously cut back, and such rockery flowers as *Arabis* and pinks. The cranesbill *Geranium cinereum* is prolific and has to be controlled, and meadow cranesbill, although grown for less than 12 years, is well-established and important.

Annuals and biennials are different from the perennials considered thus far, in that regular presence depends on either self-seeding or the whim of the gardener (decision to re-plant), and many need careful cultivation. Of the 80 annuals and biennials, only 14 have been grown in 12 or more years, whereas 35 were grown in only one or two years. Twenty-five, however, are considered to have contributed significantly to the biomass of the garden, among them self-seeding flowers, such as borage and forget-me-not, vegetables, such as white beet and runner beans, species such as *Lavatera*, repeatedly planted because they do well, and a few planted in less than six years, such as Brompton stock. Those repeatedly planted, apart from vegetables and salads, are grown for their flowers, usually because they are attractive to insects, viper's bugloss, *Lavatera* and French marigolds being in this category. Many of these are grown each year from seed harvested in the garden the previous autumn. Every year, however, it is necessary to remove many of the self-seeded borage, forget-me-nots, evening primroses, thorn apples and pot marigolds.

There are 24 garden plants grouped as bulbs or corms and their mono-cotyledonous allies, such as lily-of-the-valley (although butcher's-broom is listed as a shrub). Seventeen have been present for 12 or more years, and six contribute significantly to the garden biomass. Most are grown for their flowers, but the group also includes onions and leeks. A few scarcely survive,

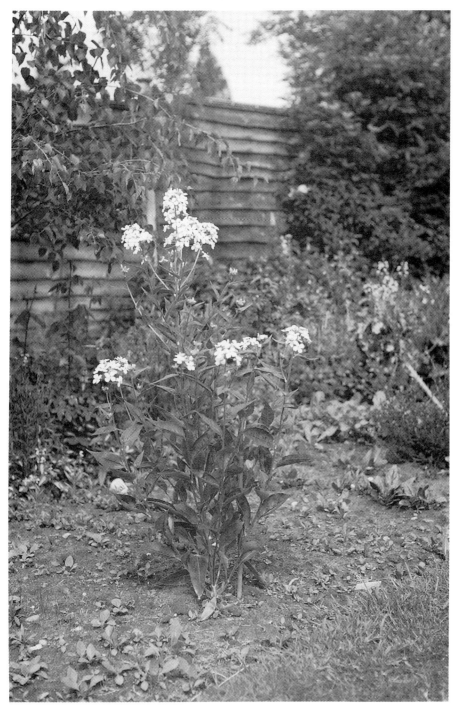

Photograph 4.2 Dame's-violet Hesperis matronalis, *one of the cultivated crucifers used as a larval foodplant by the orange-tip butterfly* Anthocaris cardamines.

such as African lily, which has been present for seven years but has hardly grown and never flowered. Others, such as lily-of-the-valley and grape hyacinths, have to be controlled.

Finally, there are plants which have been introduced to the pond. None, of course, survived the emptying of the pond in 1975, although some were reintroduced in 1984 when the pond was reinstated. The pond is small, and none of the plants is considered important in terms of biomass.

PLANTS THAT COME IN OF THEIR OWN ACCORD

For ease of expression, plants that have come into a garden of their own accord are often called weeds. To most gardeners this is what they are, although in the context of my garden it is a misleading word. A common definition of a weed is 'a plant growing in the wrong place', meaning wrong to man as manager. This is a workable definition for most purposes. A Michaelmas daisy seedling in the middle of the lawn is a weed, just as is grass in the herbaceous border. The natural flora of a garden is usually considered to include all wild or naturalized plants, whether or not these are intentionally cultivated elsewhere. Bellamy (1984) listed 245 native and naturalized plants for Buckingham Palace gardens, the inference being that these have all grown of their own accord, rather than having been planted by the gardeners. Ninety-nine of these, excluding grasses and ferns, occur in my garden, although I grow some, such as buddleia and lily-of-the-valley, intentionally, and few would class them as weeds, whatever the definition. Many, however, such as ivy-leaved speedwell, hedge bindweed and broad-leaved dock, are generally regarded as weeds.

Tolerance or even encouragement of what to most gardeners are weeds, has a long pedigree. A poem written in 1440 includes as garden plants wood avens, daisies, bugle, foxgloves, stinking iris, biting stonecrop and teasel, all of which grow in my garden, and Parkinson, the writer of *Paradisi in sole Paradisus terrestris* (1629), advocated a place for thistles in gardens (Fisher 1982). I encourage, as ground cover, so-called weeds, such as spotted deadnettle and creeping-Jenny, and have left in position until they have flowered many other plants that have come in of their own accord, either because they are attractive or in order that they might be identified. On this basis, a plant is a weed only if I decide I do not want it, and if I am tolerant, there are, by definition, no weeds in my garden. The more sensible distinction seems to be between cultivated plants, i.e. plants intentionally introduced to the garden, and plants that come in of their own accord. The latter are listed by family below. T = trees, S = shrubs, C = climbers, P = herbaceous perennials, L = low-growing perennials, A = annuals and biennials, and B = monocotyledons. 1 = present for 12 or more years, 2 = present for 6–11 years, 3 = present for 3–5 years, 4 = present for 1–2 years. * = those that make a significant contribution to the biomass.

Aceraceae

  267. Sycamore *Acer pseudoplatanus* L. – T 1

Betulaceae

*268. Silver birch *Betula pendula* Roth – T 1

Cannabaceae

*269. Hop *Humulus lupulus* L. – C 2

Caprifoliaceae

*270. Elder *Sambucus nigra* L. – S 1

Caryophyllaceae

  271. Common mouse-ear *Cerastium fontanum* Baumg. – A 3
  272. Little mouse-ear *Cerastium semidecandrum* L. – A 4
  273. Procumbent pearlwort *Sagina procumbens* L. – L 1
*274. Common chickweed *Stellaria media* (L.) Vill. – A 1

Chenopodiaceae

  275. Fat-hen *Chenopodium album* L. – A 1
  276. Fig-leaved goosefoot *Chenopodium ficifolium* Sm. – A 4

Compositae

*277. Yarrow *Achillea millefolium* L. – P 1
  278. Stinking chamomile *Anthemis cotula* L. – A 3
*279. Daisy *Bellis perennis* L. – L 1
  280. Pineappleweed *Chamomilla suavolens* (Pursh) Rydb. – A 2
  281. Creeping thistle *Cirsium arvense* (L.) Scop. – P 2
  282. Spear thistle *Cirsium vulgare* (Savi) Ten. – A 3
  283. Canadian fleabane *Conyza canadensis* (L.) Cronq. – A 1
*284. Buttonweed *Cotula coronopifolia* L. – A 1
  285. Smooth hawk's-beard *Crepis capillaris* (L.) Wallr. – A 2
  286. Scentless mayweed *Matricaria perforata* Mérat – A 3
*287. Fox-and-cubs *Pilosella aurantiaca* (L.) C. H. & F. W. Schultz – P 1
*288. Common ragwort *Senecio jacobaea* L. – A 3
  289. Heath groundsel *Senecio sylvaticus* L. – A 3
  290. Groundsel *Senecio vulgaris* L. – A 1
  291. Perennial sow-thistle *Sonchus arvensis* L. – P 4
*292. Prickly sow-thistle *Sonchus asper* (L.) Hill – A 1
*293. Smooth sow-thistle *Sonchus oleraceus* L. – A 1
*294. Feverfew *Tanacetum parthenium* (L.) Schultz Bip. – P 1
*295. Dandelion *Taraxacum officinale* Weber – P 1

296. Goat's-beard *Tragopogon pratensis* L. – A 2
*297. Colt's-foot *Tussilago farfara* L. – P 2

Convolvulaceae

*298. Hedge bindweed *Calystegia sepium* (L.) R. Br. – C 1
299. Field bindweed *Convolvulus arvensis* L. – C 4

Crassulaceae

300. Biting stonecrop *Sedum acre* L. – L 3
*301. White stonecrop *Sedum album* L. – L 1

Cruciferae

302. Garlic mustard *Alliaria petiolata* (Bieb.) Cavara & Grande – A 3
303. Shepherd's-purse *Capsella bursa-pastoris* (L.) Medicus – A 2
304. Hairy bitter-cress *Cardamine hirsuta* L. – L 2
305. Sweet Alison *Lobularia maritima* (L.) Desv. – A 3
306. Charlock *Sinapsis arvensis* L. – A 4
307. Hedge mustard *Sisymbrium officinale* (L.) Scop. – A 4

Dipsacaceae

*308. Teasel *Dipsacus fullonum* L. – P 1

Euphorbiaceae

309. Petty spurge *Euphorbia peplus* L. – A 1

Guttiferae

310. Tutsan *Hypericum androsaemum* L. – S 1

Iridaceae

311. Stinking iris *Iris foetidissima* L. – B 2

Juncaceae

312. Field wood-rush *Luzula campestris* (L.) DC. – L 1

Labiatae

313. Bugle *Ajuga reptans* L. – P 4
314. Henbit dead-nettle *Lamium amplexicaule* L. – A 3
*315. Spotted dead-nettle *Lamium maculatum* L. – L 1
316. Red dead-nettle *Lamium purpureum* L. – A 1
317. Selfheal *Prunella vulgaris* L. – L 2
*318. Hedge woundwort *Stachys sylvatica* L. – P 2

Leguminosae

  319. Common bird's-foot-trefoil *Lotus corniculatus* L. – L 1
  320. Black medick *Medicago lupulina* L. – A 3
  321. Hop trefoil *Trifolium campestre* Schreber – A 2
*322. White clover *Trifolium repens* L. – L 1
  323. Hairy tare *Vicia hirsuta* (L.) S.F. Gray – A 4
  324. Common vetch *Vicia sativa* L. – A 4

Liliaceae

  325. *Asparagus officinalis* L. – B 1

Oleaceae

*326. Ash *Fraxinus excelsior* L. – T 1
  327. Wild privet *Ligustrum vulgare* L. – S 2

Onagraceae

  328. American willowherb *Epilobium adenocaulon* Hausskn. – P 1
*329. Rosebay willowherb *Epilobium angustifolium* L. – P 1
*330. Great willowherb *Epilobium hirsutum* L. – P 1

Oxalidaceae

*331. Procumbent yellow-sorrel *Oxalis corniculata* L. – L 1
*332. Large-flowered pink-sorrel *Oxalis corymbosa* DC. – L 1

Papaveraceae

  333. Long-headed poppy *Papaver dubium* L. – A 4
  334. Common poppy *Papaver rhoeas* L. – A 4

Plantaginaceae

  335. Greater plantain *Plantago major* L. – L 1

Polygonaceae

  336. Black-bindweed *Bilderdykia convolvulus* (L.) Dumort – C 2
  337. Knotgrass *Polygonum aviculare* L. – A 2
  338. Redshank *Polygonum persicaria* L. – A 2
  339. Broad-leaved dock *Rumex obtusifolius* L. – P 2

Primulaceae

  340. Scarlet pimpernel *Anagallis arvensis* L. – A 3
*341. Creeping-Jenny *Lysimachia nummularia* L. – L 1

Ranunculaceae

*342. Creeping buttercup *Ranunculus repens* L. – P 1

Rosaceae

343. *Cotoneaster* sp. – S 2
*344. Wild strawberry *Fragaria vesca* L. – L 1
*345. Wood avens *Geum urbanum* L. – P 2
*346. Dog-rose *Rosa canina* L. – C 1
*347. Cut-leaved blackberry *Rubus laciniatus* Willd. – C 1

Rubiaceae

*348. Cleavers *Galium aparine* L. – A 1

Salicaceae

349. Sallow *Salix* sp. – S 2

Scrophulariaceae

350. Purple toadflax *Linaria purpurea* (L.) Mill. – P 3
351. Green field-speedwell *Veronica agestris* L. – A 4
*352. Ivy-leaved speedwell *Veronica hederifolia* L. – A 1
353. Thyme-leaved speedwell *Veronica serpyllifolia* L. – L 1

Solanaceae

*354. Bittersweet *Solanum dulcamara* L. – C 1
355. Black nightshade *Solanum nigrum* L. – A 2

Umbelliferae

*356. Hogweed *Heracleum sphondylium* L. – P 1

Urticaceae

357. Mind-your-own-business *Soleirolia soleirolii* (Req.) Dandy – L 2
*358. Common nettle *Urtica dioica* L. – P 1

Violaceae

359. Sweet violet *Viola odorata* L. – L 4
360. Common dog-violet *Viola riviniana* Reichenb. – L 2

The list includes three tree species, of which silver birch and ash not only grow large and so contribute significantly to plant biomass, but also seed all over the garden. There is only one sycamore, which is repeatedly cut back to ground level but continues to shoot. Of five shrubs, elder and tutsan have been present for more than 12 years, and elder, which seeds itself freely, is an important part of the biomass. Two plants of *Cotoneaster* sp., which appeared as seedlings in cracks between paving slabs, have been moved to better positions and encouraged. Of seven climbers, four have been present for 12 years or more and contribute significantly to plant biomass, as does hop. Hop, which grows in nearby gardens, first appeared in 1980 and soon had to be controlled. Hedge bindweed and, to a lesser extent, bittersweet are dominant weeds; the former is tolerated for its trumpet-like flowers and the latter for its brilliantly red, shiny berries, but both have to be controlled, the bindweed ruthlessly, lest it get completely out of hand. Some absences are more curious than presences; for instance, field bindweed was present in the back garden in the early 1970s, and then did not reappear until 1985, when a plant grew in the front garden.

Nineteen herbaceous perennials have grown of their own accord, 11 of them for 12 or more years. Ten of these are significant in terms of biomass, as are a further three species that have occurred in more than six but less than 12 years. Of these three, colt's-foot, an early successional species, was well-established in the herbaceous border in 1971 but died out in 1977, perhaps because dense soil cover and accumulation of dead plant material created conditions in which it was a poor competitor; hedge woundwort and wood avens, which appeared in 1976, were at first encouraged as interesting additions to the flora, but have since become aggressive weeds. It is curious that the pervasive creeping thistle and broad-leaved dock were not seen until 1979.

Twenty low-growing perennials contribute to the ground cover; 13 have been present for 12 or more years, and eight of these are significant in terms of biomass. Creeping-Jenny and spotted dead-nettle, with their lush growth and bright flowers, are two of the dominant and most conspicuous plants in the garden, and although both are to some extent controlled, it is impossible to think of them as weeds. Biting stonecrop is represented by one plant, which flourished for a number of years in a crack beside the house wall in the back yard; a clump of mind-your-own-business, a common accidental introduction to greenhouses, appeared in a crack between two paving slabs during the drought summer of 1976, and grew vigorously until killed by the cold winter of 1984/85. Selfheal first became apparent after one of the front lawns was dug over, common dog-violet appeared in 1980 in the middle of a vegetable patch and sweet violet at a path edge.

Not unexpectedly, the largest group of plants that have come in of their own accord, is that of annuals and biennials. Thirty-nine have appeared at some time since 1975, and 12, seven of which are considered to contribute significantly to plant biomass, have been present in 12 or more years. The most abundant of these are common chickweed, ivy-leaved speedwell and buttonweed, the latter forming a major component of the lawn. The appearance of some species can be linked with changes in land use. Thus, when part of the back lawn was dug up, stinking chamomile, scentless

Photograph 4.3. Self-sown feverfew Tanacetum parthenium *with* Begonia semperflorens, *sweet Williams* Dianthus barbatus, *the tall leaves of yellow iris* Iris pseudacorus, *old flower spikes of foxgloves* Digitalis purpurea *and many other plants.*

mayweed, pineappleweed, knotgrass, redshank and black-bindweed appeared in the newly created vegetable bed, which was sandy and lacking in humus. The soil has improved in texture and fertility with repeated cultivation, and most of these plants no longer grow there, although some have since appeared elsewhere in the garden. Long-headed poppy is a persistent weed, but common poppy has only been recorded twice. The sporadic and infrequent appearance of the larger crucifers, such as charlock, hedge mustard and garlic mustard, reflects the fact that they do not grow in any abundance in the neighbourhood. Finally, there are just two monocotyledons, asparagus, which has been present since 1971, and stinking iris, which first appeared in 1979 and is now well-established in the garden.

A number of plants that have apparently come in of their own accord are lawn 'weeds' that contribute variety to the mown turf. All are adapted to the grazing action of the lawn-mower, in that they have a low, creeping or rosette growth form. Some, such as field wood-rush, bird's-foot-trefoil, hop trefoil and buttonweed, occur nowhere else in the garden, although yarrow, selfheal, white clover, daisies, greater plantain and dandelion are present in paths, flower beds and vegetable patches, where they grow much taller.

Of the 94 plant species that have come into the garden of their own accord, 91 appear in the *Flora of Leicestershire* (Primavesi & Evans 1988); Leicestershire in this context is the pre-1974 county, excluding Rutland. Of the other three, buttonweed and large-flowered pink-sorrel, although aliens, are common as garden weeds, and *Cotoneaster* spp. are regularly cultivated. Thirteen species, including silver birch and stinking iris, are sometimes or often cultivated and hence are either probably or certainly escapes from neighbouring gardens. Mind-your-own-business is regarded as a frequent escape from gardens and greenhouses although only recorded in four tetrads (2 km × 2 km squares); 617 tetrads lie wholly or partly within the Leicestershire county boundary. Little mouse-ear (eight tetrads), heath groundsel (46 tetrads) and bugle (246 tetrads) are not typical garden weeds, and may well have been brought into the garden from elsewhere as seeds caught in mud on boots. Fig-leaved goosefoot and stinking chamomile are regarded as occasional intrusives (i.e. not constituents of a plant community) and have been recorded in 13 and 82 tetrads respectively. It is important to recognize that a record for a tetrad can equally well be one plant or many thousands.

The majority of species that have come into the garden of their own accord, i.e. 94 species, are either normal constituents of the local plant community, or frequent intrusives particularly on land disturbed or altered by man, or common garden weeds; most have been recorded in the majority of tetrads in the county. The more interesting garden occurrences in these categories are those species recorded in relatively few tetrads. Canadian fleabane (36 tetrads) mostly occurs in the vicinity of large towns, fox-and-cubs (19 tetrads) is only found on land disturbed or altered by man, and black nightshade (84 tetrads) is locally frequent in or near large towns. Three further species do not appear to be particularly widespread in the county, and probably form part of the local plant community: henbit dead-nettle is recorded for 46 tetrads, wild strawberry for 118 tetrads, and green field-speedwell for 83 tetrads. There are taxonomic difficulties with wild roses and many hybrids; the one growing in the garden is assumed to be dog-rose, but no attempt has been made to identify the intra-specific group and so county distribution cannot be cited from the *Flora of Leicestershire*. The situation is similar, although more extreme, for dandelions; there are now considered to be nearly 180 species in Britain, over half of which occur in Leicestershire, but no attempt has been made to identify the garden dandelions, which are simply listed as *Taraxacum officinale*.

ORIGINS OF GARDEN FLORA

About two-fifths of the garden's flora, excluding grasses, consists of native plants, the other 214 species originating in many different parts of the world. Some are ancient introductions, the Romans, for instance, being credited with bringing fennel, opium poppy and other aliens. Hollyhocks were brought back by the Crusaders, and garden pinks were imported with building stone by the Normans. A list of plants growing in England, published in 1548, included French and African marigolds; John Gerard's catalogue of 1596 included mock orange, pot marigolds and grape hyacinths; and among the plants growing in John Tradescant's garden in 1634 were goldenrod and the North American milkweed *Asclepias syriaca* (Fisher 1982). Other widely cultivated plants are more recent discoveries and introductions: Charles Darwin first collected *Berberis darwinii* in 1835 on the island of Chiloe off the coast of Chile (Bean 1976), and *Buddleia davidii* from China was first cultivated in Britain in 1890 (Owen, D.F. & Whiteway 1980).

The region best represented in the garden's alien flora is southern Europe, with 40 species, including many aromatic herbs such as rosemary and sage. Thirty-one species, including the common white *Arabis*, pyrethrum, spinach, peas and other crops come from southeastern Europe and Asia Minor. North America is well represented, with 31 species, such as flowering currant and many late-flowering species including Michaelmas daisies and black-eyed Susan, which are meadow plants in the wild. Twenty species, including lungwort and dame's-violet, come from central and northern parts of continental Europe. Seventeen, such as runner beans and scarlet sage, are from South America, and ten, including African and French marigolds, from Central America. Seven, including iceplant, come from China, five, including the evergreen *Aucuba japonica*, from Japan, and a further seven, such as asters, from both countries. Six, including buttonweed and *Lobelia erinus*, come from South Africa, melons and the castor oil plant come from tropical Africa, and *Chrysanthemum carinatum* comes from North Africa. Five, including cucumber and basil, come from tropical Asia, and three, including tarragon and pineappleweed, from northeastern Asia. There are marguerite daisies from the Canary Islands, a shrubby *Hypericum* from Madeira, the marvel of Peru from the West Indies, and Himalayan honeysuckle. The origins of a further three species are not known, and finally 22, such as the border pink, are hybrids of garden origin.

Grasses

In 1975, 1976 and 1977, an effort was made to identify all the grasses in the garden. To this end, those growing in flower and vegetable beds were left to flower, and areas of lawn were left uncut. The majority of the species recorded were probably introduced to the garden as lawn grasses. In all, 24 species were identified between 1975 and 1977, most of them native and four aliens. They are listed below. Hard fescue from continental Europe is a common lawn grass, and oats, barley and wheat, derived from species found in southwest Asia, grew from discarded guinea-pig food. Hubbard (1974) was used for identification, and nomenclature follows Dony, Jury & Perring (1986).

361. Velvet bent *Agrostis canina* L.
362. Common bent *Agrostis capillaris* L.
363. Black bent *Agrostis gigantea* Roth
364. Creeping bent *Agrostis stolonifera* L.
365. Sweet vernal-grass *Anthoxanthum odoratum* L.
366. Oat *Avena sativa* L.
367. Upright brome *Bromus erectus* Hudson
368. Barren brome *Bromus sterilis* L.

369. Crested dog's-tail *Cynosurus cristatus* L.
370. Cock's-foot *Dactylis glomerata* L.
371. Chewings fescue *Festuca nigrescens* Lam.
372. Hard fescue *Festuca trachyphylla* (Hackel) Krajina
373. Yorkshire fog *Holcus lanatus* L.
374. Creeping soft-grass *Holcus mollis* L.
375. Barley *Hordeum distichon* L.
376. Meadow barley *Hordeum secalinum* Schreber
377. Perennial rye-grass *Lolium perenne* L.
378. Smaller cat's-tail *Phleum pratense* L.
379. Annual meadow-grass *Poa annua* L.
380. Wood meadow-grass *Poa nemoralis* L.
381. Smooth meadow-grass *Poa pratensis* L.
382. Rough meadow-grass *Poa trivialis* L.
383. Yellow oat-grass *Trisetum flavescens* (L.) Beauv.
384. Wheat *Triticum aestivum* L.

FLOWERLESS PLANTS

The common green alga *Desmococcus olivaceus* (Pers. ex Ach.) Laundon (usually called *Pleurococcus vulgaris*), no. 385 on the plant list, grows on tree trunks, especially of holly, and an unidentified lichen (no. 386) is common on stone paving slabs. The liverwort, *Marchantia polymorpha* L. (no. 387), is abundant in and around the kitchen drain, and male fern *Dryopteris filix-mas* (L.) (no. 388) is well established in the back rockery and is increasing among the crazy-paving to the east of the house. There is a variety of different sorts of mosses on damp paving slabs and in the lawn. Thus far, nine have been identified and are listed below.

389. *Bryum capillare* Hedw.
390. *Grimmia trichophylla* Grev.
391. *Hypnum cupressiforme* Hedw. var. *resupinatum*
392. *Leptobryum pyriforme* (Hedw.) Wils.
393. *Lophocolea bidentata* (L.) Dum.
394. *Mnium hornum* Hedw.
395. *Rhynchostegium confertum* (Dicks.) Br. Eur.
396. *Rhytidiadelphus squarrosus* (Hedw.) Warnst.
397. *Tortula muralis* Hedw.

The most abundant flowerless plants are Fungi, and an effort has been made, with varying degrees of success, to identify all those that have been found. The majority have appeared in the lawn, several of them beneath a large birch tree, and it is assumed that they are associated with its roots. Thus far, 25 species have been identified, the majority of them in autumn. The large and varied list (see below) may well be a reflection of former use of the site as a mixed plantation. Kibby (1979), Lange & Hora (1981) and Phillips (1981) have been used for identification.

398. *Clavulinopsis luteo-alba* (Rea) Corner

Photograph 4.4. Shaggy parasol Lepiota rhacodes *growing in litter beneath Lawson's cypresses* Chamaecyparis lawsoniana *in June. This is one of 25 species of toadstools that have been identified in the Leicester garden.*

399. *Coriolus versicolor* (L. ex Fr.) Quél.
400. *Cortinarius evernius* Fr.
401. *Cortinarius violaceus* (L. ex Fr.) Fr.
402. *Hebeloma crustiliniforme* (Bull. ex St. Amans) Quél.
403. *Hebeloma mesophaeum* (Pers.) Quél.
404. *Hygrocybe nivea* (Scop.) Fr.
405. *Inocybe napipes* Lange
406. *Inocybe tigrina* Heim
407. *Laccaria laccata* (Scop. ex Fr.) Cke
408. *Lactarius pubescens* (Fr. ex Krombh.) Fr.
409. *Lactarius torminosus* (Schaeff. ex Fr.) S. F. Gray
410. *Lepiota rhacodes* (Vitt.) Quél.
411. *Lepista nuda* (Bull. ex Fr.) Cooke
412. *Lyophyllum decastes* (Fr. ex Fr.) Sing.
413. *Mycaena fibula* (Bull. ex Fr.) Kuhner

414. *Mycaena flavo-alba* (Fr.) Quél.
415. *Mycaena* sp.
416. *Paxillus involutus* (Fr.) Fr.
417. *Peziza vesiculosa* Bull. ex St. Amans
418. *Pluteus* sp.
419. *Psathyrella candolleana* (Fr.) Maire
420. *Psilocybe semilanceata* (Fr. ex Secr.) Kummer
421. *Tarzetta cupularis* (L. ex Fr.) Lamb.
422. *Thelephora terrestris* (Ehrh.) Fr.

SEASONS

The quantity of vegetation in the garden varies considerably during the year, with the abundance of lush, green growth usually reaching a peak in late June and early July. However, there is green vegetation present throughout the year, not only evergreens, such as holly, but also perennials, such as spotted dead-nettle and foxglove. In mild winters, privet retains its leaves and many plants continue growing, albeit slowly, although in hard winters, privet leaves shrivel and herbaceous plants and buddleia bushes die.

An amazing number of plants may be in flower in the autumn or even during winter. The reason that gardens are so colourful in late summer and autumn is the preponderance of North American meadow plants, such as Michaelmas daisy, black-eyed Susan and goldenrod, which, in Britain's cooler summers, flower later than in their country of origin. These flowers, together with those of iceplant, attract many flower-feeding insects into gardens at a time of year when there are few flowers in hedgerow, woodland and meadow. One October day in 1979 there were 27 plants in flower: African marigolds, asters, aubrieta, borage, buddleia, dwarf convolvulus, evening primrose, fennel, feverfew, flowering tobacco, foxglove, fuchsia, goldenrod, honeysuckle, iceplant, larkspur, lavender, marguerite daisies, Michaelmas daisies, montbretia, pansy, perennial cornflower, roses, snapdragon, spiked speedwell, spotted dead-nettle and sweet pea. But in the particularly mild autumn of 1978, even more plants were in flower on 23 November: African marigolds, asparagus pea, *Berberis darwinii*, black-eyed Susan, blackberry, cabbage, dwarf convolvulus, fennel, feverfew, flowering tobacco, French marigolds, fuchsia, *Geranium cinereum*, hydrangea, lavender cotton, lobelia, marguerite daisies, mezereon, Oxford ragwort, pansy, perennial cornflower, primrose, rose, sage, scarlet sage, smooth sow-thistle, spiked speedwell, spotted dead-nettle, sweet pea and wallflower. And, in the exceptionally mild Christmas of 1974, chrysanthemums, groundsel, lavender cotton, marguerite daisies, mezereon, petty spurge, pot marigold, primulas, pyrethrum, roses, spotted dead-nettle and wallflower were in flower on 25 December. The impression is that in a garden, despite the seasonal progression of leaf growth, flowering, fruiting and dormancy, leaves and flowers are available as food throughout the year. This undoubtedly has an effect on animal life, particularly insects.

The flora of my garden is rich and varied, with up to 268 species of flowering plants, other than grasses, recorded in any one year, and an average of 240 a year. In addition, there are mosses, many Fungi, and other flowerless plants, as well as an unexpected variety of grasses. More than 73 per cent of the 360 species of flowering plants (other than grasses) growing during the 12-year period 1975–86 were intentionally grown, the others came in of their own accord. Nearly 60 per cent are of alien origin, from many different parts of the world, the rest being native to the British Isles. The composition of the flora to some extent changed from year to year – that is the essence of gardening, but in terms of the quantity of vegetation and in its extraordinary variety, the flora consistently provided a broad, firm base to support, directly or indirectly, the many animals that are the subject of the following chapters.

5

Butterflies

Any garden, and certainly my garden, is dominated in terms of wildlife by two groups of animals, insects and birds, and among insects, the most familiar, most welcome and best known group are undoubtedly butterflies. They are conspicuous, readily identified and attractive, and consequently are popularly regarded as indicating the health of a habitat. This is quite a reasonable view, since butterflies are typical insects and, as they have herbivorous larvae, are directly dependent on the vegetation. That most people recognize and remark on butterflies is undoubtedly because they are large relative to most British insects and are active by day. Indeed, many people seem to accord butterflies the status of honorary birds, thus admitting them to the 'fur and feathers' category of animal, which is all that some people recognize as wildlife.

Butterflies have been the object of interest and investigation for a long time, the first account of British butterflies by Christopher Merrett having been written in 1666, and there are good records of their past distribution and status, particularly for the last 150 years (Heath, Pollard & Thomas 1984). Interest has taken a new direction in the last couple of decades with the fear that many British butterflies are on the decline and, since 1976, the Institute of Terrestrial Ecology has been organizing a national scheme for monitoring the abundance of butterflies, in the hope of detecting trends which may affect their status (Pollard, Hall & Bibby 1986).

Butterflies, together with moths, form the insect order Lepidoptera, characterized by the presence of scales on the wings, body and legs. The mouthparts almost always form a long, suctorial proboscis, coiled when not in use, and the larval form is a caterpillar, most of which are herbivorous. Two superfamilies, Hesperioidea (skippers) and Papilionoidea are popularly known as butterflies, but, although there is confident consensus about recognition of a butterfly as such, there is no single, absolutely trustworthy taxonomic character by which butterflies can be separated from moths. Most, but not all, moths have one or more bristles, the frenulum, on the hindwing which locks into the forewing, whereas in all butterflies the expanded edge of the hindwing holds the wings together. Butterflies have a club-like swelling at or near the tip of thread-like antennae; moths have antennae of various forms, but all those that have antennae with club-like swellings at the tip also have a frenulum. Most moths rest with the wings folded flat over the back, whereas butterflies tend to rest with the wings held upright, but there are exceptions to both. Moths are typically drab in colour and nocturnal or crepuscular, whereas butterflies are usually diurnal and tend to be brightly coloured. There are, however, brilliantly coloured diurnal moths, and dull coloured butterflies that fly at dusk, in overcast weather and in the shade. Many moths weave a silken cocoon in which they pupate, whereas butterfly pupae are naked and often brightly ornamented. However, moths that pupate underground do not usually form a cocoon, and skippers regularly make a loose cocoon. Skippers are indeed rather moth-like and different from typical butterflies, as is recognized by their classification in a separate superfamily.

THE GARDEN SPECIES OF BUTTERFLIES: THEIR NATURAL HISTORY

Apart from observation, butterflies have been monitored in the garden by two methods: hand-netting, marking and releasing from 1972 to 1979 inclusive, and Malaise trapping from 1972 onwards (see Chapter 3). In all, 21 species have been recorded. The frequencies with which each of the species was hand-netted or caught in the Malaise trap during the eight-year period 1972–79 are shown in Table 5.1 (p. 93).

Butterflies are better known than any other group of insects. There are a large number of books detailing the life histories and habits of British butterflies so, rather than repeating much easily obtainable information, I shall briefly describe only the species that have been recorded in my garden. Less than half of these are regularly and particularly associated with gardens: brimstone, large white, small white, green-veined white, orange tip, red admiral, painted lady, small tortoiseshell and peacock. The others are infrequent, occasional or rare visitors, but describing all 21 gives reasonable coverage of the range of life styles in the group.

Hesperiidae

1. The small skipper *Thymelicus sylvestris* (Poda) is a butterfly of rough grasslands, on the wing in July and August, and has one generation a year. The caterpillars feed mainly on Yorkshire fog *Holcus lanatus* and also on other grasses, and hibernate when newly hatched. Pupation is in a silken cocoon, covered with grass leaves, at the base of the foodplant.
2. The large skipper *Ochlodes venata* (Turati) is found on grassland, on grassy rides in woods, along hedgerows and in similar sites. There is one generation a year, adults flying from June to August. Caterpillars feed mainly on cock's-foot *Dactylis glomerata* and on other grasses. They hibernate as larvae, complete their growth the following spring, and pupate in cocoons spun in 'tents' of grass blades.

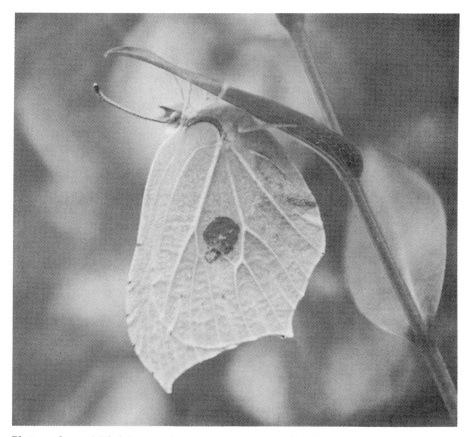

Photograph 5.1. *Male brimstone* Gonepteryx rhamni, *which has been marked on the underside of the hindwing with a dot of waterproof ink from a felt-tip marker.*

Pieridae

3. The brimstone *Gonepteryx rhamni* (L.) is on the wing from March until September or October; they fly in woods, scrub and wetlands, as well as in built-up areas. They hibernate as adults, which is why they are seen so early and late in the year, but tend to be scarce in June and July when the single generation is at the caterpillar and pupal stages. In the winter of 1976/77, I found two male brimstones hibernating in a bramble patch in an abandoned railway cutting not far from the garden. On a cold day in January, the butterflies' wings were covered with hoar frost and larger beads of ice, and at a distance they were hard to distinguish from the leaves that still remained on the plant; they were still in the same position in March. The caterpillars are restricted to alder buckthorn *Frangula alnus* and purging buckthorn *Rhamnus catharticus*, but pupate among low-growing vegetation. Brimstones are mobile butterflies, and make local movements from egg-laying to feeding and wintering sites.

4. The large white *Pieris brassicae* (L.) is a widespread and highly mobile butterfly particularly associated with gardens and agricultural land where *Brassica* spp.,

the caterpillar foodplants, are cultivated. On the wing from April to October, two or, in some years, three generations are completed. Large whites pass the winter as pupae, formed on tree trunks or walls. It has long been believed that they are migratory, and that the British population is boosted each year by immigrants from continental Europe, but the butterfly monitoring scheme has produced no evidence of this (Pollard *et al.* 1986).

5. Small whites *Pieris rapae* (L.) are also widespread, mobile and particularly associated with gardens and fields where brassicas, a common larval foodplant, are cultivated, but caterpillars also feed on other crucifers, such as garlic mustard *Alliaria petiolata* and hedge mustard *Sisymbrium officinale*. They fly from April to October, hibernating as pupae. There are two or sometimes three generations a year, the earlier ones tending to pupate on the foodplant, the later ones on tree trunks and walls. Migratory flights and directional movements of small whites are often reported, but there is no consensus as to whether or not there is regular, directional movement.

6. The green-veined white *Pieris napi* Verity is another widespread and mobile butterfly. It is on the wing from April to October, passing the winter as a pupa. The caterpillars feed on cuckooflower *Cardamine pratensis*, on garlic and hedge mustard, and on other crucifers, and pupate on nearby vegetation. There are two or three generations a year, but some of the first-generation pupae persist until the following spring before emergence of the adult.

7. The orange tip *Anthocaris cardamines* Verity is widespread, particularly in open woods and lanes, and is becoming increasingly associated with gardens. Adults are on the wing in May and June, and there is only one generation a year, hibernation being as a pupa. Caterpillars feed on the flowers and seed pods of cuckooflower, garlic mustard and other hedgerow crucifers, and there are many reports of them using garden arabis *Arabis caucasica* and dame's-violet *Hesperis matronalis*. The angular, well-camouflaged pupae are formed on the foodplant or on nearby plant stems.

Lycaenidae

8. The white-letter hairstreak *Strymonidia w-album* (Knoch) is usually very local around elms *Ulmus* spp., on which the caterpillars feed. Adults fly in July and August and there is a single generation. They hibernate as eggs, and the larvae feed first on elm flowers, later on the leaves, and pupate on the elm twigs.

9. Small coppers *Lycaena phlaeas* (Fab.) are found from May to October in open situations in woods or along hedgerows and in similar sites. Caterpillars feed on common sorrel *Rumex acetosa* and sheep's sorrel *R. acetosella*, and pupate among low vegetation. There are two or, sometimes, three generations a year, hibernation being as larvae.

10. The common blue *Polyommatus icarus* (Rottemburg) flies on open grassland and in large woodland glades from May to September. There are two generations a year, and few adults are seen in July. Caterpillars, the hibernating stage, feed on bird's-foot-trefoil *Lotus corniculatus* and pupate on the ground.

Nymphalidae

11. Red admirals *Vanessa atalanta* (L.) are widespread, flying in a range of habitats from April to October. Some individuals hibernate as adults, and these may be

seen on the wing in early April, but most red admirals seen early in the summer are migrants from the south. The British butterflies are part of a widespread population extending from northern Europe to the Mediterranean and North Africa, which shifts northwards in spring and early summer and retreats southwards in late summer and autumn. Two generations are passed in Britain, the caterpillars feeding on common nettle *Urtica dioica* and spinning the leaves together as a shelter for pupation.

12.   Painted ladies *Cynthia cardui* (L.) are similarly widespread, and may be seen in a variety of habitats from May to September. Unlike red admirals, however, they seem unable to survive the British winter in any form, and their appearance in early summer is dependent on immigration. They, too, are part of a population extending to North Africa, which ebbs and flows, southwards in autumn and northwards in early summer. The abundance of painted ladies in Britain varies greatly from year to year, depending on breeding conditions to the south. One or two generations are produced in the British summer, using thistles, *Cirsium* spp. and *Carduus* spp., as larval foodplants, and pupating on the foodplants or nearby.

13.   The small tortoiseshell *Aglais urticae* (L.), a butterfly widespread in open situations, from March to October, hibernates as an adult. There are two generations a year, the caterpillars feeding on common nettle and pupating on the foodplant, on tree trunks or on walls.

14.   Peacocks *Inachis io* (L.), though mobile and widespread, are predominantly woodland butterflies. They fly from March until October, and hibernate as adults. One generation a year is produced using common nettle as a foodplant and pupating on tree trunks.

15.   Commas *Polygonia c-album* (L.) also use common nettle as a caterpillar foodplant, although they sometimes feed on elm. They fly from March to October in woodland, among shady scrub and along hedgerows. Pupation is on the foodplant, and they hibernate as adults. Although there are two generations a year, not all adults produced from the first generation of larvae breed immediately but instead go into early hibernation until the following year.

16.   The silver-washed fritillary *Argynnis paphia* (L.) is a woodland butterfly that flies in July and August. Caterpillars feed on violets *Viola* spp. and pupate on any convenient object. There is one generation a year, overwintering as larvae.

Satyridae

17.   Wall butterflies *Lasiommata megera* (L.) fly from May to September on unimproved grassland, on roadsides and along woodland edge. There are two, sometimes three, generations a year, and adults are scarce in July. Caterpillars feed on tor-grass *Brachypodium pinnatum*, false brome *B. sylvaticum*, cock's-foot and other grasses, and pupate on grass stems. They hibernate as larvae.

18.   The marbled white *Melanargia galathea* Verity flies on rough grassland from June to August. Caterpillars feed on sheep's-fescue *Festuca ovina*, tor-grass and other grasses, and pupate on the ground at the base of the foodplant. There is one generation a year which, as in other satyrids (browns), hibernate as larvae.

19.   Hedge browns *Pyronia tithonus* Verity are on the wing from July to September along hedges, lanes and woodland edge and among scrub. Various fine-leaved

grasses are used as larval foodplants, and pupae are formed on grass stems. There is one generation a year, hibernating as larvae.

20.   Meadow browns *Maniola jurtina* (L.) fly from June to September on grasslands and in other grassy areas. They, too, feed as caterpillars on various fine-leaved grasses and pupate on grass stems. The single generation pass the winter as larvae.

21.   The small heath *Coenonympha pamphilus* (L.) is on the wing on unimproved grasslands and other grassy areas from June to September. Two generations are produced a year, using various fine-leaved grasses as food and pupating on grass stems. The second generation overwinter as larvae.

SPECIES DIVERSITY OF BUTTERFLIES

Conservationists and naturalists are usually interested in seeing a variety of species, and put a high value on habitats with a large species list. It has become clear that hand-netting and Malaise trapping tend to catch species in different proportions. The relative ranking by abundance of species caught over the eight-year period 1972–79 (hand-netting was discontinued after 1979), is shown in Table 5.1, which indicates differences in the effectiveness of hand-netting and Malaise trapping for catching different species.

Small tortoiseshells, peacocks and red admirals are more likely to be hand-netted than to enter the Malaise trap, and the painted lady, comma, common blue, small heath, white-letter hairstreak, silver-washed fritillary, marbled white and hedge brown were not taken at all in the trap, although this is a function of their relative rarity and the great difference between the total sample sizes. The common blue and the hedge brown have been caught in the Malaise trap since 1979, when hand-netting was discontinued. Meadow browns and walls, however, are particularly likely to enter the Malaise trap.

A further indication of the relative contribution of different species to catches by the two methods, can be gained by ranking them in terms of the number of years in which each was caught. The small white was caught in the Malaise trap in every year (15), the green-veined white in every year but one (14), large white in 12 years, small tortoiseshell and meadow brown in seven, orange tip and wall in six, small copper and peacock in five, brimstone and red admiral in three, large skipper in two, and the small skipper, common blue and hedge brown were each caught in just one year out of 15. However, so much more is captured with a hand-net, that the representation of species in different years is much higher. Six species – brimstone, large white, small white, green-veined white, small tortoiseshell and peacock – were caught in all eight years; orange tip, red admiral and meadow brown in seven years; painted lady and wall in six years; small copper and comma in five years; small skipper in four years; large skipper, common blue and small heath in two years; and white-letter hairstreak, silver-washed fritillary, marbled white and hedge brown in one year.

The differences between species in representation in the two types of samples and in the number of years in which they were trapped or hand-netted,

Table 5.1 *Ranking by abundance of butterfly species hand-netted and captured in a Malaise trap, 1972–79*

Hand-netting		Malaise trap	
Small white	8577	Small white	336
Small tortoiseshell	2860	Green-veined white	116
Large white	2469	Large white	52
Green-veined white	1135	Small tortoiseshell	27
Peacock	939	Meadow brown	13
Red admiral	331	Wall	12
Meadow brown	78	Peacock	6
Wall	64	Orange tip	4
Orange tip	54	Red admiral	4
Brimstone	46	Large skipper	2
Small copper	16	Brimstone	2
Painted lady	16	Small skipper	1
Small skipper	6	Small copper	1
Comma	5	Total	576
Large skipper	2		
Common blue	2		
Small heath	2		
White-letter hairstreak	1		
Silver-washed fritillary	1		
Marbled white	1		
Hedge brown	1		
Total	16 606		

are a consequence of behavioural differences. The preference for shade and the weak flight of meadow browns and walls make them more likely to enter the Malaise trap and then less likely to find a way out than active, strong fliers such as painted ladies and red admirals. Satyrids are probably over-represented in the Malaise trap catch, and the red admiral and the painted lady (which has never been trapped), under-represented. One advantage in employing two different methods to compile a list of garden butterflies, evident particularly in the short term, is precisely that different species are likely to be caught. The large skipper had been trapped for two years and the meadow brown for one, before they were seen and hand-netted. On the other hand, Malaise trapping is unlikely to produce such rare vagrants as white-letter hairstreaks or silver-washed fritillaries.

Species richness alone is not a very satisfactory way of describing diversity; ecologists prefer some measure that also takes into account the relative frequency of the different species. This can be expressed quantitatively as an Index of Diversity which characterizes the sample, and is assumed to be a property of the community sampled. Diversity indices, such as α of Fisher, Corbet & Williams (1943), which has proved an efficient statistic for moth diversity in large, random, light-trap samples (Taylor 1978), are parametric

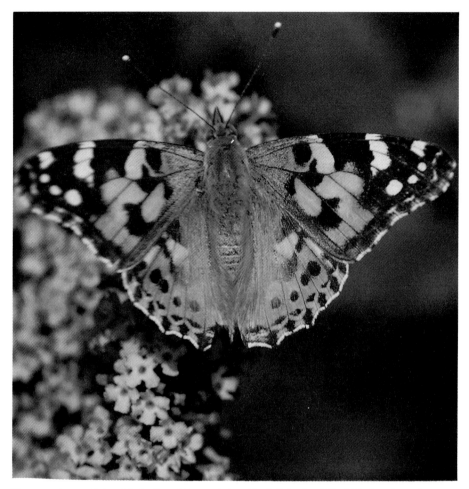

Photograph 5.2. Painted lady Cynthia cardui *on* Buddleia davidii. *This is a long-distance migrant from the south, which is unable to survive the British winter in any form.*

statistics which assume that the distribution of abundance among species fits a particular model – a logarithmic series in the case of α. Indices based on the proportional abundance of species, which take both evenness and species richness into account, make no assumptions about the shape of the underlying species abundance distribution, and Southwood (1978) has referred to them as non-parametric indices. One of the most useful is Simpson's index, considered by Routledge (1979) to be the best single measure of diversity. Simpson's index, λ, when used for species diversity, is an estimate of the probability that any two individuals taken at random will be of the same species. Low values of λ indicate a high diversity and high values a low diversity. In calculating diversity of garden butterflies for 1972, 1973 and 1974, Owen, D. F. (1975a) modified the index to $1 - \lambda$, expressed as β, an estimate of the probability that an individual sampled at random will be different from the previous individual sampled. The advantage of β is that high values indicate a high diversity and low values a low diversity.

Diversity indices are most frequently employed to discriminate between sites, and Magurran (1988) has compared the value for this purpose of a number of indices. She describes Simpson's index as having moderate discriminant ability, having low sensitivity to sample size, and responding mainly to dominance. In other words, it is more affected by the abundance of the commonest species than by species richness. Its overriding advantage, however, in marked contrast to most diversity indices, is that it is quite clear what it means: when Simpson's index expressed as β is high, there is a greater probability that two individuals taken at random will be different than there is when β is low. Use of Simpson's index expressed as β, together with S (number of species), usefully describes the composition of a sample, although Usher (1983) concluded that it is best to quote an equitability index, such as $1/\lambda S$, as well.

The modified Simpson index, β, has been calculated for all annual samples of butterflies. The β-diversity values for the series of 15 Malaise trap samples are given in Table 5.2. This series is markedly heterogeneous ($\chi^2 = 43.3950$, df = 14, p<0.001), and so each β-diversity value can legitimately be compared with that of the preceding year. Diversity was significantly higher in 1979 than in 1978, and in 1984 than in 1983, both at the 5 per cent level of probability; 1986, when only one species was trapped, axiomatically had a significantly lower diversity than 1985. Neither 1979 nor 1984 were particularly good butterfly years in terms of numbers trapped, and both were years when abundance decreased over the preceding year. The increase in β-diversity over the preceding year simply means that the probability that an individual sampled at random would be different from the previous individual sampled, was greater.

Values of β-diversity for the eight hand-netted samples were similarly calculated and compared (Table 5.3); the series is markedly heterogeneous ($\chi^2 = 472.9377$, df = 7, p<0.001). The increases in β-diversity from the preceding year in 1973, 1975, 1976 and 1979, and the decreases in 1977 and 1978, were all highly significant at the 1 per cent level of probability. Diversity was lower than usual in 1972 and 1978, and higher than usual in 1979 and, particularly, in 1976; the value of β for 1976 is significantly higher than that for 1979, at the 1 per cent probability level. In both 1972 and 1978, low diversity was at least partly a consequence of the high relative frequency of small whites; high diversity in 1976 was a direct consequence of the large number of species, and in 1979 a consequence of relatively high numbers of several species, although the species total was the same as in 1978. This confirms that β-diversity is a useful index of the interest or 'health' of a habitat, because a large number of species, and relatively high numbers of many of them, are exactly what a conservationist or naturalist likes to see.

The high numbers of butterflies hand-netted over eight years in the garden are themselves an indication of the richness, or at any rate potential, of gardens as butterfly habitats. It is pertinent, therefore, to ask what so many

Table 5.2 *Number of individuals (N) and species (S), and diversity (β) of annual Malaise trap samples of butterflies*

	Individuals (N)	Species (S)	Diversity (β)	Standard error
1972	41	8	0.659	±0.0652
1973	172	10	0.524	±0.0413
1974	77	5	0.631	±0.0404
1975	105	7	0.593	±0.0046
1976	79	9	0.651	±0.0514
1977	14	5	0.725	±0.0858
1978	61	4	0.547	±0.0320
1979	27	4	0.675	±0.0533
1980	27	5	0.764	±0.0472
1981	4	2	0.667	±0.2041
1982	63	7	0.642	±0.0544
1983	74	8	0.660	±0.0365
1984	52	9	0.784	±0.0346
1985	13	4	0.692	±0.1154
1986	9	1	0.00	

Note: The figures for 1972–74 differ slightly from those given in Owen, D. F. (1975a) due to earlier errors in tabulation.

Table 5.3 *Number of individuals (N) and species (S), and diversity (β) of annual hand-netted samples of butterflies*

	Individuals (N)	Species (S)	Diversity (β)	Standard error
1972	640	9	0.552	±0.0203
1973	2667	13	0.650	±0.0059
1974	932	12	0.625	±0.0146
1975	4438	14	0.692	±0.0048
1976	1827	19	0.741	±0.0073
1977	967	14	0.614	±0.0145
1978	3162	12	0.515	±0.0098
1979	1973	12	0.712	±0.0063

Note: The figures for 1972–74 differ slightly from those given in Owen D. F. (1975a) due to earlier errors in tabulation.

butterflies were doing in a fairly average suburban garden. On no more than three or four occasions during the period 1972–86, was the garden alive with butterflies. The more usual situation on a stroll around, was to see perhaps one or two cabbage whites, a peacock and a small tortoiseshell, and on repeating the stroll a couple of hours later, to see the same. The reasonable assumption is that the same butterflies were seen on each occasion, but marking-and-releasing has shown that this was not necessarily so. Few individuals remain in the garden for any length of time, and most are away

Photograph 5.3. Small tortoiseshell Aglais urticae, *one of the commonest garden butterflies, feeding on Michaelmas daisies* Erigeron sp. *in September, when there are few flowers available in the wild.*

again as soon as they have re-fuelled at the many nectar-rich garden flowers. This is why so many butterflies were caught and marked over eight years. Confirmation of this has come from a mark-and-release programme followed during summer 1984 in the suburban West Midlands by a Leicester Polytechnic student. Marked butterflies, however, tend to fly away on release, so that hand-netting acts in the same way as removal-trapping in creating a space which is quickly filled by an immigrant individual. The reasons for hand-netting large numbers may lie in the very process of capture and release. Although this may have boosted the numbers caught actually in the garden, those butterflies must have already been present in the immediate vicinity. It is evident that the butterflies of suburbia belong to vast mobile populations, constantly on the move seeking mates, egg-laying sites, and, above all, food. The gaudy colours and sweet perfumes of a flower garden attract them down for just as long as it takes to feed, then off they go again. This is why so many different individual butterflies may be caught in a garden and may also go some way to explaining how unlikely rarities, such as hedge browns and marbled whites, come to drop in.

ANNUAL FLUCTUATIONS OF BUTTERFLIES

The results of 15 years of Malaise trapping, 1972–86, are shown in Fig. 5.1 as histograms of species (*S*) and of individuals (*N*). This shows clearly that there was a run of good butterfly years from 1973 to 1976, a good year in 1978, and another run of good years from 1982 to 1984. 1973 was far and away the best year, followed by 1975, *N*=172 and *N*=105, respectively. 1977, 1981, 1985 and 1986 were particularly poor years, with only 14, 4, 13 and 9 individuals, respectively, caught. However, annual fluctuations in number of species were not nearly so marked, although most (ten) were caught in 1973 and fewest (one) in 1986. This is because the average number of individuals per species (*N/S*) tended to be highest in years of greatest abundance, and lowest in poor years: *N/S* was 17.2 in 1973, and 2 in 1981. This means that in good butterfly years, abundance is in large part a consequence of increases in numbers of the species already there, rather than addition of new species. In every year but one, the small white was the commonest butterfly, forming up to 66.86 per cent (in 1973) of the total catch; in 1977, the green-veined white was the commonest.

The report of the first ten years (1976–85) of the butterfly monitoring scheme organized by the Institute of Terrestrial Ecology (Pollard *et al.* 1986) indicates national fluctuations in numbers echoed by the Malaise trap figures. Thus, numbers were high in 1976 and fell sharply in 1977; in the warm summers of 1982–84 numbers were again high, followed by a decline in many species in 1985. The organizers are at some pains to point out that their data are not accurate national figures, as the sites where monitoring takes place are not representative, incorporating a disproportionate number of nature reserves. However, they believe that for many species, their information gives a good indication of national trends. They conclude that weather has an important effect on short-term trends in abundance, although they are unprepared to establish any causal connections on only ten years' data, especially as different species respond differently to weather. They are, however, confident that high summer temperatures are associated with good butterfly years. For instance, on their collated index for all 116 sites regularly monitored, five of the 21 species recorded in the garden increased significantly in 1978 and 18 in 1982; 12 decreased significantly in 1977, seven in 1981, and 12 in 1985.

While recognizing that annual variations in total Malaise trap catch mask differences in response between species, I have attempted to correlate the size of the butterfly catch with weather conditions in the current and preceding year. Inspection of the figures suggested that rainfall and minimum temperature in the preceding winter, spring rainfall and maximum temperature, and summer rainfall and maximum temperature might be important. Values for each of these factors for every year were calculated using the daily weather data produced by the Meteorological Office for Newtown Linford (see Chapter 2). Winter was taken as December, January and February, spring as March, April and May, and summer as June, July and

Annual fluctuations of butterflies

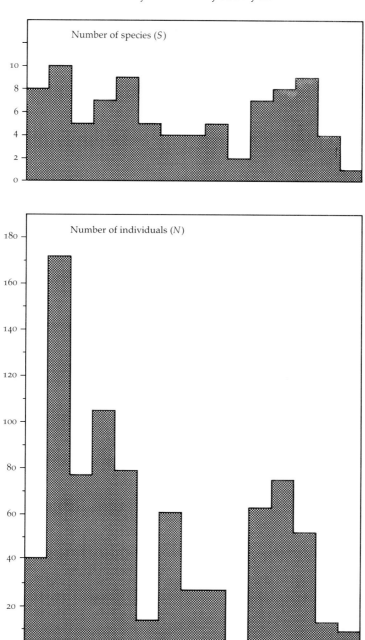

Fig. 5.1 Annual fluctuations in numbers of individuals and species of butterflies caught in Malaise trap, 1972–86.

August. The values calculated were average monthly rainfall for each season, and mean maximum and minimum temperatures.

Using a regression equation, a computer program selected the mean winter minimum temperature as the best explanation of variations in butterfly abundance in the Malaise trap catch, and no significant improvement was made by adding any other variable. When all six variables were put into a regression equation and backward elimination done by computer to select, progressively, variables that make least difference to the equation, the elimination stopped when only average monthly summer rain and mean maximum summer temperature were still in the equation. It is unclear why the winter mean minimum temperature was not one of this pair, but it was the last variable to be removed. Straight correlation of each of the weather variables with abundance ranks mean winter minimum temperature as the most important (0.564), followed by the mean summer maximum temperature (0.486) and by a negative correlation with winter rainfall (−0.427).

This means that two of the most important effects of weather on butterfly abundance are that numbers go up in a warm summer and after a mild winter, especially if that winter was fairly dry. The correlations of weather with the Malaise trap figures are not ideal, because total butterfly numbers have been used (because of low figures), despite the fact that different species almost certainly respond in different ways. However, the general association is shown in Fig. 5.2 where plots of winter minimum and summer maximum temperatures are superimposed on a histogram of butterfly numbers by year.

The figures for hand-netting in the period 1972–79 (Fig. 5.3) accord well with those for the Malaise trap, with peaks in 1973, 1975 and 1978, and a decline in 1977, although this latter was not so marked as for the Malaise trap. Hand-netting tends to be more subjective, and this is one reason why the Malaise trap figures were preferred for correlations with weather. Annual fluctuations in species caught may have been particularly affected by the human element involved, the peak year being 1976, when 19 species, including four rarities (for the garden) were captured. As with the Malaise trap catch, the average number of individuals per species (N/S) was highest in years of peak abundance, showing again that increased overall abundance reflected increases in the common species, but also suggesting that hand-netting was not as far biased towards the unusual as might have been expected.

General observation reinforces the hand-netting figures, and of course has been particularly important since the hand-netting programme ended. My garden diary records that butterflies were very common in August 1975 in the hottest weather for several years, and that in 1975 many small coppers were seen (seven were caught). In 1976, what looked like a holly blue *Celastrina argiolus* was seen flying high over the garden, and there was a clouded yellow *Colias croceus* in a nearby garden. Painted ladies were recorded as common in 1980, and the garden diary calls 1982 'a good comma year'. Since painted ladies, in particular, and also red admirals, are long-distance migrants, high

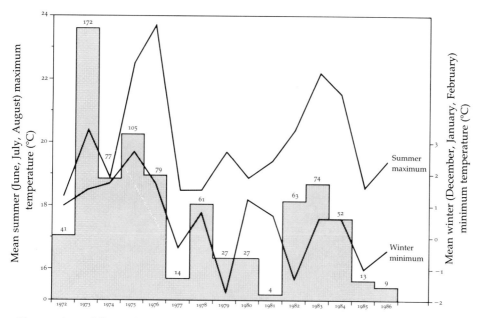

Fig. 5.2 Annual fluctuations in Malaise trap catch of butterflies, and mean winter minimum and summer maximum temperatures, 1972–86.

numbers probably reflect weather and breeding conditions in southern Europe and the Mediterranean rather than in England. Similarly, conditions far away may account for absences of painted ladies, as in 1984.

Among the aims of the Institute of Terrestrial Ecology's butterfly monitoring scheme is to detect trends in the abundance of butterflies which may affect their status. The ten-year report (Pollard *et al.* 1986) reaches no conclusions as to long-term trends, but does include national figures and graphs showing fluctuations. Of the 21 species recorded in my garden, only the silver-washed fritillary showed a downward trend over the ten years, attributed to the increased shade conditions in many woodlands because of a decline in coppice management. (No figures are given for the white-letter hairstreak because of the impossibility of monitoring this butterfly as it flies in the tree canopy.) The small skipper, large skipper, green-veined white, peacock, comma and marbled white showed an upward trend, and the other 13 species were more or less stable. Even after 15 years, it is not possible to make any firm statement about trends in butterfly numbers in the garden, as the greatest changes are the short-term ones caused by weather. It may, however, be significant that numbers in the good summers of 1983 and 1984 were nowhere near as high as in the period 1973–76.

SEASONALITY OF BUTTERFLIES

Of necessity, British butterflies are strictly seasonal in their flight activity, but even within the flying season, there are definite patterns of activity. Capture

Butterflies

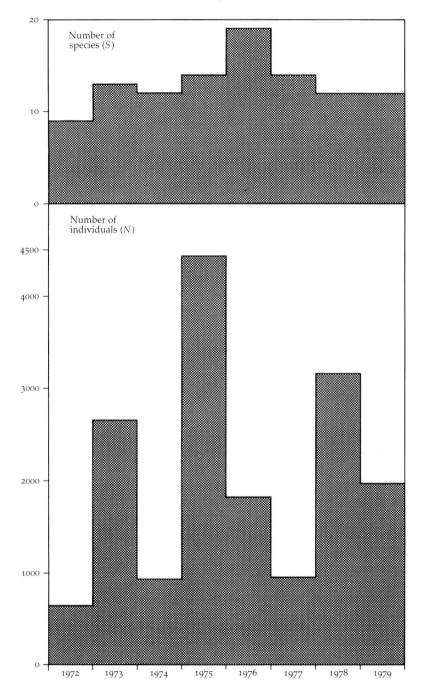

Fig. 5.3 Annual fluctuations in numbers of individuals and species of butterflies hand-netted, 1972–79.

of individuals by both Malaise trapping and hand-netting reached a marked peak in August (Figs. 5.4 & 5.5). July and August were the best months for species; 15 species were hand-netted in July and in August, and ten and 11 were caught in July and August, respectively, in the Malaise trap. The smoother seasonal increases and rapid declines in numbers of individuals and species caught in the Malaise trap (Fig. 5.4) probably give a more accurate picture of seasonality than do the figures for hand-netting (Fig. 5.5), which are affected by variations in time and effort, often dependent on weather.

The compilation of years to give overall monthly figures obscures differences in seasonality between years. Observation suggested that garden butterflies were scarce in July 1974, in June and July 1975, in August and September 1976, and in July 1977. This is mostly borne out by trapping figures, although these indicate that June and July 1975 were in fact better than they appeared. Furthermore, single individuals of four species – white-letter hairstreak, silver-washed fritillary, marbled white and hedge brown – were only hand-netted in 1976, the former three in July, the latter in August. (However, the hedge brown has since been seen in the garden and caught in the Malaise trap.) 1976 was the drought summer when plant growth and insect activity reached an early peak, and by August the countryside and many gardens were parched and brown. All four species are typically butterflies of the countryside, not expected in gardens, and I assumed that in the arid conditions of 1976 they wandered far from their normal haunts seeking scarce nectar sources.

The monthly occurrence of different butterfly species, combining hand-netting and Malaise trapping, shows clearly their differing seasonality. For instance, brimstones and peacocks, which hibernate as adults, were caught from March to May and from July to October, with none in June, when the single generation is at the caterpillar or pupal stages. Large, small and green-veined whites, which hibernate as pupae and have two or more generations a year, were caught from April to October, and small tortoiseshells, which hibernate as adults, from March to October. Orange tips, which hibernate as pupae and produce a single generation a year, were only caught in May and June, and red admirals and painted ladies, long-distance migrants, from July onwards.

BREEDING OF BUTTERFLIES IN GARDENS

The only butterflies definitely to have bred in the garden are pierids (whites): the large white on cultivated brassicas and nasturtium *Tropaeolum majus*; the small white on cultivated brassicas, wild cabbage, and dame's-violet; the green-veined white on *Aubrieta deltoidea*, honesty *Lunaria annua*, dame's-violet and candytuft *Iberis sempervirens*; and the orange tip on garden arabis and dame's-violet. In addition, a small white laid an egg on radish *Raphanus sativus*, and a meadow brown laid eggs on lawn grass.

That none of the other species recorded in the garden has bred there, bears out the importance of location and aspect of egg-laying site because, at one

Butterflies

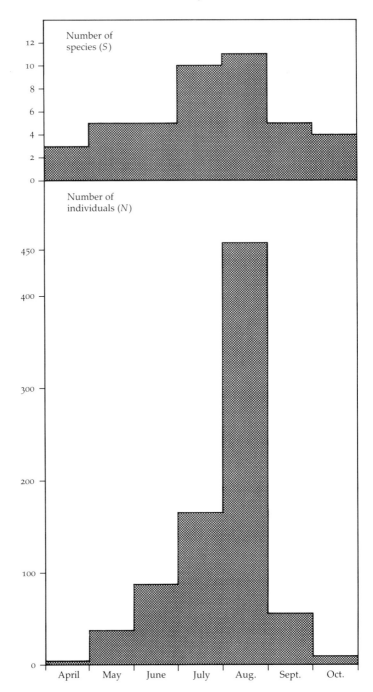

Fig. 5.4 Numbers of individuals and species of butterflies caught in Malaise trap, 1972–86, by month.

time or another, there have been foodplants in the garden of all but the brimstone. For instance, there are three patches of nettles, but not even the abundant tortoiseshells and peacocks have used them for egg-laying. This is almost certainly because the nettles are not growing in sunny enough or open enough situations. Butterflies whose caterpillars feed on nettles often do find breeding sites in gardens, but it is not enough simply to provide the foodplant for these or for other species – the location must be right as well.

STATUS OF BUTTERFLIES IN LEICESTERSHIRE

Concern has been expressed since the 1950s over the apparent decline in British butterflies; there seem to be fewer than people remember seeing in the past, and many species are no longer found where they once flew commonly. Recognition in the 1970s and 1980s of the destructive effect that modern farming methods have had since World War II on the countryside has focused the attention of conservationists, politicians and the concerned public on changes in the range and abundance of British wildlife in general. Because butterflies are attractive, conspicuous and easily identified, they have come in for particular attention. An atlas of British butterflies published by the Biological Records Centre (Heath *et al.* 1984) details changes in range and, by implication, abundance, allowing some conclusions to be drawn.

Butterflies are predominantly a tropical group, and are on the northern edge of their range in the British Isles. They are thus particularly susceptible in Britain to short-term changes in the climate and other environmental factors, which produce fluctuations in numbers and range, but these are genuine fluctuations, and a decrease is eventually followed by an increase. There is, however, evidence of overwhelming net decline in butterfly ranges, and consequently in abundance.

Over the last 150 years, four British species have become extinct, and 18 (including the silver-washed fritillary and marbled white) have suffered a major contraction of range. Of six species that declined in the latter part of the nineteenth century, three, including the comma and wall, have again spread to much of their former range, and three, including the orange tip and peacock, have now exceeded their former range. The other 34 species have remained more or less stable in range for 150 years, but several, including the white-letter hairstreak, have disappeared from many areas they once frequented (Heath *et al.* 1984).

The general decline in range and abundance cannot be attributed to poor climatic conditions that might have restricted adult activity and prolonged development, thus increasing predation, because species such as the cabbage whites that benefit from man's activities, and some local species in protected habitats, have prospered. The main cause of butterfly decline seems to be the way that man has altered habitats by such procedures as draining of wetlands, cultivating and 'improving' permanent pasture, heath and moorland, planting conifers on uplands or in place of hardwoods, and neglecting traditional woodland management. In particular, there has been conspicuous loss

Butterflies

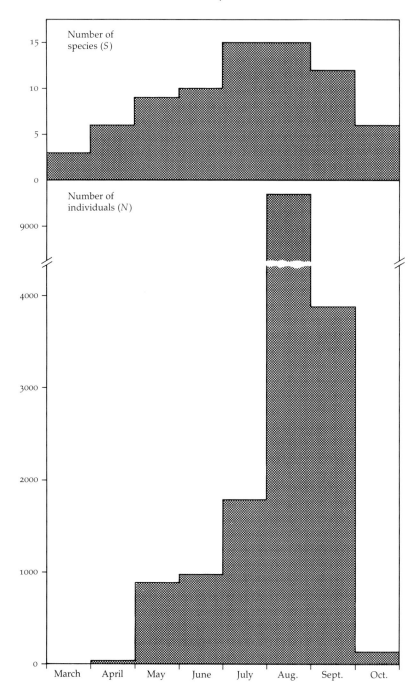

Fig. 5.5 Numbers of individuals and species of butterflies hand-netted, 1972–79, by month.

of coppice woodland, and of chalk or limestone grassland. The cumulative effect has been the disappearance of butterflies, either completely or at the edges of their range, northern species having tended to contract to the north and west, and southern species to the south and west, to areas of less intensive land use. Some species, such as the large and small skippers, green-veined white, small tortoiseshell, hedge brown and small heath, can survive in uncultivated corners in all but the most sterile of arable prairies, but these are, in general, exceptions (Heath *et al.* 1984).

What is the situation in Leicestershire? How many butterfly species are there, and, more particularly, are there any that might be added to the garden list? As well as the garden's 21 species, six further species – grizzled skipper *Pyrgus malvae*, holly blue, small pearl-bordered fritillary *Boloria selene*, pearl-bordered fritillary *B. euphrosyne*, speckled wood *Pararge aegeria* and clouded yellow – were recorded elsewhere in Leicester city during 1983–87 (Lomax 1987). Heath *et al.* (1984) give additional records for the area: in the 10-kilometre square, large tortoiseshell *Nymphalis polychloros* and ringlet *Aphantopus hyperantus* pre-1940; and in the 30-kilometre square (the garden's 10-kilometre square and all adjacent ones), dingy skipper *Erynnis tages*, green hairstreak *Callophrys rubi*, purple hairstreak *Quercusia quercus*, small blue *Cupido minimus*, dark green fritillary *Argynnis aglaja* and marsh fritillary *Eurodryas aurinia* post-1940, and brown argus *Aricia agestis*, Duke of Burgundy *Hamearis lucina*, purple emperor *Apatura iris* and high brown fritillary *Argynnis adippe* pre-1940. Leicestershire Museums Service, which collates all county records, has provided information on the status in the county of all 39 species listed above and of five additional species – chequered skipper *Carterocephalus palaemon*, Essex skipper *Thymelicus lineola*, brown hairstreak *Thecla betulae*, black hairstreak *Strymonidia pruni*, and white admiral *Ladoga camilla*.

Of the 21 garden species, the large white, small white, green-veined white, red admiral, small tortoiseshell and peacock are generally abundant in the county, and the meadow brown locally abundant. The brimstone, orange tip, small copper and wall are widespread, the small and large skippers widespread in rough grassland, the common blue widespread but local, the comma widespread since a resurgence in the 1950s, and the hedge brown widespread since a marked increase in the 1940s. The painted lady, whose presence depends on immigration, is frequent, as is the small heath. The white-letter hairstreak has declined since the death of mature elms from Dutch elm disease, and the silver-washed fritillary appears to be extinct as a breeding species, although occasionally recorded. There is an established colony of marbled whites in east Leicestershire, and a new record for south Leicestershire in 1985.

Of the six species recently recorded elsewhere in the city, the grizzled skipper was frequent on disused railway lines in the 1960s, but has since declined, having been recorded from only six tetrads (2 km × 2 km squares) in central and north Leicestershire in the 1980s. The clouded yellow is only seen in acknowledged migration years, so only turns up in the county at irregular intervals. The holly blue occurred in 14 tetrads in the period 1980–85, all but

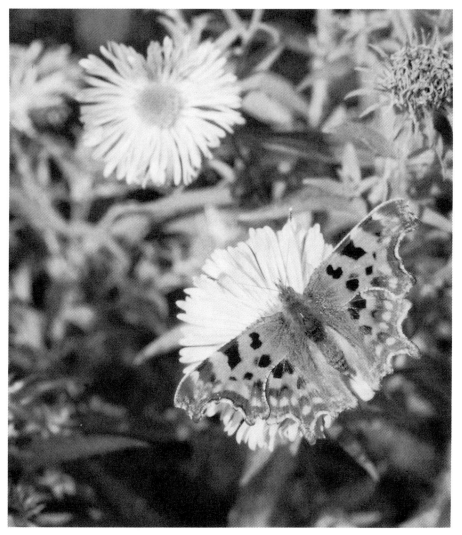

Photograph 5.4. Comma Polygonia c-album, *an occasional garden visitor. This species of butterfly is widespread in Leicestershire since a resurgence in the 1950s.*

one of them being in the western half of the county. The small pearl-bordered fritillary is thought to be extinct in the county, and there is an element of doubt over the city record. In 1954, the pearl-bordered fritillary, though local, was fairly abundant; in the 1970s there were no records, but in 1986 there were two reports at sites other than those recorded in 1954 or in the nineteenth century. There are two strongholds of the speckled wood, one in the northwest, the other in southern Rutland; it has been reported from 12 tetrads in the 1980s and seems to be on the increase.

The last English sighting of the chequered skipper was in Rutland in 1975, and it is now considered extinct in England, although thriving in the west of Scotland. The Essex skipper was reported from only two tetrads in Leicestershire in the 1980s, but it is probably overlooked because of its resemblance to the small skipper. The dingy skipper was reported from seven tetrads in the 1970s and from a different seven in the 1980s; it is thought to be maintaining its populations, but to be very vulnerable to land-use changes.

The green hairstreak, a very local species, was abundant in the mid-1950s, but may have declined since, having been reported from five tetrads, four of them in Rutland, in the 1980s and having been introduced to another. There is only a single 1983 record of the brown hairstreak *Thecla betulae*, but the purple hairstreak seems to be expanding its range, and now occurs in seven tetrads, mainly on Charnwood Forest to the northwest of Leicester. The black hairstreak is well-established at one locality in Rutland, and there is a single 1974 record for the south of the county. According to the County Records Office, the small blue has never been recorded. The brown argus, which used to occur on limestone grassland in the northeast of the county, has not been seen for about 30 years.

There is only a single record for 1972 of the Duke of Burgundy, but several for the white admiral. The latter was not known in Leicestershire when the *Victoria County History* was published in 1907, but was present in 1954, although described as rare; there are no records for 1970–79, but in 1986 it was recorded in three tetrads widely spaced. The purple emperor was last seen in the county more than 100 years ago, the large tortoiseshell has not been seen for more than 30 years, and the high brown fritillary was last seen in a deer park near Leicester more than 20 years ago; all three are almost certainly extinct in the county. The dark green fritillary was common at the same deer park near Leicester in the nineteenth century, but by 1907 was declared extinct; it was recorded in one tetrad in 1976, and in two tetrads in the south of the county in the 1980s, although these are considered as vagrants or possibly released specimens. The only recent record of the marsh fritillary is a doubtful one from an abandoned railway cutting quite close to my garden. The ringlet is common in south and east Leicestershire, and is spreading, although it is more or less absent from the north and west.

On the basis of this review of the county's butterflies, and evidence of the attractiveness of a well-stocked garden to wandering individuals, it is possible to predict some additions to the garden butterfly list. The holly blue has not been included in the garden list because it was only a possible sighting, but undoubtedly it will visit soon. If, as seems likely, the Essex skipper is much more common than is reported, this may well be recorded too. Sooner or later, in a good year for immigration of clouded yellows, one or more will be attracted by the abundance of flowers, and as the speckled wood increases its range and abundance, it is bound eventually to be recorded. It is likely, too, that in a warm summer, dingy and grizzled skippers or a ringlet will visit. On the basis of 1976, however, even less probable visitors cannot be ruled out, and another hot, dry summer will bring the possibility of yet more rarities.

6

Moths

To the general public, moths are less familiar than butterflies, simply because the vast majority of them are nocturnal. There are, however, many more British moths than butterflies, and they are taxonomically more variable. Moreover, a great deal is known about British moths, as there have been dedicated and knowledgeable moth collectors for as long as there have been butterfly collectors. Collection of moths (and the now more generally acceptable study of them and their population dynamics without killing them) was made far easier, and indeed revolutionized, with the invention of the mercury-vapour light trap in the late 1940s. Mercury-vapour light does not attract all species of moths equally (Taylor & Carter 1961), but it draws in vast numbers of many species. To a large extent, lepidopterists still concentrate on the macro-Lepidoptera, the families of larger moths, but there is increasing interest in the families of smaller moths, the micro-Lepidoptera. Most moth books concentrate on macro-Lepidoptera, considering only the larger and more familiar micro-Lepidoptera, despite the fact that the latter constitute the majority of British moths, which number in total well over 2000 species. The distinction between 'macros' and 'micros' is neither scientific nor biological but only convenient and traditional, and in the checklist of British insects (Kloet & Hincks 1972) the families are not separated in this way. Inevitably, much of this chapter is about macro-moths, partly because they are more familiar and easier to identify, but also because they have been intensively investigated in the garden; micro-Lepidoptera are less well known but all those that have been recorded are included.

Natural history of moths

The natural history of moths is varied, depending on habitat, the way in which caterpillars exploit the foodplant, dispersal and mating behaviour, mode of hibernation, and whether or not they are migratory. All, however, have the same basic life history of egg, caterpillar as a feeding and growing stage, immobile pupa in which massive re-organization of tissues occurs, and mobile, usually winged, adult. A moth without wings sounds rather a contradiction of terms, but females of a number of species are wingless.

Eggs are laid singly or in clusters on the foodplant, or in some cases broadcast in flight over suitable vegetation, as in the ghost moth *Hepialus humuli*. Caterpillars of many garden moths feed by chewing leaves, some feed on roots, others on flower buds, and a number bore into stems of plants, including trees, or, in the case of micros, mine burrows within the blades of leaves. Pupae tend to be concealed or protected in some way; many of those of macros are formed in the surface soil, often in a cocoon, while those of micros tend to be formed in the leaf-mines, stem-tunnels, or wherever else the larva has been feeding. Newly emerged female moths produce pheromones which males detect using their antennae, which are complex batteries of chemoreceptors, and males of at least some species, such as the angle shades *Phlogophora meticulosa*, have elaborate brush organs on the abdomen which disseminate an aphrodisiac as they approach the female (Birch 1970). A number of species are migratory, the prime example being the silver-Y *Autographa gamma*, which arrives in Britain from southern Europe in late spring and summer, the numbers varying from year to year. Migration of the silver-Y is a large-scale population drift north and northwest early in the year, and a southwards retreat in the autumn.

THE GARDEN SPECIES OF MOTHS AND THEIR RELATIVE ABUNDANCE

The garden list, 1972–86, includes 343 species of moths, 263 macro-Lepidoptera and 80 micro-Lepidoptera. The majority were attracted to a mercury-vapour light trap, many entered the Malaise trap, and some were attracted to a trap baited with fermenting apple. The Malaise trap is in operation continuously every year and so the trapping figures can be used to illustrate annual and seasonal fluctuations. Use of the mercury-vapour trap is irregular, and so the trapping results are qualitative not quantitative. The baited trap has only been used occasionally, mostly in 1975.

In this listing of the species, the macros will be described first, the order of families being that of the British checklist, and then the micros, again in checklist order of families (Kloet & Hincks 1972). However, the better to describe the garden moth fauna, the macros have been divided into four categories: common garden moths – those that are common or quite common; uncommon garden moths – those that are uncommon, even if regular, and those that are irregular; scarce garden moths – recorded less than ten times in the 15-year period; and moths recorded only once in the 15-year period. On this basis, there are 115 common macro-moth species, 29 uncommon, 78 scarce and 41 recorded once only. The micros have not been divided into categories but are treated in taxonomic order.

Macro-lepidoptera

The scientific names are as used in the British checklist (Kloet & Hincks 1972). Foodplants and overwintering stages are taken from Skinner (1984), with some additional information from Stokoe & Stovin (1948), and foodplants

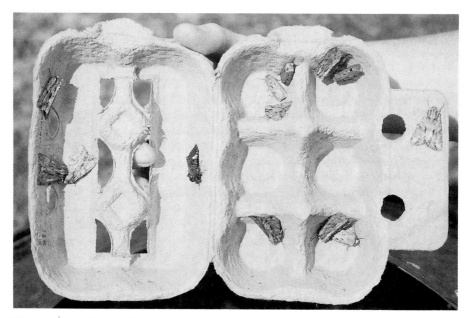

Photograph 6.1. Part of the catch of moths from a mercury-vapour light trap. Papier mâché egg cartons packed in the collecting box of the trap provide resting places for moths, which can be examined and identified without handling.

used in the garden are given. The majority of moths were identified by D. F. Owen. MT indicates that the species has been taken in the Malaise trap; BT indicates that it has been taken in a trap baited with fermenting apple.

COMMON GARDEN MOTHS

Hepialidae

1. Orange swift *Hepialus sylvina* (L.). Common, July–September. Larvae on roots of wide variety of herbaceous plants. Overwinter as larvae. MT
2. Common swift *Hepialus lupulinus* (L.). Common, especially in June, but straggling on through summer. Larvae on roots of grasses and occasionally herbaceous plants. Overwinter as larvae. MT

Thyatiridae

3. Buff arches *Habrosyne pyritoides* (Hufnagel). Common, June–August. Forty-eight in seven nights in mercury-vapour trap in July 1983. Larvae on *Rubus fruticosus*. Overwinter as pupae. MT
4. Figure of eighty *Tethea ocularis* (L.). Common, June and July. Larvae on *Populus* spp. Overwinter as pupae.

Geometridae

5. March moth *Alsophila aescularia* (D. & S.). Common, March and April. Larvae on wide variety of deciduous trees. Overwinter as pupae. MT

6. Riband wave *Idaea aversata* (L.). Common, June–September. Polymorphic. Larvae on variety of herbaceous plants. Overwinter as larvae. MT
7. Dark-barred twin-spot carpet *Xanthorhoe ferrugata* (Clerck). Common, July and August. Larvae on variety of herbaceous plants. Overwinter as pupae. MT
8. Silver-ground carpet *Xanthorhoe montanata* (D. & S.). Quite common in June; two in July 1980. Larvae on variety of herbaceous plants. Overwinter as larvae. MT
9. Garden carpet *Xanthorhoe fluctuata* (L.). Common, May–September in two broods. One each in October 1979 and 1986. Larvae on range of Cruciferae; on *Alyssum saxatile*, *Arabis caucasica*, *Brassica oleracea* and (especially) *Iberis sempervirens* in garden. Overwinter as pupae. MT, BT
10. Common carpet *Epirrhoe alternata* (Müller). Common, July and August. Larvae on various species of *Galium*. Overwinter as pupae. MT
11. Yellow shell *Camptogramma bilineata* (L.). Common, July and August, especially in 1980s. Larvae on variety of herbaceous plants; on *Arabis caucasica*, *Aubrieta deltoidea*, *Origanum majorana* and *Taraxacum officinale* in garden. Overwinter as larvae. MT
12. Spinach *Eulithis mellinata* (Fab.). Quite common, June and July. Larvae on *Ribes nigrum* and *R. rubrum*; on *R. uva-crispa* in garden. Overwinter as eggs. MT
13. Common marbled carpet *Chloroclysta truncata* (Hufnagel). Common, June–September in two broods. Almost all melanics. Larvae on variety of trees, shrubs and herbaceous plants; on *Althaea rosea*, *Buddleia davidii*, *Fragaria virginiana* × *chiloensis*, *Geranium cinereum*, *Hypericum androsaemum*, *Paeonia officinalis*, *Ribes sanguineum*, *Rubus fruticosus* and *Saxifraga* sp. in garden. Overwinter as larvae. MT, BT
14. Grey pine carpet *Thera obeliscata* (Hübner). Common, June and September–October in two broods. Many melanics. Larvae on variety of conifers. Overwinter as larvae. MT
15. July highflyer *Hydriomena furcata* (Thunberg). Common, July and August. Larvae on various species of *Calluna*, *Salix*, *Vaccinium*, and on *Corylus avellana*. Overwinter as eggs.
16. November moth *Epirrita dilutata* (D. & S.). Common, October and November. Larvae on wide variety of trees and shrubs; on *Prunus cerasifera* in garden. Overwinter as eggs.
17. Mottled pug *Eupithecia exiguata* (Hübner). Common, May and June. Larvae on *Acer pseudoplatanus*, *Crataegus* sp., *Prunus spinosa* and *Sorbus aucuparia*; on *Buddleia davidii*, *Crataegus monogyna*, *Fraxinus excelsior*, *Iberis sempervirens*, *Paeonia officinalis*, *Populus nigra*, *Potentilla fruticosa*, *Ribes uva-crispa*, *Rosa* sp., *Sambucus nigra*, *Spiraea* × *vanhouttei* and *Weigela florida* in garden. Overwinter as pupae. MT
18. Lime-speck pug *Eupithecia centaureata* (D. & S.). Common, July and August. Larvae on flowers of many herbaceous plants. Overwinter as pupae. MT
19. Common pug *Eupithecia vulgata* (Haworth). Common, May–July. Larvae on leaves and flowers of a variety of trees, shrubs and herbaceous plants; on *Ribes sanguineum* in garden. Overwinter as pupae.
20. Tawny speckled pug *Eupithecia icterata* (de Villers). Common, August and September. Larvae on *Achillea millefolium*. Overwinter as pupae. MT
21. Green pug *Chloroclysta rectangulata* (L.). Common, June and July. Larvae on flowers of various Rosaceae. Overwinter as eggs.
22. Brimstone *Opisthograptis luteolata* (L.). Common, May and June, and second

brood in August and September. Larvae on variety of trees in Rosaceae; on
Chaenomeles sp., *Crataegus monogyna*, *Prunus cerasifera* and *Sorbus aucuparia* in
garden. The *P. cerasifera* in the garden is the purplish variety *atropurpurea*;
luteolata caterpillars on it are mostly purplish, matching the leaves, and the
few green caterpillars are conspicuous. Overwinter as larvae or pupae. MT, BT

23. Canary-shouldered thorn *Ennomos alniaria* (L.). Common, August and
September. Larvae on a variety of trees. Overwinter as eggs. MT

24. Dusky thorn *Ennomos fuscantaria* (Haworth). Common, August and September.
Larvae on *Fraxinus excelsior*. Overwinter as eggs.

25. Early thorn *Selenia dentaria* (Fab.). Common, March–May, and second brood in
July and August. Larvae on variety of trees and shrubs. Overwinter as pupae.
MT

26. Scalloped hazel *Odontopera bidentata* (Clerck). Common, May and June.
Occasional melanics. Larvae on variety of trees; on *Buddleia davidii, Malus
sylvestris, Philadelphus coronarius, Potentilla fruticosa, Prunus cerasifera, Spiraea* ×
vanhouttei and *Weigela florida* in garden. Overwinter as pupae. MT

27. Scalloped oak *Crocallis elinguaria* (L.). Common, July and August. Larvae on
most deciduous trees and shrubs; on *Buddleia davidii* in garden. Overwinter as
eggs. MT

28. Swallow-tailed moth *Ourapteryx sambucaria* (L.). Common, June and July.
Larvae on *Hedera helix* and a variety of trees and shrubs, mainly Rosaceae; on
Ligustrum ovalifolium and *Ribes sanguineum* in garden. Overwinter as larvae. MT

29. Brindled beauty *Lycia hirtaria* (Clerck.). Common, April and May. Larvae on
variety of deciduous trees; eggs on *Lonicera periclymenum* in garden.
Overwinter as pupae.

30. Peppered moth *Biston betularia* (L.). Common, May–July. Larvae on wide
variety of trees and herbaceous plants; on *Buddleia davidii, Mahonia aquifolium,
Malus sylvestris, Potentilla fruticosa* and *Salix fragilis* in garden. Overwinter as
pupae. The widespread phenomenon of industrial melanism has been most
intensively investigated in *B. betularia* (see e.g. Kettlewell 1973). In the 12-year
period 1972–83, the catch of *B. betularia* in the garden comprised 97 *carbonaria*
(melanic), two *insularia* (melanic) and 19 *typica* (the normal, pale form); none
was caught in 1984–86. The relative frequency of *typica* seems to have
increased slightly from 1978 onwards, suggesting that Clean Air policies have
been effective locally.

31. Willow beauty *Peribatodes rhomboidaria* (D. & S.). Abundant, July–September.
Larvae on wide variety of trees and other plants; on *Cytisus scoparius* in
garden. Overwinter as larvae. MT

32. Mottled beauty *Alcis repandata* (L.). Common, July–September. Larvae on a
wide variety of trees and other plants; on *Crataegus monogyna* in garden.
Overwinter as larvae. MT, BT

33. Light emerald *Campaea margaritata* (L.). Common, June and July, and one in
August 1981. Larvae on wide variety of deciduous trees. Overwinter as larvae.
MT

Sphingidae

34. Lime hawk *Mimas tiliae* (L.). Common, May–July; may have been less common
in 1980s than in 1970s. Larvae on *Alnus* spp., *Betula* spp., *Tilia* spp. and *Ulmus*
spp. Overwinter as pupae.

35. Eyed hawk *Smerinthus ocellata* (L.). Quite common, June and July. Larvae on *Malus sylvestris*, *Populus* spp. and *Salix* spp.; on *Salix fragilis* in garden. Overwinter as pupae.
36. Poplar hawk *Laothoe populi* (L.). Common, May–August. Larvae on various species of Salicaceae; on *Populus nigra* and *Salix fragilis* in garden. Overwinter as pupae. MT

Notodontidae

37. Swallow prominent *Pheosia tremula* (Clerck). Quite common, May and June, and second brood in August and September. Larvae on various species of Salicaceae. Overwinter as pupae.
38. Pale prominent *Pterostoma palpina* (Clerck). Quite common, May and June, and one in July 1979. Larvae on various species of Salicaceae. Overwinter as pupae.

Lymantriidae

39. Vapourer *Orygia antiqua* (L.). Males common, September and October, and one in August 1976, the drought year; females are wingless. Larvae on most deciduous trees and shrubs. Overwinter as eggs. MT
40. Yellow-tail *Euproctis similis* (Fuessly). Common, July and August; less common in 1980s than in 1970s. Larvae on variety of trees; on flower of *Rosa* sp. and on *Crataegus monogyna* in garden. Overwinter as larvae. MT

Arctiidae

41. Common footman *Eilema lurideola* (Zincken). Common, June–August, although not recorded until 1975. Larvae on lichens. Overwinter as larvae.
42. Garden tiger *Arctia caja* (L.). Common, July and August. Larvae on wide variety of herbaceous plants. Overwinter as larvae.
43. Buff ermine *Spilosoma luteum* (Hufnagel). Common, June and July, and one in August 1980. Larvae on variety of herbaceous plants and occasionally on trees and shrubs. Overwinter as pupae. MT

Noctuidae

44. Garden dart *Euxoa nigricans* (L.). Common, July and August. Larvae on variety of herbaceous plants. Overwinter as eggs.
45. Turnip *Agrotis segetum* (D. & S.). Common, June–October. Larvae on roots and lower stems of variety of herbaceous plants. Overwinter as larvae. MT, BT
46. Heart and dart *Agrotis exclamationis* (L.). One of the commonest moths from May to August. Larvae on wide variety of herbaceous plants. Overwinter as larvae. MT
47. Shuttle-shaped dart *Agrotis puta* (Hübner). Common in June with second brood in August, and twice recorded in September. Larvae on wide variety of herbaceous plants; one larva in garden in 1983 but foodplant not known. Overwinter as larvae. MT
48. Flame *Axylia putris* (L.). Common, June and July, and two in August 1980. Larvae on wide variety of herbaceous plants. Overwinter as pupae. MT

49. Flame shoulder *Ochropleura plecta* (L.). Common in June, and three in August 1981. Larvae on wide variety of herbaceous plants; on *Malva sylvestris* in garden. Overwinter as pupae. MT
50. Large yellow underwing *Noctua pronuba* (L.). The commonest garden moth, June–October, as it is everywhere in England. 230 in mercury-vapour trap on night of 13 August 1980. Annual Malaise trap catches fluctuated considerably from lows of five in both 1983 and 1984 to high of 112 in 1975. Larvae on a wide range of herbaceous plants and grasses; on *Aubrieta deltoidea, Buddleia davidii, Epilobium adenocaulon* and *Lamium maculatum* in garden. Overwinter as larvae. MT, BT
51. Lesser yellow underwing *Noctua comes* (Hübner). Common, August and September, and one in July 1981. Larvae on wide variety of trees, shrubs and herbaceous plants; on *Aubrieta deltoidea, Chrysanthemum frutescens, Dianthus barbatus, Digitalis purpurea, Hedera helix, Iberis sempervirens, Phlox douglasii* and, especially, on *Lamium maculatum* in garden. Overwinter as larvae. MT, BT
52. Lesser broad-bordered yellow underwing *Noctua janthina* (D. & S.). Common, August, with a few in September. Larvae on wide variety of shrubs and herbaceous plants. Overwinter as larvae. MT
53. Least yellow underwing *Noctua interjecta* (Hübner). Common, July and August. Larvae on various grasses and herbaceous plants. Overwinter as larvae. MT
54. Stout dart *Spaelotis ravida* (D. & S.). Quite common, July–September. Larval foodplant(s) unknown. Overwinter as larvae.
55. Ingrailed clay *Diarsia mendica* (Fab.). Common, June–August. Larvae on variety of shrubs and herbaceous plants. Overwinter as larvae. MT, BT
56. Small square-spot *Diarsia rubi* (Vieweg). Common, June–September. Larvae on wide variety of herbaceous plants. Overwinter as larvae. MT
57. Setaceous Hebrew character *Xestia c-nigrum* (L.). Common, June–October in two broods. Larvae on many herbaceous plants; on *Digitalis purpurea* in garden. Overwinter as larvae. MT, BT
58. Double square-spot *Xestia triangulum* (Hufnagel). Common, June and July. Larvae on wide variety of trees, shrubs and herbaceous plants; on *Lamium maculatum* in garden. Overwinter as larvae.
59. Square-spot rustic *Xestia xanthographa* (D. & S.). Common, June–September. Larvae on variety of grasses and herbaceous plants. Overwinter as larvae. MT, BT
60. Gothic *Naenia typica* (L.). Common, July and August; more often taken in Malaise trap than at light. Larvae on wide variety of trees, shrubs and herbaceous plants. Overwinter as larvae. MT
61. Nutmeg *Dicestra trifolii* (Hufnagel). Common, July and August. Larvae on several species of family Chenopodiaceae. Overwinter as pupae. MT, BT
62. Cabbage *Mamestra brassicae* (L.). Common, June–October. Larvae on extraordinary range of herbaceous plants and occasionally shrubs; on 35 species of plants in garden and three species of pot plants indoors (see Table 6.3, p. 139). Usually overwinter as pupae, sometimes as larvae. MT, BT
63. Dot *Melanchra persicariae* (L.). Common, June–August. Larvae on extraordinary range of herbaceous plants and also trees and shrubs; on 38 species of plants in garden (see Table 6.4, p. 140). Overwinter as pupae. MT
64. Pale-shouldered brocade *Lacanobia thalassina* (Hufnagel). Quite common, June and July. Larvae on variety of trees and shrubs. Overwinter as pupae. MT
65. Bright-line brown-eye *Lacanobia oleracea* (L.). Very common, May–July, and on

into autumn in small numbers. Unusually early record on 14 April 1979. Larvae on extraordinary range of shrubs and herbaceous plants; on 36 species of plants in garden (see Table 6.5, p. 141). Overwinter as pupae. MT, BT

66. Broom *Ceramica pisi* (L.). Common, June and July, occasionally in August. Larvae on wide variety of trees, shrubs and herbaceous plants. Overwinter as pupae.

67. Campion *Hadena rivularis* (Fab.). Common, June and July. Larvae on ripening seeds of many Caryophyllaceae. Overwinter as pupae. MT

68. Lychnis *Hadena bicruris* (Hufnagel). Quite common, June and August. Larvae on various species of Caryophyllaceae. Overwinter as pupae.

69. Powdered quaker *Orthosia gracilis* (D. & S.). Fairly common, April and May. Larvae on variety of trees, shrubs and herbaceous plants. Overwinter as pupae. MT

70. Common quaker *Orthosia stabilis* (D. & S.). Common, March–June. Larvae on variety of trees; on *Buddleia davidii*, *Prunus cerasifera* and *Salix fragilis* in garden. Overwinter as pupae. MT

71. Clouded drab *Orthosia incerta* (Hufnagel). Common, March–June. Larvae on most trees and shrubs. Overwinter as pupae. MT

72. Hebrew character *Orthosia gothica* (L.). Common, April and May. Larvae on variety of trees, shrubs and herbaceous plants; on *Potentilla fruticosa* in garden. Overwinter as pupae. MT

73. Brown-line bright-eye *Mythimna conigera* (D. & S.). Quite common, July and August. Larvae on many grasses. Overwinter as larvae.

74. Clay *Mythimna ferrago* (Fab.). Common, July and August. Larvae mainly on grasses, occasionally on herbaceous plants. Overwinter as larvae. MT

75. Smoky wainscot *Mythimna impura* (Hübner). Common, July and August. Larvae on many grasses. Overwinter as larvae. MT

76. Common wainscot *Mythimna pallens* (L.). Common, June–August. Larvae on many grasses. Overwinter as larvae. MT, BT

77. Shark *Cucullia umbratica* (L.). Quite common, June and July, and one in May 1976, the drought year. Larvae on various species of *Sonchus* and *Lactuca* (Compositae). Overwinter as pupae.

78. Early grey *Xylocampa areola* (Esper). Common, April–June. Larvae on *Lonicera periclymenum*; on *L. periclymenum* in garden. Overwinter as pupae. MT

79. Brown-spot pinion *Agrochola litura* (L.). Common, August and September. Larvae on various trees, shrubs and herbaceous plants; on *Salix fragilis* in garden. Overwinter as eggs. MT, BT

80. Beaded chestnut *Agrochola lychnidis* (D. & S.). Common, September and October. Larvae on grasses and various herbaceous plants. Overwinter as eggs. MT, BT

81. Lunar underwing *Omphaloscelis lunosa* (Howarth). Quite common in September. Larvae on grasses. Overwinter as larvae.

82. Sallow *Xanthia icteritia* (Hufnagel). Common, August and September. Larvae on *Salix* catkins when young, later on leaves or on herbaceous plants. Overwinter as eggs. MT

83. Poplar grey *Acronicta megacephala* (D. & S.). Common, May–August. Larvae on various Salicaceae. Overwinter as pupae.

84. Grey dagger *Acronicta psi* (L.). Common, May–July, and one in August 1981. Larvae on variety of trees; on *Betula pendula*, *Crataegus monogyna* and *Rosa* sp. in garden. Overwinter as pupae. MT

85. Marbled beauty *Cryphia domestica* (Hufnagel). Common, June–August. Larvae on lichens. Overwinter as larvae. MT

86. Mouse *Amphipyra tragopoginis* (Clerck). Common, July–September. Larvae on variety of shrubs and herbaceous plants. Overwinter as eggs. MT, BT

87. Straw underwing *Thalpophila matura* (Hufnagel). Common, July and August. Larvae on various grasses. Overwinter as larvae. MT

88. Small angle shades *Euplexia lucipara* (L.). Common, June–August. Larvae on variety of trees, shrubs, herbaceous plants and, especially, ferns; in garden, on *Althaea rosea*, *Buddleia davidii*, *Calystegia sepium*, *Epilobium angustifolium*, *E. hirsutum*, *Lupinus polyphyllus*, *Lycopersicon esculentum*, *Malva sylvestris*, *Oenothera biennis*, *Paeonia officinalis*, *Solanum dulcamara*, *Spinacia oleracea*, *Weigela florida* and, especially, male fern *Dryopteris filix-mas*. Overwinter as pupae. MT

89. Angle shades *Phlogophora meticulosa* (L.). Common, almost all months of year, especially September–November. Larvae on extraordinary range of herbaceous plants, and also trees and shrubs; on 31 species of plants in garden and one pot plant indoors (see Table 6.6, p. 142). Overwinter as larvae. MT, BT

90. Dun-bar *Cosmia trapezina* (L.). Quite common, July–September. Larvae on wide variety of trees and shrubs; on *Buddleia davidii* in garden. Overwinter as eggs.

91. Dark arches *Apamea monoglypha* (Hufnagel). Abundant with occasional melanics, June–August and two in September 1980. Larvae on various grasses. Overwinter as larvae. MT, BT

92. Clouded-bordered brindle *Apamea crenata* (Hufnagel). Common, May and June. Dimorphic. Larvae on various grasses. Overwinter as larvae. MT

93. Clouded brindle *Apamea characterea* (Clerck). Common, June and July. Larvae on various grasses. Overwinter as larvae. MT

94. Dusky brocade *Apamea remissa* (Hübner). Common, June–September. Larvae on various grasses. Overwinter as larvae. MT

95. Small clouded brindle *Apamea unanimis* (Hübner). Fairly common, July and August. Larvae on various grasses. Overwinter as larvae. MT

96. Rustic shoulder-knot *Apamea sordens* (Hufnagel). Common, May–July. Larvae on various grasses. Overwinter as larvae. MT

97. Marbled minor *Oligia strigilis* (L.). Common, June–August. Larvae on various grasses. Overwinter as larvae. MT

98. Tawny marbled minor *Oligia latruncula* (D. & S.). Common, June–August. Larvae on various grasses. Overwinter as larvae. MT

99. Middle-barred minor *Oligia fasciuncula* (Haworth). Common, June. Larvae on various grasses. Overwinter as larvae. MT

100. Rosy minor *Mesoligia literosa* (Haworth). Common, June–August. Larvae on various grasses. Overwinter as larvae. MT

101. Common rustic *Mesapamea secalis* (L.). Common and extremely polymorphic, June–August. Exceptionally abundant in 1975 and 1976, when 149 and 154, respectively, were taken in Malaise trap. A total of 244 in 15 nights in mercury-vapour trap in August 1979. Larvae on various grasses. Overwinter as larvae. MT, BT

102. Flounced rustic *Luperina testacea* (D. & S.). Common, August and September. Larvae on various grasses. Overwinter as larvae. MT

103. Rosy rustic *Hydraecia micacea* (Esper). Common, August and September, in most years. Larvae on variety of herbaceous plants. Overwinter as eggs. MT

104. Uncertain *Hoplodrina alsines* (Brahm). Common, June and July, and one in August 1972. Larvae on variety of herbaceous plants. Overwinter as larvae.

105. Rustic *Hoplodrina blanda* (D. & S.). Common, July and August. Larvae on variety of herbaceous plants. Overwinter as larvae.
106. Vine's rustic *Hoplodrina ambigua* (D. & S.). Possibly common and frequently overlooked, August and September. Larvae on variety of herbaceous plants. Overwinter as larvae.
107. Mottled rustic *Caradrina morpheus* (Hufnagel). Common, June–August. Larvae on variety of herbaceous plants; on *Aster amellus*, *Lamium maculatum* and *Lavatera trimestris* in garden. Overwinter as larvae. MT
108. Pale mottled willow *Caradrina clavipalpis* (Scopoli). Common, July–September. Larvae on various grasses and on variety of herbaceous plants. Overwinter as larvae. MT
109. Burnished brass *Diachrysia chrysitis* (L.). Common, June–September in two broods. Larvae on variety of herbaceous plants; on *Aster novi-belgii*, *Mentha spicata*, *Origanum majorana*, *Petroselinum crispum* and *Pisum sativum* in garden. Overwinter as larvae. MT
110. Silver-Y *Autographa gamma* (L.). Common, May–November, as immigrant. Usually arrive in June, earliest record was 22 May 1976. During the period 1972–86, two good years, 1973 and 1975, with 233 and 155, respectively, in Malaise trap; otherwise annual Malaise trap catches fluctuated from 4 in 1981 to 92 in 1985. Larvae on variety of herbaceous plants; on *Borago officinalis*, *Lactuca sativa*, *Nepeta mussinnii* and *Petroselinum crispum* in garden. MT, BT
111. Beautiful golden Y *Autographa pulchrina* (Haworth). Common, June. Larvae on variety of herbaceous plants. Overwinter as larvae.
112. Plain golden Y *Autographa jota* (L.). Common, June and July. Larvae on variety of shrubs and herbaceous plants. Overwinter as larvae. MT
113. Spectacle *Abrostola triplasia* (L.). Common, May–August, in two broods. Larvae on *Urtica dioica*. Overwinter as pupae. MT
114. Snout *Hypena proboscidalis* (L.). Common, July and August, and one in May 1976, the drought year. Larvae on *Urtica dioica*; on *U. dioica* in garden. Overwinter as larvae. MT
115. Fan-foot *Polypogon tarsipennalis* (Treitschke). Common, July and August, and one in May 1976, the drought year. Larvae on various trees and shrubs. Overwinter as larvae. MT

UNCOMMON GARDEN MOTHS: THOSE THAT ARE UNCOMMON, EVEN IF REGULAR, AND THOSE THAT ARE IRREGULAR

Hepialidae

116. Ghost moth *Hepialus humuli* (L.). Quite common in June and July, but irregular. Larvae on roots of many herbaceous plants and grasses; on roots of *Chrysanthemum frutescens* in garden. Overwinter as larvae.

Cossidae

117. Wood leopard *Zeuzera pyrina* (L.). Twelve records, June–September. Larvae in wood of variety of trees and shrubs. Overwinter as larvae.

Lasiocampidae

118. Lackey *Malacosoma neustria* (L.). Ten records, nine in July and one in August 1980. Larvae on variety of trees and shrubs. Overwinter as eggs.

Drepanidae

119. Chinese character *Cilix glaucata* (Scopoli). Uncommon, May–July and second brood in August. Larvae on various trees and shrubs. Overwinter as pupae. MT

Geometridae

120. Common emerald *Hemithea aestivaria* (Hübner). Not seen until 1976, when four taken in June; thereafter common in July and August. Larvae on variety of trees and shrubs. Overwinter as larvae. MT
121. Blood-vein *Timandra griseata* (Petersen). Uncommon, June–August, and one each in September 1981 and 1986. Larvae on various Chenopodiaceae and Polygonaceae. Overwinter as larvae. MT
122. Single-dotted wave *Idaea dimidiata* (Hufnagel). Infrequent, July and August. Larvae on *Anthriscus sylvestris* and *Pimpinella saxifraga*, both Umbelliferae. Overwinter as larvae.
123. Small rivulet *Perizoma alchemillata* (L.). Rather infrequent, July and August. Larvae on *Galeopsis* spp. Overwinter as pupae.
124. Bordered pug *Eupithecia succenturiata* (L.). Infrequent, July–September. Larvae on *Artemisia vulgaris*. Overwinter as pupae. MT
125. Magpie *Abraxas grossulariata* (L.). Common in 1970s, June–August, but much scarcer in 1980s. Larvae on variety of shrubs; on *Ribes uva-crispa* and *R. sanguineum* in garden. Overwinter as larvae. MT
126. Feathered thorn *Colotois pennaria* (L.). Ten in October and November. Larvae on variety of trees and shrubs; on *Potentilla fruticosa* in garden. Overwinter as eggs.

Sphingidae

127. Elephant hawk *Deilephila elpenor* (L.). Irregular, June and July. Larvae on *Epilobium* spp., *Fuchsia* spp., *Galium* spp., and other plants; on *E. angustifolium* in garden. Overwinter as pupae. Last instar larvae may be green or brown. In 1975, a female confined on *E. angustifolium* laid many eggs which hatched successfully. Forty were reared singly in jars; 60 were reared in 12 batches of five. Three of 38 successfully reared singly turned brown at the 4th instar, the rest were green; four were still green in the 5th and final instar. Not all the crowded larvae survived in every jar, but 36 of the surviving 55 larvae were brown in the 4th instar, and all 12 jars had at least one brown 4th instar larva; all the 5th instar larvae were brown. The large 5th instar larvae are presumably brown when crowded because this gives greater camouflage as they rest by day on the soil surface below the foodplant. One green larvae may resemble a fallen leaf, but several are less likely to evade detection.

Notodontidae

128. Coxcomb prominent *Ptilodon capucina* (L.). Ten records, June–August. Larvae on variety of deciduous trees. Overwinter as pupae.

Photograph 6.2. Fully grown caterpillars of the elephant hawk moth Deilephila elpenor *may remain green if reared singly, but crowded caterpillars become brown usually by the 4th instar and always by the 5th and final instar.*

Arctiidae

129. Muslin *Diaphora mendica* (Clerck). Common in early 1970s but not seen since 1976. Larvae on variety of herbaceous plants. Overwinter as pupae.
130. Ruby tiger *Phragmatobia fuliginosa* (L.). Rather uncommon in early 1970s, July and August, and not seen since 1975. Larvae on variety of herbaceous plants. Overwinter as larvae.

Nolidae

131. Short-cloaked *Nola cucullatella* (L.). Regular but not common in July, and two in August 1978. Larvae on various trees in Rosaceae. Overwinter as larvae. MT

Noctuidae

132. Broad-bordered yellow underwing *Noctua fimbriata* (Schreber). Twelve records, ten in August and two in July. Larvae on variety of shrubs and herbaceous plants. Overwinter as larvae.
133. Double dart *Graphiphora augur* (Fab.). Frequent, June–August. Larvae on various trees and shrubs, and occasionally on herbaceous plants. Overwinter as larvae.
134. Six-striped rustic *Xestia sexstrigata* (Haworth). Eleven records. Larvae on variety of herbaceous plants. Overwinter as larvae.
135. Shoulder-striped wainscot *Mythimna comma* (L.). Quite common until 1976, then only four records up to 1986, all June and July. Larvae on various grasses. Overwinter as larvae.
136. Green-brindled crescent *Allophyes oxycanthae* (L.). One in Malaise trap, October 1973, was only record until October 1978 and 1979, when it was common with more melanics than typicals; then not until October 1986, two melanics. Larvae on *Crataegus* spp. and *Prunus spinosa*. Overwinter as eggs.
137. Red-line quaker *Agrochola lota* (Clerck). Not seen until October 1977, when common; common in October 1978 and 1979, and one in November 1978; not

seen then until one each in October 1985 and 1986. Larvae on *Salix* spp. Overwinter as eggs.

138. Centre-barred sallow *Atethmia centrago* (Haworth). Thirteen records in August and September. Larvae on *Fraxinus excelsior*. Overwinter as eggs.

139. Copper underwing *Amphipyra pyramidea* (L.). Not seen until 1978, when one in August; thereafter, quite common, August and September. Larvae on various trees and shrubs. Overwinter as eggs.

140. Light arches *Apamea lithoxylaea* (D. & S.). Frequent, but never common, July and August. Larvae on grasses. Overwinter as larvae. MT

141. Small dotted buff *Photedes minima* (Haworth). Twelve records in July and August. Larvae on tufted hair-grass *Deschampsia caespitosa*. Overwinter as larvae.

142. Golden plusia *Polychrysia moneta* (Fab.). Eighteen records, July–September. A recent immigrant to Britain. Larvae on *Aconitum* spp. and *Delphinium* spp. Overwinter as larvae.

143. Herald *Scoliopteryx libatrix* (L.). Frequent, August–October, and after hibernation in May and June. Larvae on various Salicaceae. Overwinter as adults. MT

144. Small fan-foot *Polypogon nemoralis* (Fab.). Uncommon, June and July. Larvae on *Alnus* spp. and *Quercus* spp. Overwinter as pupae.

SCARCE GARDEN MOTHS: THOSE RECORDED MORE THAN ONCE BUT LESS THAN TEN TIMES, 1972–86

Drepanidae

145. Oak hook-tip *Drepana binaria* (Hufnagel)

Thyatiridae

146. Peach blossom *Thyatira batis* (L.)

Geometridae

147. Small blood-vein *Scopula imitaria* (Hübner) MT
148. Small dusty wave *Idaea seriata* (Schrank)
149. Streamer *Anticlea derivata* (D. & S.)
150. Barred straw *Eulithis pyraliata* (D. & S.)
151. Small phoenix *Ecliptopera silaceata* (D. & S.)
152. Barred yellow *Cidaria fulvata* (Forster)
153. Juniper carpet *Thera juniperata* (L.)
154. Broken-barred carpet *Electrophaes corylata* (Thunberg)
155. Green carpet *Colostygia pectinataria* (Knoch)
156. Tissue *Triphosa dubitata* (L.)
157. Winter *Operophtera brumata* (L.). Larvae in garden on *Potentilla fruticosa*, *Ribes sanguineum*, *Salix fragilis* and *Ulmus glabra*.
158. Twin-spot carpet *Perizoma didymata* (L.)
159. Toadflax pug *Eupithecia linariata* (D. & S.)
160. Wormwood pug *Eupithecia absinthiata* (Clerck)

161. Currant pug *Eupithecia assimilata* (Doubleday)
162. V-pug *Chloroclystis v-ata* (Haworth)
163. Double-striped pug *Gymnoscelis rufifasciata* (Haworth). Larvae abundant in garden on flowers of *Buddleia davidii* in 1974, 1975 and 1986.
164. Seraphim *Lobophora halterata* (Hufnagel)
165. Clouded border *Lomaspilis marginata* (L.)
166. Tawny-barred angle *Semiothisa liturata* (Clerck)
167. Latticed heath *Semiothisa clathrata* (L.)
168. V-moth *Semiothisa wavaria* (L.)
169. Brown silver-line *Petrophora chlorosata* (Scopoli)
170. Scorched wing *Plagodis dolabraria* (L.)
171. Lilac beauty *Apeira syringaria* (L.) MT
172. September thorn *Ennomos erosaria* (D. & S.)
173. Lunar thorn *Selenia lunularia* (Hübner)
174. Pale brindled beauty *Apocheima pilosaria* (D. & S.)
175. Dotted border *Agriopis marginaria* (Fab.)
176. Waved umber *Menophra abruptaria* (Thunberg)
177. Bordered white *Bupalus piniaria* (L.)
178. Common white wave *Cabera pusaria* (L.)
179. Clouded silver *Lomographa temerata* (D. & S.). Pupa found in garden in May 1975.
180. Barred red *Hylaea fasciaria* (L.)

Sphingidae

181. Humming-bird hawk *Macroglossum stellatarum* (L.)

Notodontidae

182. Buff-tip *Phalera bucephala* (L.)
183. Puss *Cerura vinula* (L.)
184. Sallow kitten *Harpyia furcula* (Clerck)
185. Iron prominent *Notodonta dromedarius* (L.)
186. Figure of eight *Diloba caeruleocephala* (L.)

Lymantriidae

187. Pale tussock *Dasychira pudibunda* (L.)
188. White satin *Leucoma salicis* (L.)

Arctiidae

189. White ermine *Spilosoma lubricipeda* (L.)
190. Cinnabar *Tyria jacobaeae* (L.)

Noctuidae

191. Dark sword-grass *Agrotis ipsilon* (Hufnagel), MT
192. Dotted rustic *Rhyacia simulans* (Hufnagel)
193. Pearly underwing *Peridroma saucia* (Hübner)

194. Red chestnut *Cerastis rubricosa* (D. & S.)
195. Light brocade *Lacanobia w-latinum* (Hufnagel)
196. Broad-barred white *Hecatera bicolorata* (Hufnagel)
197. Tawny shears *Hadena perplexa* (D. & S.)
198. Varied coronet *Hadena compta* (D. & S.). MT
199. Antler *Cerapteryx graminis* (L.)
200. Feathered gothic *Tholera decimalis* (Poda)
201. Small quaker *Orthosia cruda* (D. & S.)
202. Chamomile shark *Cucullia chamomillae* (D. & S.)
203. Brindled green *Dryobotodes eremita* (Fab.)
204. Grey chi *Antitype chi* (L.)
205. Dark chestnut *Conistra ligula* (Esper)
206. Dusky-lemon sallow *Xanthia gilvago* (D. & S.)
207. Alder *Acronicta alni* (L.)
208. Knot grass *Acronicta rumicis* (L.)
209. Svensson's copper underwing *Amphipyra berbera* (Fletcher)
210. Old lady *Mormo maura* (L.). MT
211. Bird's wing *Dypterygia scabriuscula* (L.). MT
212. Olive *Ipimorpha subtusa* (D. & S.)
213. Dingy shears *Enargia ypsillon* (D. & S.)
214. Lesser-spotted pinion *Cosmia affinis* (L.)
215. Double-lobed *Apamea ophiogramma* (Esper). MT
216. Cloaked minor *Mesoligia furuncula* (D. & S.)
217. Frosted orange *Gortyna flavago* (D. & S.)
218. Brown-veined wainscot *Archanara dissoluta* (Treitschke)
219. Large wainscot *Rhizedra lutosa* (Hübner)
220. Scarce silver-lines *Bena prasinana* (L.)
221. Dark spectacle *Abrostola trigemina* (Werneburg)
222. Red underwing *Catocala nupta* (L.)

MOTHS RECORDED ONCE ONLY, 1972–86

Sesiidae

223. Currant clearwing *Synanthedon salmachus* (L.) 1976 MT
224. Yellow-legged clearwing *Synanthedon vespiformis* (L.) 1976 MT
225. Red-belted clearwing *Conopia myapaeformis* (Borkhausen) 1976 MT

Lasiocampidae

226. Drinker *Philudoria potatoria* (L.) 1975

Geometridae

227. Grass emerald *Pseudoterpna pruinata* (Hufnagel) 1977
228. Shaded broad-bar *Scotopteryx chenopodiata* (L.) 1980
229. Dark umber *Philereme transversata* (Hufnagel) 1975
230. Pinion-spotted pug *Eupithecia insigniata* (Hübner) 1980
231. White-spotted pug *Eupithecia tripunctaria* (Herrich-Schäffer) 1974

232. Grey pug *Eupithecia subfuscata* (Haworth) 1974
233. Plain pug *Eupithecia simpliciata* (Haworth) 1974
234. Ochreous pug *Eupithecia indigata* (Hübner) 1976
235. Ash pug *Eupithecia fraxinata* (Crewe) 1981
236. Juniper pug *Eupithecia pusillata* (D. & S.) 1975
237. Chimney sweeper *Odezia atrata* (L.) 1975 MT
238. Bordered beauty *Epione repandaria* (Hufnagel) 1981
239. Small brindled beauty *Apocheima hispidaria* (D. & S.) 1981
240. Oak beauty *Biston strataria* (Hufnagel) 1978
241. Scarce umber *Agriopis aurantiaria* (Hübner) 1977
242. Mottled umber *Erannis defoliaria* (Clerck). One larva on *Prunus cerasifera* produced a moth in November 1978.
243. Early thorn *Theria rupicapraria* (D. & S.) 1980

Notodontidae

244. Pebble prominent *Eligmodonta ziczac* (L.) 1978

Noctuidae

245. True lover's knot *Lycophotia porphyrea* (D. & S.) 1981
246. Dotted clay *Xestia baja* (D. & S.) 1982
247. Great brocade *Eurois occulta* (L.) 1976
248. Grey arches *Polia nebulosa* (Hufnagel) 1974
249. Marbled coronet *Hadena confusa* (Hufnagel) 1973
250. Twin-spotted quaker *Orthosia munda* (D. & S.) 1973
251. Obscure wainscot *Mythimna obsoleta* (Hübner) 1976
252. Wormwood shark *Cucullia absinthii* (L.) 1974
253. Mullein shark *Cucullia verbasci* (L.). 1974. Four larvae on *Buddleia davidii* in 1977.
254. Deep-brown dart *Aporophyla lutulenta* (D. & S.) 1976
255. Merveille du jour *Dichonia aprilina* (L.) 1979
256. Brick *Agrochola circellaris* (Hufnagel) 1986
257. Barred sallow *Xanthia aurago* (D. & S.) 1979
258. Sycamore *Acronicta aceris* (L.) 1979
259. Miller *Acronicta leporina* (L.) 1978
260. Brown rustic *Rusina ferruginea* (Esper) 1979
261. Dusky sallow *Emerobia ochroleuca* (D. & S.) 1976
262. Gold spangle *Autographa bractea* (D. & S.) 1973
263. Beautiful hook-tip *Laspeyria flexula* (D. & S.) 1972

Micro-Lepidoptera

Names and foodplants from Emmet (1988), with foodplants recorded in garden. Number of records given.

Incurvariidae

264. *Incurvaria masculella* (D. & S.). Two. Larvae in leaves of *Crataegus* or *Rosa* spp.

Tineidae

265. *Tinea trinotella* (Thunberg). One. Larvae in birds' nests.

Yponomeutidae

266. *Yponomeuta cagnagella* (Hübner). Abundant. Larvae on *Euonymus* spp.; on *Crataegus monogyna* and *Cytisus scoparius* in garden.
267. *Ypsolopha dentella* (Fab.). Two. Larvae on *Lonicera* spp.
268. *Ypsolopha scabrella* (L.). Three. Larvae on *Cotoneaster, Crataegus* or *Malus* spp.
269. *Orthotaelia sparganella* (Thunberg). One. Larvae in leaves and stems of *Glyceria maxima, Iris* and *Sparganium* spp.

Oecophoridae

270. *Hofmannophila pseudospretella* (Stainton). Two. On variety of dried organic material.
271. *Carcina quercana* (Fab.). Common. Larvae on leaves of variety of trees; on *Prunus cerasifera* in garden.
272. *Agonopterix arenella* (D. & S.). One. Larvae on leaves of thistles and similar Compositae; on *Centaurea montana* in garden.
273. *Agonopterix nervosa* (Haworth). Larvae on various shrubs of family Leguminosae; common on *Cytisus × kewensis* in garden.

Gelechiidae

274. *Mirificarma mulinella* (Zeller). Larvae on flowers of various Leguminosae; on *Cytisus scoparius* in garden.
275. *Dichomeris marginella* (Fab.). Two. Larvae in needles of *Juniperus* spp.
276. *Brachmia rufescens* (Haworth). One. Larvae on leaves of various grasses.

Tortricidae

277. *Agapeta hamana* (L.). Common. Larvae in roots of *Carduus* and *Cirsium* spp.
278. *Pandemis corylana* (Fab.). Common, August and September. Larvae on variety of trees and shrubs.
279. *Pandemis cerasana* (Hübner). Three. Larvae on variety of trees and herbaceous plants; on *Epilobium angustifolium* and *Rosa* spp. in garden.
280. *Pandemis heparana* (D. & S.). Two. Larvae on variety of trees and shrubs; on *Ligustrum ovalifolium, Potentilla fruticosa* and *Prunus cerasifera* in garden.
281. *Archips podana* (Scopoli). Common, June and July, and one each in August and September. Larvae on variety of trees and shrubs; on *Buddleia davidii* in garden.
282. *Aphelia paleana* (Hübner). Two. Larvae on variety of trees, herbaceous plants and grasses.
283. *Clepsis spectrana* (Treitschke). Three. Larvae on variety of trees and herbaceous plants.
284. *Lozotaenioides formosana* (Geyer). Four. Larvae on *Pinus sylvestris*.
285. *Lozotaenia forsterana* (Fab.). Two. Larvae on variety of trees, shrubs and herbaceous plants.

286. *Cnephasia stephensiana* (Doubleday). Larvae on variety of herbaceous plants; on *Centaurea montana* in garden.
287. *Cnephasia* sp.1. One larva on *Buddleia davidii*.
288. *Cnephasia* sp.2. One.
289. *Tortrix viridana* (L.). Common. Larvae on *Quercus* spp., occasionally on other trees.
290. *Croesia forsskaleana* (L.). Common, July and August. Larvae on *Acer* spp.
291. *Acleris sparsana* (D. & S.). Three. Larvae on various trees.
292. *Acleris rhombana* (D. & S.). One. Larvae on various trees, usually of family Rosaceae; on *Prunus cerasifera* in garden.
293. *Acleris variegana* (D. & S.). Common, July–September. Larvae on variety of trees and shrubs, usually Rosaceae; on *Rosa* sp. in garden.
294. *Celypha striana* (D. & S.). Two. Larvae on *Taraxacum* spp.
295. *Olethreutes lacunana* (D. & S.). Four. Larvae on most species of herbaceous plants, occasionally on trees or shrubs.
296. *Apotomis betuletana* (Haworth). One. Larvae on *Betula* spp.
297. *Epinotia bilunana* (Haworth). One. Larvae on *Betula* spp.
298. *Epiblema cynosbatella* (L.). Three. Larvae on *Rosa* or *Rubus* spp.
299. *Epiblema foenella* (L.). Four. Larvae in roots of *Artemisia* spp.
300. *Eucosma pupillana* (Clerck). One. Larvae in stem and roots of *Artemisia absinthium*.
301. *Cydia nigricana* (Fab.). One. Larvae in pods of *Lathyrus*, *Pisum* and *Vicia* spp.
302. *Cydia splendana* (Hübner). One. Larvae in nuts of *Castanea*, *Juglans* or *Quercus* spp.
303. *Cydia pomonella* (L.). Common. June–September. Larvae in fruits of various trees; in apples *Malus sylvestris* every year.
304. *Dichrorampha alpinana* (Treitschke) or *flavidorsana* Knaggs. Two. Larvae in roots of *Leucanthemum vulgare* (*alpinana*) or *Tanacetum vulgare* (*flavidorsana*).

Alucitidae

305. *Alucita hexadactyla* (L.). Common. Larvae on *Lonicera periclymenum* flowerbuds and leaves; on *L. periclymenum* in garden.

Pyralidae

306. *Chrysoteuchia culmella* (L.). Abundant. June–August. Larvae in stem bases of grasses.
307. *Crambus lathoniellus* (Zincken). Two. Larvae on stem bases of grasses, especially *Deschampsia caespitosa*.
308. *Crambus perlella* (Scopoli). Six. Larvae on stem bases of grasses, especially *Deschampsia* and *Festuca* spp.
309. *Agriphila selasella* (Hübner). One. Larvae on variety of grasses.
310. *Agriphila straminella* (D. & S.). Four. Larvae on variety of grasses.
311. *Agriphila tristella* (D. & S.). Abundant, July and August. Larvae on variety of grasses.
312. *Catoptria pinella* (L.). Two. Larvae on variety of grasses and sedges.
313. *Eudonia truncicolella* (Stainton). One. Larvae on mosses.
314. *Eudonia angustea* (Curtis). One. Larvae on mosses.
315. *Eudonia mercurella* (L.). Four. Larvae on mosses.

316. *Acentria ephemerella* (D. & S.). One. Larvae on submerged leaves of water plants.
317. *Evergestis forficalis* (L.). Common. Larvae on *Brassica*, *Raphanus* and other species of Cruciferae; on wild cabbage *Brassica oleraceus* and cultivars in garden.
318. *Evergestis pallidata* (Hufnagel). Four. Larvae on various species of Cruciferae.
319. *Pyrausta purpuralis* (L.). One. Larvae on *Mentha arvensis* and *Thymus drucei*.
320. *Eurrhypara hortulata* (L.). Common. Larvae on *Urtica* and other herbaceous plants; on *Urtica dioica* in garden.
321. *Phlyctaenia coronata* (Hufnagel). Common. Larvae on various shrubs.
322. *Udea prunalis* (D. & S.). One. Larvae on variety of trees, shrubs and herbaceous plants, particularly Labiatae.
323. *Udea olivalis* (D. & S.). Common, June and July. Larvae on variety of shrubs and herbaceous plants, particularly Labiatae.
324. *Udea ferrugalis* (Hübner). One. Larvae on variety of herbaceous plants.
325. *Nommophila noctuella* (D. & S.). One. Larvae on various herbaceous plants.
326. *Pleuroptya ruralis* (Scopoli). Common. Larvae on variety of herbaceous plants; on *Urtica dioica* in garden.
327. *Sceliodes laisalis* (Walker). One on 29 July 1979, the second of four British records. An African species feeding on tomato fruits (Owen, D.F. 1986a).
328. *Hypsopygia costalis* (Fab.). Common. Larvae on clover- and grass-hay, in squirrels' dreys, and probably in thatch and dead leaves.
329. *Orthopygia glaucinalis* (L.). Frequent. Larvae on dead and decaying vegetable matter of all types.
330. *Pyralis farinalis* (L.). One. Larvae on stored cereals and cereal refuse.
331. *Aglossa pinguinalis* (L.). One. Larvae on chaff and refuse of cereals, seeds, hay and sheep dung.
332. *Aphomia sociella* (L.). Four. Larvae in nests of social wasps and bumblebees.
333. *Pyla fusca* (Haworth). Three. Larvae on *Erica* and *Vaccinium* spp.
334. *Acrobasis ?repandana* (Fab.). One. Larvae on *Quercus* spp.
335. *Numonia advenella* (Zincken). Two. Larvae on *Crataegus* spp. and *Sorbus aucuparia*.
336. *Myelois cribrella* (Hübner). One. Larvae on various thistles and burdock (Compositae).
337. *Euzophera pinguis* (Haworth). One. Larvae in inner bark of *Fraxinus*.
338. *Phycitodes binaevella* (Hübner). Four. Larvae on thistles (Compositae).

Pterophoridae

339. *Platyptilia gonodactyla* (D. & S.). One. Larvae on *Tussilago* and possibly *Petasites*.
340. *Platyptilia ochrodactyla* (D. & S.). Two. Larvae in stems of *Tanacetum vulgare*.
341. *Pterophorus pentadactyla* (L.). Common, July. Larvae on *Calystegia* and *Convolvulus* spp.
342. *Oidaematophorus lithodactyla* (Treitschke). Larvae on *Inula conyza* and *Pulicaria dysenterica*; one pupa found on *Populus* in garden.
343. *Emmelina monodactyla* (L.). Seven. Larvae on *Calystegia* or *Convolvulus* spp.; on *Calystegia sepium* in garden.

ANNUAL FLUCTUATIONS IN NUMBERS OF MOTHS

Mercury-vapour light trapping has not been done regularly or consistently, so the results can only be used qualitatively for describing variations from year to year in the abundance of moths. The Malaise trap, however, although not an ideal way of catching moths, has been run continuously from 1 April to 31 October for every year, 1972–86. Moreover, since it uses no attractant, it is non-selective in the species that it intercepts in their flight. Tough, fast, active moths, such as hawk moths, may be less likely to be caught in a Malaise trap, but they constitute only a small proportion of the garden moth fauna. And if that argument had much foundation, silver-Ys *Autographa gamma* might be expected to evade capture, whereas they are by far the commonest species in the total 15-year sample, outnumbering the next commonest, the large yellow underwing *Noctua pronuba*, by more than two to one (N = 921 and 396, respectively). Annual fluctuations are, therefore, represented here by changes from year to year of the Malaise trap catch. This is for macro-moths only; the figures and identifications are incomplete for micros. The composition, in terms of species, of the annual catches is known completely only for 1974 onwards, largely because of initial problems in identifying moths caught in alcohol.

Annual fluctuations in moth numbers are shown in Fig. 6.1. By far the best years were 1973 and 1975; 1974 and 1976 were also quite good. The large catches for the period 1973–76 have not been matched since, although there was a slight improvement in 1984–86. The period 1973–76 was also good for butterflies (Fig. 5.1, p. 99), although butterfly numbers peaked in 1973, moth numbers in 1975. Moth numbers showed only a slight improvement in 1978, compared with that of butterflies, and butterflies showed a marked increase in 1982–84, which was not matched by moth numbers. The agreement between Figs. 5.1 and 6.1 is in the exceptionally poor figures for 1981, and this was also the low point in annual hoverfly catches (Fig. 7.2, p. 162); evidently the dull but dry summer of 1981, following a dull, wet spring, was as unfavourable for moths as for butterflies and hoverflies. Otherwise, there is no clear correlation between moth numbers and weather factors. All that can be said with confidence, is that warm, sunny summers are good for insects in general, including moths.

Autographa gamma was the commonest moth in ten of the 15 annual Malaise trap samples, forming 15.0 to 44.9 per cent of the annual catch (Table 6.1). Although it is an immigrant every year, *A. gamma* breeds in the garden, thus augmenting the population. High numbers of *A. gamma* are a reflection of favourable conditions in the garden as well as of the extent of immigration. That is why the two peaks in moth catches included 233 *A. gamma* in 1973 and 156 in 1975, the next best being 1985 with 92. Both 1973 and 1975 were years when *A. gamma* was abundant throughout the country (Bretherton 1982). Removal of *A. gamma* from the annual totals does not appreciably alter the 15-year pattern of fluctuations. Similarly, *Noctua pronuba*, a resident, was common in 1973 and 1975, 49 and 112, respectively, although it was also common

Table 6.1 *Annual fluctuations in catch of macro-moths in Malaise trap, 1972–86*

	No. of individuals (N)	No. of species (S)	Commonest species	Commonest as %
1972	98	>15	*Autographa gamma*	44.9
1973	622	>16	*A. gamma*	37.5
1974	339	47	*Noctua pronuba*	18.6
1975	693	57	*A. gamma*	22.5
1976	425	48	*Mesapamea secalis*	36.2
1977	153	30	*A. gamma*	24.2
1978	205	38	*N. pronuba*	21.5
1979	149	38	*Lacanobia oleracea*	19.5
1980	160	33	*A. gamma*	15.0
1981	73	29	*N. pronuba*	13.7
1982	200	43	*A. gamma*	27.0
1983	141	36	*A. gamma*	19.1
1984	270	38	*A. gamma*	23.0
1985	285	35	*A. gamma*	32.3
1986	237	37	*A. gamma*	32.1
1972–86	4050	>103	*A. gamma*	22.7

Note: Malaise trap specimens are collected in 70 per cent alcohol, not ideal for Lepidoptera, and skill in identifying the wet specimens had to be acquired. Consequently, the number of species given for 1972 and 1973 is a minimum. Thereafter, identifications of the trap catch were complete.

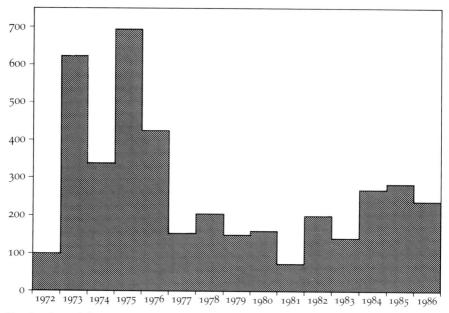

Fig. 6.1 *Annual fluctuations in total number of macro-moths caught in Malaise trap, 1972–86.*

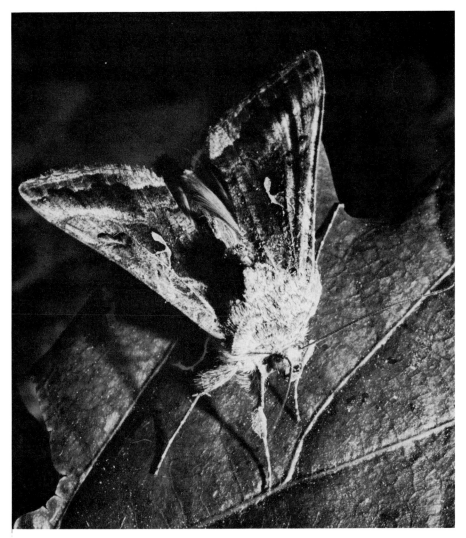

Photograph 6.3. The silver-Y Autographa gamma *is a migratory moth which breeds in the Leicester garden. In ten of the fifteen years 1972–86, it was the commonest species in annual Malaise trap samples of moths. Roger Whiteway.*

in 1974 (63) and again in 1978 (44), in both of which years it was the com-monest moth. *Mesapamea secalis* and *Lacanobia oleracea* were each the com-monest in one year, 1976 and 1979, respectively. *M. secalis* was exceptionally abundant in 1975 and in 1976, when 149 and 154, respectively, entered the Malaise trap.

Eleven species other than *A. gamma* and *N. pronuba* contributed more than 50 to the total of 4050 macro-moths in the Malaise trap: *Mesapamea secalis* (387), *Lacanobia oleracea* (260), *Xanthorhoe fluctuata* (169), *Noctua comes* (100), *Agrotis exclamationis* (93), *Caradrina morpheus* (85), *Hepialus lupulinus* (72), *Phlogophora meticulosa* (62), *Abrostola triplasia* (58), *Odontopera bidentata* (55) and *Idaea aversata* (51). Not all were present every year, however. *N. comes* was absent from the trap sample in 1980 and 1983; *A. exclamationis* in 1974; *H. lupulinus* in 1973, 1979 and 1981; *P. meticulosa* in 1981; *A. triplasia* in 1977 and 1978; and *O. bidentata* in 1977 and 1982.

The greatest annual number of species, 57, was taken in the Malaise trap in 1975, indicating that high catches are caused as much by addition of species as by increase in abundance of species. A total of 103 species was trapped. However, 27 of the species judged by mercury-vapour light trapping to be common in the garden, and listed as such in the previous section, were not caught in the Malaise trap. In the case of some, such as *M. tiliae* and *S. ocellata*, this may be a reflection of behaviour, but in most cases it is no more than chance and a small sample, reflecting the paucity of Malaise trap catches of moths relative to catches at light.

In the previous section, the 41 species recorded only once in the period 1972–86, usually by mercury-vapour light trapping, are listed. Thirty-two of these were caught before 1980, only nine in the 1980s. The peak year for rarities was 1976, when eight were recorded, including three clearwings (Sesiidae) all taken in the Malaise trap. 1976 was the drought year, when many nectar-feeding insects wandered far from their usual haunts in search of relatively scarce food, and in the same year several unusual butterfly species were caught in the garden.

One question that is inevitably raised by the lack of good moth years since the mid-1970s, is whether there might have been a reduction in moth population sizes caused by increasing urbanization and disturbance. One of the factors believed by many to affect moths is the increase in outdoor lighting evident in such densely populated and urbanized countries as England. A consideration of the impact of outdoor lighting on moths (Frank 1988) acknowledged that it has many potentially adverse effects, but draws no firm conclusions on its effect on moth numbers. It would be premature to interpret Fig. 6.1 as indicating an irreversible decline in moth populations until we have another run of years climatically like the period 1973–76.

SEASONAL FLUCTUATIONS IN NUMBERS OF MOTHS

For the 144 commoner garden moths, an indication of their seasonality has been given (by species) in the earlier section listing the species. This, however, indicates only their presence by month and not frequency, and for a quantitative view of seasonality it is necessary to look again at the Malaise trap catches. In Fig. 6.2 the monthly catches of moths are shown as percentages of the total for that year. The Malaise trap jar is changed weekly, for convenience on a Sunday, and monthly catches have been achieved by summing weeks to give the nearest fit to calendar months; this inevitably means that some months contain four weeks, some five, and the pattern changes from year to year. However, to judge from results obtained when hoverfly catches are summed in the same way, this should not make any marked

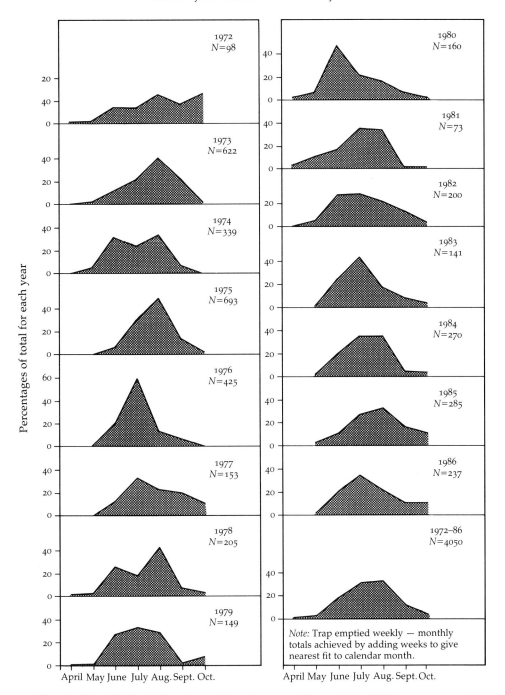

Fig. 6.2 *Monthly fluctuations in numbers of macro-moths caught in Malaise trap, 1972–86, expressed as percentages of annual total.*

difference to the seasonal pattern that is obtained. The striking feature of Fig. 6.2 is the lack of consistency, and the considerable differences in pattern between years.

In seven years, moth numbers reached a peak in July, in six years in August, in 1984 the July and August figures were equally high, and in 1980 the peak was in June. In seven years, there was no catch in April, and in 1977, when spring and summer were cool, there was none in May either; other April and May catches were so small, they scarcely appear on Fig. 6.2. In 1974, there was no catch in October, but in 1972, more than a quarter of the year's total, comprising 26 moths including 24 *Autographa gamma*, was caught in October; in all years when the October catch was relatively large, it consisted mainly of *A. gamma*. These marked differences from year to year result in the 15-year seasonal pattern differing from almost all the annual patterns.

When the catches for the same calendar month in different years are compared, it becomes clear why the shapes of the seasonal patterns differ in different years. The May 1981 catch was relatively large, 11 per cent of the annual total, and consisted of seven *Odontopera bidentata* and only one other moth. June catches were relatively large in 1974 and 1980, and small in 1975; the June 1974 and 1980 catches were diverse, and indeed it was the 1980 peak, but in 1975, what turned out later to be a good year had a slow start. The catch in July 1976 was relatively large, and it formed the pronounced July peak characteristic for many insects in the drought summer. In July 1972, however, the catch was relatively small, no better than that of June, and was associated, perhaps, with the cool, dry, dull summer. The August catches of 1973, 1975 and 1978 were relatively large, and formed the summer peak in all three years; 1973 and 1975 were good moth years, and August was hot, but in 1978, August picked up very much after a cool July when the catch was relatively poor. In 1976, 1980 and 1983, the August catch was relatively small because the season was already declining after a peak earlier in the summer. In both 1973 and 1977, September catches were relatively large because of prolonged summer weather, the catches consisting, respectively, of 138 moths including 121 *A. gamma*, and 31 moths including 16 *A. gamma*. In 1979, 1981 and 1984, the September catches were relatively small, in each case associated with the weather; 1979 was a generally cool summer, 1981 was cool earlier although August and September were warm, and in 1984 the previously hot summer had finished.

When running a mercury-vapour light trap, it is easy to get the impression that July and August are always good for moths, because the light is such a good attractant. Looking at the smaller, but perhaps less biased Malaise trap catch, however, it becomes clear that there are striking differences between years in the seasonal pattern of moth flight.

LARVAL FOODPLANTS OF MOTHS

Of the 343 moths (macros and micros) recorded in the garden, the caterpillars of 97 feed only on herbaceous plants and those of a further 53 feed sometimes on herbaceous plants. Seventy-seven feed on trees, 22 on shrubs, 42 on trees or shrubs and a further 44 feed sometimes on trees or shrubs. Of the remainder, 37 have caterpillars that feed on grass and a further 12 sometimes feed on grass. In addition, there are a handful of species whose caterpillars eat other things, such as lichen, moss or dried vegetable material. In summary, the caterpillars of 150 feed only or sometimes on herbaceous plants, the caterpillars of 148 feed only or sometimes on trees, and the caterpillars of 103 feed only or sometimes on shrubs. There is no clear bias towards one or another growth form of foodplant among the garden moth fauna, and the numbers feeding only on woody or only on non-woody plants are almost equal.

Sixty-eight species of moths (macros and micros) have been found breeding in the garden at one time or another on a total of 115 species of plants in 35 different families. These moths are listed by family in Table 6.2, and their foodplants can be found where the species are listed and described in an earlier section. Examination of the degree of specialization on foodplants of the 144 commoner garden moths, shows that 14 are specialists (restricted to a species or genus of plants), 40 are moderately specialized (restricted to a family of plants), and 89 are polyphagous, indiscriminate as to plant family. (The foodplant of one is unknown.) In comparison, of the 49 macros that breed in the garden, three are specialists, five are moderately specialized, and 41 are more or less polyphagous. Among macros, the three specialists that breed in the garden are *Eulithis mellinata* on *Ribes uva-crispa*, *Hypena proboscidalis* on *Urtica dioica*, and *Xylocampa areola* on *Lonicera periclymenum*. Moderately specialized are *Xanthorhoe fluctuata* on Cruciferae, *Opisthograptis luteolata* on Rosaceae, *Laothoe populi* on Salicaceae, *Cucullia verbasci* on Scrophulariaceae (and on Loganiaceae), and *Scoliopteryx libatrix* on Salicaceae. The group that is conspicuous by its absence from the garden breeding records is the one that uses grasses as foodplants.

Thus, only just over 16 per cent of those moths that breed in the garden are moderately or highly specialized, whereas over 37 per cent of the commoner garden moths are to some extent specialized. This suggests that it is those species with restricted larval food requirements that are unlikely to breed in the garden. Conversely, over 83 per cent of the species that breed in the garden are polyphagous, compared with only just over 60 per cent of the commoner garden moths, implying that polyphages, more or less indiscriminate as to foodplant, are particularly successful in a garden. Similarly, among the micros that breed in the garden, 11 are polyphagous, seven are moderately specialized, and only one is specialized (on *Lonicera periclymenum*).

Cucullia verbasci, however, is particularly interesting, having adopted the alien *Buddleia davidii* as an alternative foodplant in gardens. *Buddleia* belongs to a family, Loganiaceae, unrepresented in the native flora of northwestern Europe, and yet it is apparently acceptable as a foodplant to a specialist species. The reason for this appears to be a chemical resemblance between Loganiaceae and Scrophulariaceae, which has led to the moth's acceptance of

Table 6.2 *Moth larvae feeding in the garden on native and alien species of plants*

	Native	Alien	Total
HEPIALIDAE			
Hepialus humuli	—	1	1
YPONOMEUTIDAE			
Yponomeuta cagnagella	2	—	2
OECOPHORIDAE			
Carcina quercana	—	1	1
Agonopterix arenella	—	1	1
A. nervosa	1	—	1
GELECHIIDAE			
Mirificarma mulinella	1	—	1
TORTRICIDAE			
Pandemis cerasana	2	—	2
P. heparana	1	2	3
Archips podana	—	1	1
Cnephasia stephensiana	—	1	1
Cnephasia sp.	—	1	1
Acleris rhombana	—	1	1
A. variegana	1	—	1
Cydia pomonella	1	—	1
ALUCITIDAE			
Alucita hexadactyla	1	—	1
PYRALIDAE			
Evergestis forficalis	1	—	1
Eurrhypara hortulata	1	—	1
Pleuroptya ruralis	1	—	1
PTEROPHORIDAE			
Oidaematophorus lithodactyla	1	—	1
Emmelina monodactyla	1	—	1
GEOMETRIDAE			
Xanthorhoe fluctuata	1	3	4
Camptogramma bilineata	1	3	4
Eulithis mellinata	1	—	1
Chloroclysta truncata	2	7	9
Epirrita dilutata	—	1	1
Operophtera brumata	3	1	4
Eupithecia exiguata	7	5	12
E. vulgata	—	1	1
Gymnoscelis rufifasciata	—	1	1
Abraxas grossulariata	1	1	2
Opisthograptis luteolata	2	2	4
Odontopera bidentata	3	5	8
Crocallis elinguaria	—	1	1
Ourapteryx sambucaria	—	2	2
Colotois pennaria	1	—	1
Lycia hirtaria	2	—	2
Biston betularia	3	2	5
Agriopsis marginaria	1	—	1

Table 6.2 *Continued*

	Native	Alien	Total
Erannis defoliaria	—	1	1
Peribatodes rhomboidaria	1	—	1
Alcis repandata	1	—	1
SPHINGIDAE			
Smerinthus ocellata	1	—	1
Laothoe populi	2	—	2
Deilephila elpenor	1	—	1
LYMANTRIIDAE			
Euproctis similis	2	—	2
NOCTUIDAE			
Ochropleura plecta	1	—	1
Noctua pronuba	—	4	4
N. comes	2	6	8
Xestia c-nigrum	1	—	1
X. triangulum	—	1	1
Naenia typica	—	3	3
Mamestra brassicae	8	30	38
Melanchra persicariae	15	23	38
Lacanobia oleracea	7	29	36
Orthosia stabilis	1	2	3
O. gothica	1	—	1
Cucullia verbasci	—	1	1
Xylocampa areola	1	—	1
Agrochola litura	1	—	1
Acronicta psi	3	—	3
Euplexia lucipara	6	8	14
Phlogophora meticulosa	6	26	32
Cosmia trapezina	—	1	1
Caradrina morpheus	—	3	3
Diachrysia chrysitis	1	4	5
Autographa gamma	—	4	4
Scoliopteryx libatrix	2	—	2
Hypena proboscidalis	1	—	1

Note: Foodplants are listed with the description of the moths' garden status. Based on Owen, D. F. (1986b) with additions.

the alien family as suitable food (Owen, D. F. 1984). Eighteen moths, including two micros, have bred on *Buddleia* in the garden, but all the others are more or less polyphagous.

The four most polyphagous moths in the garden have fed on *Buddleia* at some time or another. Lists of foodplants for *Mamestra brassicae*, *Melanchra persicariae*, *Lacanobia oleracea* and *Phlogophora meticulosa* are given in Tables 6.3, 6.4, 6.5 and 6.6; they feed, respectively, on 38, 38, 36 and 32 plant species of many different families. This means that they are exposed to a whole range of

Photograph 6.4. The angle shades Phlogophora meticulosa *is one of the four most polyphagous moth species in the Leicester garden. Its caterpillars have been recorded on 31 species of garden plants, six native and 25 alien, and on one pot plant indoors. Roger Whiteway.*

plant chemistry, particularly with regard to secondary plant substances. These are chemicals that play no part in a plant's life processes but have probably evolved as part of its defensive strategy against herbivores; they are its characteristic chemical signature. Species that are able to accommodate such chemically diverse foods are probably able to do so because of the enzyme complement of their digestive systems. Brattsten, Wilkinson & Eisner (1977) showed that the noctuid moth *Spodoptera eridania* can adjust to new foodplants because of rapid induction of mixed function oxidases in response to small quantities of secondary plant substances; these enzymes degrade toxic compounds and, following induction, a larva is less susceptible to dietary poisoning. This is probably the secret to feeding on a wide variety of different foodplants, and the level of mixed function oxidases in polyphagous species may be 15 times greater than in monophagous species (Krieger, Feeny & Wilkinson 1971).

Why then should not all moth species be polyphagous as larvae? The answer has to be that different moth species have taken different evolutionary options. There are obvious drawbacks to specialization in that it sets restrictions on where and when a moth can find larval foodplants, but there are

Table 6.3 *Foodplants of* Mamestra brassicae
(N=native, A=alien)

A	*Thunbergia alata* Acanthaceae (pot plant indoors)
A	*Acer pseudoplatanus* Aceraceae
A	*Mahonia aquifolium* Berberidaceae
A	*Echium plantagineum* Boraginaceae
A	*Weigela florida* Caprifoliaceae
A	*Spinacia oleracea* Chenopodiaceae
A	*Artemisia dracunculus* Compositae
A	*Aster novi-belgii* Compositae
A	*Calendula officinalis* Compositae
A	*Callistephus chinensis* Compositae
A	*Erigeron* sp. Compositae
A	*Lactuca sativa* Compositae
A	*Tagetes patula* Compositae
N	*Calystegia sepium* Convolvulaceae
A	*Bryophyllum* sp. Crassulaceae (pot plant indoors)
N	*Brassica oleracea* Cruciferae
A	*Sinningia speciosa* Gesneriaceae (pot plant indoors)
A	*Ribes sanguineum* Grossulariaceae
N	*Ribes uva-crispa* Grossulariaceae
A	*Crocosmia crocosmiiflora* Iridaceae
A	*Lamium maculatum* Labiatae
N	*Mentha spicata* Labiatae
A	*Ocimum basilicum* Labiatae
A	*Origanum majorana* Labiatae
A	*Salvia splendens* Labiatae
N	*Stachys sylvatica* Labiatae
A	*Lathyrus odoratus* Leguminosae
A	*Pisum sativum* Leguminosae
A	*Vicia faba* Leguminosae
A	*Buddleia davidii* Loganiaceae
N	*Epilobium hirsutum* Onagraceae
A	*Papaver somniferum* Papaveraceae
A	*Delphinium* sp. Ranunculaceae
A	*Prunus cerasifera* Rosaceae
N	*Atropa bella-donna* Solanaceae
A	*Datura stramonium* Solanaceae
A	*Petroselinum crispum* Umbelliferae
N	*Urtica dioica* Urticaceae

almost certainly physiological advantages in the efficiency that comes from being highly adapted to processing one type of food. If, however, intra-specific competition between herbivores is strong, there is selective advantage in switching to an unrelated foodplant, whose insect fauna is taxonomically different and which has a different complex of predators and parasitoids associated with it (Holloway & Hebert 1979). One of the advantages of

Table 6.4 *Foodplants of* Melanchra
persicariae *(N=native, A=alien)*

N	*Betula pendula* Betulaceae
N	*Myosotis alpestris* Boraginaceae
N	*Humulus lupulus* Cannabaceae
N	*Lonicera periclymenum* Caprifoliaceae
N	*Sambucus nigra* Caprifoliaceae
A	*Spinacia oleracea* Chenopodiaceae
A	*Aster amellus* Compositae
A	*Aster novi-belgii* Compositae
A	*Calendula officinalis* Compositae
A	*Centaurea montana* Compositae
A	*Chrysanthemum frutescens* Compositae
A	*Solidago canadensis* Compositae
A	*Tagetes erecta* Compositae
N	*Tanacetum vulgare* Compositae
A	*Lunaria annua* Cruciferae
A	*Ribes sanguineum* Grossulariaceae
N	*Ribes uva-crispa* Grossulariaceae
N	*Hypericum androsaemum* Guttiferae
A	*Crocosmia crocosmiiflora* Iridaceae
A	*Lamium maculatum* Labiatae
A	*Melissa officinalis* Labiatae
N	*Mentha spicata* Labiatae
A	*Origanum majorana* Labiatae
A	*Salvia officinalis* Labiatae
A	*Lupinus polyphyllus* Leguminosae
A	*Buddleia davidii* Loganiaceae
A	*Lavatera trimestris* Malvaceae
N	*Malva sylvestris* Malvaceae
A	*Forsythia intermedia* Oleaceae
N	*Aquilegia vulgaris* Ranunculaceae
A	*Aruncus dioicus* Rosaceae
N	*Malus sylvestris* Rosaceae
A	*Potentilla fruticosa* Rosaceae
N	*Salix fragilis* Salicaceae
A	*Philadelphus coronarius* Saxifragaceae
A	*Lycopersicon esculentum* Solanaceae
N	*Urtica dioica* Urticaceae
also on fern N	*Dryopteris filix-mas*

extreme polyphagy in the Leicester garden is that foodplants are available all the year round. *Lamium maculatum*, for instance, has green leaves throughout the year, and such moth caterpillars as those of *Phlogophora meticulosa* can be found on it in almost every month of the year. Specialists tend to be restricted to a short breeding season, but polyphages can breed at any time and hence have fewer restraints on population growth.

Table 6.5 *Foodplants of* Lacanobia
oleracea *(N=Native, A=alien)*

A	*Mahonia aquifolium* Berberidaceae
N	*Betula pendula* Betulaceae
A	*Symphytum* × *uplandicum* Boraginaceae
A	*Dianthus barbatus* Caryophyllaceae
A	*Spinacia oleracea* Chenopodiaceae
A	*Artemisia dracunculus* Compositae
A	*Aster amellus* Compositae
A	*Aster novi-belgii* Compositae
A	*Callistephus chinensis* Compositae
A	*Chrysanthemum frutescens* Compositae
A	*Lactuca sativa* Compositae
A	*Tagetes erecta* Compositae
A	*Tagetes patula* Compositae
N	*Calystegia sepium* Convolvulaceae
A	*Convolvulus minor* Convolvulaceae
A	*Iberis sempervirens* Cruciferae
A	*Matthiola incana* Cruciferae
N	*Dipsacus fullonum* Dipsacaceae
A	*Geranium cinereum* Geraniaceae
N	*Ribes nigrum* Grossulariaceae
N	*Ribes uva-crispa* Grossulariaceae
A	*Crocosmia crocosmiiflora* Iridaceae
A	*Melissa officinalis* Labiatae
A	*Origanum majorana* Labiatae
A	*Lathyrus odoratus* Leguminosae
A	*Pisum sativum* Leguminosae
A	*Buddleia davidii* Loganiaceae
A	*Oxalis corymbosa* Oxalidaceae
N	*Rumex acetosa* Polygonaceae
A	*Paeonia officinalis* Ranunculaceae
A	*Aruncus dioicus* Rosaceae
A	*Potentilla fruticosa* Rosaceae
A	*Lycopersicon esculentum* Solanaceae
A	*Nicotiana alata* Solanaceae
N	*Solanum dulcamara* Solanaceae
A	*Petroselinum crispum* Umbelliferae

Buddleia davidii is the most widely used plant species in the garden, despite its alien origin, with moth larvae of 18 species feeding on it. Next most widely used are *Potentilla fruticosa* and *Prunus cerasifera* with nine each, *Rosa* sp. and *Lamium maculatum* with eight, *Salix fragilis*, *Ribes sanguineum* and *Crataegus monogyna* with seven, *Chrysanthemum frutescens*, *Ribes uva-crispa* and *Origanum majorana* with six, and *Aster novi-belgii*, *Iberis sempervirens*, *Urtica dioica* and *Petroselinum crispum* with five.

Moths

Table 6.6 *Foodplants of* Phlogophora
meticulosa *(N=native, A=alien)*

A	*Borago officinalis*	Boraginaceae
N	*Myosotis alpestris*	Boraginaceae
A	*Symphytym × uplandicum*	Boraginaceae
A	*Dianthus × allwoodii*	Caryophyllaceae
A	*Aster novi-belgii*	Compositae
A	*Centaurea montana*	Compositae
A	*Chrysanthemum frutescens*	Compositae
A	*Lactuca sativa*	Compositae
A	*Rudbeckia hirta*	Compositae
A	*Solidago canadensis*	Compositae
A	*Tagetes patula*	Compositae
A	*Convolvulus minor*	Convolvulaceae
N	*Sedum acre*	Crassulaceae
A	*Alyssum saxatile*	Cruciferae
A	*Arabis caucasica*	Cruciferae
A	*Aubrieta deltoidea*	Cruciferae
N	*Brassica oleracea*	Cruciferae
A	*Hesperis matronalis*	Cruciferae
A	*Iberis sempervirens*	Cruciferae
A	*Coleus* sp.	Labiatae (pot plant indoors)
A	*Lamium maculatum*	Labiatae
A	*Origanum majorana*	Labiatae
A	*Salvia officinalis*	Labiatae
A	*Buddleia davidii*	Loganiaceae
A	*Althaea rosea*	Malvaceae
A	*Dicentra formosa*	Papaveraceae
N	*Rumex acetosa*	Polygonaceae
N	*Geum urbanum*	Rosaceae
A	*Saxifraga* sp.	Saxifragaceae
A	*Antirrhinum majus*	Scrophulariaceae
N	*Digitalis purpurea*	Scrophulariaceae
A	*Petroselinum crispum*	Umbelliferae

FEEDING ON ALIENS

Of the most widely used foodplants, on each of which caterpillars of five or more species feed, nine are aliens, and only six are native. This is despite the widely held belief that 'Introduced plants, such as garden forms, often have no herbivores at all and therefore no food-chains exploiting them' (Elton 1966). In large measure this belief arises because of the way that books on moths and on foodplants, such as Stokoe & Stovin (1948) and Allan (1949), present data on foodplants, and the way in which others have used this information. These books were written for naturalists and collectors; they have, therefore, concentrated on plants, such as blackthorn *Prunus spinosa*, likely to yield rarities, and have largely ignored alien plants. Furthermore,

142

they tend to single out tree and shrub species, and are much more general about herbaceous plants; the only exceptions are when citing the herbaceous foodplants of specialist-feeders or rarities. Little specific information is given about polyphages, especially those on herbaceous plants. Unfortunately, it is just such books which have been used as sources by people such as Leather (1986), in compiling lists of herbivorous species utilizing particular plants. For instance, Leather does not record any Lepidoptera using *Potentilla fruticosa* as a foodplant, whereas, in this garden alone, nine moth species feed on it. In large part, this arises because *P. fruticosa* is rare in the wild and because few look at garden plants. However, it is widely grown in gardens, and, moreover, in a form that differs little, if at all, from the wild-type in foliage characteristics.

It has become generally accepted among urban conservation groups that 'native is best' when planting areas for habitat creation. The origins of this belief seem to be investigations by Southwood and others of the insect fauna associated with particular trees. They found that native willows *Salix* spp., oaks *Quercus* spp. and birches *Betula* spp. support many insect species, whereas such common aliens as sycamore *Acer pseudoplatanus* and horse chestnut *Aesculus hippocastanum*, although introduced several centuries ago, support relatively few (Kennedy & Southwood 1984, Southwood 1961). Tree species abundance, number of years that the species has been present in Britain, and whether it is evergreen or coniferous were found to account for 74 per cent of the variation in insect species richness between British trees. It is assumed that the longer a tree species has been present, the more time herbivores have had to evolve adaptations to overcome its physical and chemical defences. Aliens are assumed to have defences that are unfamiliar and hence initially impregnable. This may well be true of trees, which are long-lived and therefore might be expected to have evolved and to develop strong defence, but there are at best tenuous grounds for extending this theory to herbaceous plants. Yet this is exactly what has happened, in that advice that 'native is best' when planting for wildlife has been extended to all plants.

Analysis of the use by herbivorous moth larvae of native and alien plants in the Leicester garden has shown that extensive planting of aliens is by no means a disaster for moths (Owen, D.F. 1986b). I have updated the table in Owen's paper, and present the information on moths in Table 6.2. Sixty-eight species of moths use 115 plant species of 35 families in the garden. Forty-six species of moths use 40 native plants and 38 species of moths use 75 alien plants. Over the period of investigation (1972–86), the garden has included 146 native and 214 alien plant species of 78 families. Consequently, moths have used more than 27 per cent of the native plants, and 35 per cent of the aliens, suggesting that aliens are more heavily utilized, although the difference is not statistically significant.

The four most polyphagous of the moths that breed in the garden illustrate well this dependence on aliens. *Mamestra brassicae* has fed on eight native and 30 alien species, *Melanchra persicariae* on 15 native and 23 alien, *Lacanobia*

143

oleracea on seven native and 29 alien, and *Phlogophora meticulosa* on six native and 26 alien species (Tables 6.3, 6.4, 6.5 and 6.6). These four feed mostly on herbaceous plants, but what of species that utilize trees and shrubs? *Odontopera bidentata* has been found on five shrubs (one native, four alien), two trees (one native, one alien) and a herbaceous plant (native). *Ourapteryx sambucaria*, which is reported as feeding on a variety of trees and shrubs, mainly Rosaceae, occurs in the garden on *Ribes sanguineum* and *Ligustrum ovalifolium*, both aliens, and neither of them in the family Rosaceae.

Many of the native plants grown intentionally in the garden are cultivars, but they have usually been changed in either flower or fruit character and not in any way that would affect leaf-feeders. Thus, the cultivars of *Potentilla fruticosa* and *Rosa* sp. that are grown are probably indistinguishable from wild forms to a herbivore. Since *P. fruticosa* is rare in the wild, it is scarcely surprising that it is rarely mentioned as a foodplant, but, in the garden, no less than nine species of moth use it: *Pandemis heparana, Operophtera brumata, Eupithecia exiguata, Odontopera bidentata, Colotois pennaria, Biston betularia, Melanchra persicariae, Lacanobia oleracea* and *Orthosia gothica*. *Brassica oleracea* has been changed by cultivation into what is probably a more palatable form, and numbers of at least two of the four species that feed on it can be something of a problem in the garden.

There are more moth species on native than on alien plants, despite the fact that more alien than native plants are used. Twenty-five families of garden plants that are widely utilized by moth larvae include 98 native and 151 alien plants (Table 6.7). Thirty-three of the native and 64 of the alien plants are used as foodplants by a total of 68 species of moths, 46 on native and 38 on alien plants. The most heavily used plant family is the Rosaceae, of which seven natives and five aliens are used by a total of 27 moths. Second to this, however, is the alien family Loganiaceae, with just the one species, *Buddleia davidii*, used by 18 species of moths. In contrast, Compositae, the largest family of garden plants, with 18 natives and 40 aliens, is used by 13 moth species, only two on natives and all 13 on aliens. The families Caprifoliaceae, Grossulariaceae and Salicaceae are heavily used relative to their representation in the garden. Two native and one alien Caprifoliaceae (out of a total of four) are used by eight moths; two native and one alien Grossulariaceae (out of a total of four) are used by eight moth species; and the three native Salicaceae are used by ten species of moths (Table 6.7).

There is not strong evidence here that 'native is best'. On the contrary, there is every indication that alien plant species are widely acceptable as food for moth caterpillars. Some families of plants, such as Labiatae, Grossulariaceae, Rosaceae and Salicaceae, are evidently good as foodplants for moth caterpillars, but in general terms, the advice for making a garden attractive as a breeding site for moths is to diversify the vegetation.

Status of moths in Leicestershire

Table 6.7 *Use of alien and native species in different plant families as foodplants by moth larvae*

	Garden species		Use as foodplants		Moth species involved		
	Native	Alien	Native	Alien	On native	On alien	Total
Berberidaceae	0	2	—	1	—	3	3
Betulaceae	2	0	1	—	4	—	4
Boraginaceae	1	6	1	3	2	4	5
Caprifoliaceae	2	2	2	1	5	4	8
Caryophyllaceae	6	6	0	2	—	3	3
Chenopodiaceae	3	1	0	1	—	4	4
Compositae	18	40	2	13	2	13	13
Convolvulaceae	2	1	1	1	4	2	4
Cruciferae	8	13	1	7	4	8	10
Grossulariaceae	3	1	2	1	5	7	8
Guttiferae	1	2	1	0	2	—	2
Labiatae	9	15	2	8	3	12	12
Leguminosae	7	8	1	5	3	6	8
Loganiaceae	0	1	—	1	—	18	18
Malvaceae	1	2	1	2	3	6	6
Oleaceae	2	3	1	2	1	3	4
Onagraceae	2	5	2	2	4	2	5
Ranunculaceae	2	8	1	2	1	5	6
Rosaceae	10	11	7	5	21	13	27
Salicaceae	3	0	3	—	10	—	10
Saxifragaceae	0	4	—	2	—	4	4
Scrophulariaceae	6	5	1	1	3	1	3
Solanaceae	4	10	2	3	3	4	4
Umbelliferae	5	4	0	1	—	5	5
Urticaceae	1	1	1	0	5	—	5

STATUS OF MOTHS IN LEICESTERSHIRE

All but one of the moths listed as common or uncommon in the garden (144 species), are common and widespread in the county. The exception is *Hoplodrina ambigua* (listed as no. 106), which I have described as 'possibly common and frequently overlooked'. Leicestershire Museums hold four records of this species for 1986, two from other gardens in the city of Leicester and two from places west of the city, and one from just north of the city for 1976. However, as has been suggested, it is possible that this species is frequently overlooked, and it may well be common.

Two *Archanara dissoluta* (no. 218) were caught on 8 August 1975, and single specimens of *Rhizedra lutosa* (no. 219) were taken in September 1973 and October 1978; both are listed as scarce garden moths. These two species are characteristic of wet places, the larvae feeding on reeds *Phragmites australis*, and are not the sort of moth expected in a suburban garden. There are a few

Leicestershire records of each for the 1980s and also for the 1970s, mostly from wet places, such as Narborough Bog.

Single individuals of each of the clearwings, *Synanthedon salmachus* (no. 223), *S. vespiformis* (no. 224) and *Conopia myapaeformis* (no. 225), were taken in the Malaise trap in July 1976, the drought year. *S. salmachus* was recorded from a short distance north of the garden in 1976, and *S. vespiformis* was taken three times in the early 1960s at the same place in the east of the county, but there is no other recent record of *C. myapaeformis*. Since they fly by day, they are not normally caught in light traps, and are probably more common than the paucity of records suggests.

A single specimen of *Apocheima hispidaria* (no. 239) was caught in March 1981; the only other county record is from Ulverscroft, on the Charnwood Forest, in 1976. One *Hadena confusa* (no. 249) was taken in June 1973; there are two county records for 1985, from different gardens in Leicester, and five for the 1970s from widely dispersed localities in the county. The only specimen of *Autographa bractea* (no. 262) was caught in the garden in July 1973; there are five county records for the 1980s (three of them from Leicester in 1982, 1983 and 1986), four others, widely dispersed, for the 1970s and two for the 1960s. These three species are undoubtedly scarce to rare in the county, and their capture in a suburban garden is remarkable.

Recording of macro-Lepidoptera in the county has been good over the last 20 or 30 years because of a number of keen amateur and professional lepidopterists. There are, however, fewer keen and competent micro-Lepidopterists, and this may be why county records of micros are much sparser. Four of the 80 garden micros have not otherwise been recorded in the county during the past 30 years: *Cydia nigricana* (no. 301), *Sceliodes laisalis* (no. 327), *Platyptilia ochrodactyla* (no. 340) and *Oidaematophorus lithodactyla* (no. 342). One *C. nigricana* was caught in June 1974, and one *S. laisalis* in July 1979, the latter being chronologically the second of four British records (Owen, D.F. 1986a). *P. ochrodactyla* was taken once in July and once in August 1975, and *O. lithodactyla* is known only from a pupa found on *Populus nigra* in the garden. For three other micros taken in the garden, the only other county record is one in the 1980s: *Orthotaelia sparganella* (no. 269) taken once in the garden in July 1975, *Eucosma pupillana* (no. 300) taken once in July 1981, and *Evergestis pallidata* (no. 318), single individuals of which were taken in July, in August and in September 1974, and in August 1975. The identification of no. 304, two of which were caught in August 1981, is not certain; it could be either *Dichrorampha flavodorsana*, which has not been recorded in the county, or *D. alpinana*, which has been recorded once, in the south of the county in 1985. *Mirificarma mulinella* (no. 274), which was bred in July 1975 from a larva found on *Cytisus scoparius* in the garden, is otherwise only known in the county from records in four different years in the 1980s from another garden in Leicester. *Agriphila selasella* (no. 309), recorded once in the garden in August 1975, was otherwise recorded once in the county in 1980 and five times at widely dispersed localities in the 1970s. *Aglossa pinguinalis* (no. 331), recorded once in the garden in July 1974, has only otherwise been recorded twice in the county, in 1980 and in 1983 at sites on the outskirts of Leicester. These ten micros are undoubtedly scarce in the county, *Sceliodes laisalis* being nationally a rarity, but there is still much to discover about Leicestershire's micro-Lepidoptera.

Trapping of moths and rearing of larvae found even in a suburban garden is a fascinating and productive enterprise. Despite a general interest among naturalists in Lepidoptera, monitoring the garden fauna has produced a considerable body of new information for the county. The information on foodplants, gained from rearing all larvae found in the garden, has helped resolve the controversy over use of alien foodplants, and goes some way to explaining why gardens are such good moth habitats.

7

Hoverflies

Of the many flower-dependent groups of insects that abound in gardens, one of the most abundant and diverse is the family Syrphidae, the hoverflies. Moreover, few discrete, readily recognizable groups are as curious, as interesting, and so much an indicator of habitat diversity. Here, however, I must admit to some partiality, hoverflies being the garden group in which I am most interested, and on which I have spent most time. Those most often noticed in gardens are the larger species, many of which are brightly coloured, but from April until October a variety of small, slim, dark hoverflies can also usually be found feeding at flowers or hovering in sunny spots or dappled shade. Most hoverflies are glossy bodied, and many have yellow bands or pairs of yellow spots or lunules on the black abdomen; some are robust, others elongate in form; a few bear a striking resemblance to solitary or to social wasps; some resemble honeybees; and several furry, polymorphic species are mimics of bumblebees. Hoverflies, nevertheless, constitute a family, in contrast to butterflies and moths, which encompass, respectively, several and many families, i.e. hoverflies are taxonomically less variable.

The behavioural feature of the family Syrphidae which leads to them being called hoverflies is the ability to hover seemingly motionless, the transparent wings a shimmering blur, then dart away so fast they seem to vanish, only to re-appear, often in the exact spot they left. They are predominantly nectar- and pollen-feeders, hence their American name of flower flies. Anatomically, members of the family can be distinguished from other Diptera by the *vena spuria*, a thickening of the wing membrane along the central axis of the wing, looking like a vein but unconnected to any others, and by well-developed marginal cross-veins which are more conspicuous than the flimsy edge of the wing.

NATURAL HISTORY OF HOVERFLIES

So varied are the life histories of hoverflies that it is difficult, and would be misleading, to generalize. The common denominator is the dependence of adults on energy-rich fluids, usually in the form of nectar. However, some smaller species, such as *Melanostoma* and *Episyrphus*, seem to satisfy their energy requirements by eating pollen, and *Xylota* spp., whose fidgety behaviour reinforces their resemblance to solitary wasps, apparently collect pollen and other substances from leaf surfaces, and rarely visit flowers. I often see *Eupoedes* (=*Metasyrphus*) *luniger* and *Syrphus ribesii* lapping honeydew from leaf surfaces on aphid-infested garden plants, and suspect that hoverflies, like many other nectar-feeding insects, are indiscriminate as to the source of fluids rich in sugars, salts and nutrients. I have caught *Platycheirus scutatus, Episyrphus balteatus, Eupoedes corollae, E. luniger, Scaeva selenitica, Syrphus ribesii, S. vitripennis, Ferdinandea cuprea, Helophilus hybridus* and *Volucella pellucens* in traps baited with rotting fruit, and *S. ribesii* and *S. vitripennis* in traps baited with fish offal.

Different species take nectar and pollen in different proportions when feeding at flowers, and have correspondingly adapted mouthparts. *Melanostoma scalare, E. balteatus, Meliscaeva auricollis* and *S. ribesii*, which take more than 90 per cent pollen, have short proposces with broad, fleshy labella (terminal lobes); *Platycheirus peltatus, Sphaerophoria scripta, Rhingia campestris, Eristalis* spp. and *Syritta pipiens*, which more often take nectar, have long proposces with slim, elongate labella; and species such as *Platycheirus cyaneus* and *Eupoedes corollae*, which take nectar and pollen in more or less equal proportions, have mouthparts intermediate in length and structure (Gilbert 1981). It seems that, in all species, the protein component of pollen is necessary for maturation of eggs and probably also of sperm. Small hoverflies, when feeding, pollinate plantains once thought to be wind-pollinated (Stelleman 1978), and it is highly likely that the larger, furry species play a major part in pollination, as pollen readily adheres to their hairs. Certainly, hoverflies can be trained to visit artificial flowers of particular colours by provision of a reward in the form of sugar solution; such behaviour is a necessary prerequisite for a pollinator.

Since adults differing taxonomically and in size have similar food requirements, it is no wonder that gardens, with their abundant cultivated flowers, are good sites for hoverflies. Indeed males of some species use flowers as courtship territories. There is, however, great divergence in larval habits and hence in habitat typically frequented by adults, particularly egg-laying females. Colyer & Hammond (1968) recognized five categories of larval habits: aphidivorous, frequenters of decaying organic matter, frequenters of sap from tree-wounds or of rotting wood, frequenters of plant stems and roots, or fungi, and dwellers in nests of bees, wasps and ants. I adopted these five categories and expressed them in terms of the trophic level at which larvae operate (Owen, J. 1981a). The larval stages of many hoverflies are unknown, however, and such categorization depends on the assumption that taxonomically similar species have similar life histories. Furthermore, there is some overlap between the trophic levels of Colyer & Hammond's five categories, and it may be more realistic to follow Gilbert (1986) in recognizing four groups: larvae that eat plant tissues and plant products (including tree sap), larvae that feed on decaying organic material (including wood in rot-holes in trees), larvae that live in the nests of social insects, mainly as scavengers, and carnivorous larvae, most of which eat aphids.

The details of larval life are best known for the common aphid-feeding species and for those, such as *Eristalis tenax*, which feed on decaying organic material, often in water. It is clear that some species are facultative to some extent in their feeding: all *Platycheirus* (except *P. scutatus*), *Melanostoma mellinum* and *M. scalare* can complete their development on rotting plant material if no aphids are available (Gilbert 1986); and *E. tenax* is found in fully aquatic habitats, in sewage or dung, and in rotting matter of various sorts. Indeed, Samson's riddle in the Book of Judges (14:14) 'Out of the eater came forth meat, and out of the strong came forth sweetness' is often assumed to arise from misidentification of *E. tenax* breeding in putrid carcasses, as honeybees. The one major habitat absent from suburban gardens is stands of old wood, and as a consequence, those hoverflies that breed in sap runs or rot-holes in trees are under-represented in the garden fauna. The group with herbivorous larvae is represented in gardens mainly by the narcissus fly *Merodon equestris* and the lesser bulb flies *Eumerus strigatus* and *tuberculatus*. In more rural localities, the large genus *Cheilosia* would be better represented in this category. *Merodon* and *Eumerus* are the only hoverflies of economic importance, inflicting considerable damage on commercial bulb-growing, but, although fairly common, their activities in gardens are rarely evident enough for them to rank as pests. Indeed, *Merodon* is a beautiful fly, and adds considerable interest to the garden fauna as it is highly polymorphic, the different colour forms mimicking different species of bumblebees *Bombus*.

DIVERSITY AND ABUNDANCE OF HOVERFLIES

Ninety-one species of hoverflies were recorded in the garden during the period 1972–86. All appeared in Malaise trap samples, although many have never been observed alive. Conversely, *Melangyna umbellatarum*, although hand-netted in 1972, was not captured in the Malaise trap until 1986. A recent book (Stubbs & Falk 1983) has clarified some problems of identification; on the other hand, distinguishing characters given in Coe (1953) for females of some species are now discounted, and identification of *Baccha*, *Neocnemodon*, and of most species of *Paragus* and *Sphaerophoria* is valid for males only. Stubbs & Falk (1983) recognized 256 British species of hoverflies, six of them as yet unnamed, and include in their checklist a further nine forms of uncertain status. The following account of garden species follows the taxonomy and classification of this checklist, except where noted. The numbers given for each species refer to Malaise trap captures for 1972–86 in a total sample of 43 749.

Subfamily: Syrphinae

Baccha. Occasionally seen in shady, overgrown parts of garden. Larvae feed on aphids. Determinations prior to 1982 followed Coe (1953) recognizing two species:

subsequently, difficulties in identification of females led to amalgamation of records.

 1. *B. elongata* (Fab.) 1972–81   92
 2. *B. obscuripennis* Meigen 1972–81   36
  Baccha spp. 1972–86   153

Melanostoma.   Adults frequent at flowers. Larvae are facultative aphid-feeders, able to complete their development on rotting vegetation; both *mellinum* and *scalare* breed in garden.

 3. *M. mellinum* (L.) 3841
 4. *M. scalare* (Fab.) 883

Platycheirus.   Adults of many species are frequent at flowers or basking on leaves. Larvae of all but *P. scutatus* are facultative aphid-feeders, able to complete their development on rotting vegetation: *P. scutatus* is thought to be an obligate aphid-feeder. *P. cyaneus* and *P. scutatus* breed in garden; *P. scutatus* has been found feeding on *Myzus cerasi*, *Brevicoryne brassicae* and other aphids. *P. cyaneus* is called *P. albimanus* in Stubbs & Falk (1983); the name change is due to Torp (1984).

 5. *P. ambiguus* (Fallén)   426
 6. *P. cyaneus* (Müller) 5391
 7. *P. angustatus* (Zett.)   427
 8. *P. clypeatus* (Meigen) 2342
 9. *P. immarginatus* (Zett.)   11
10. *P. manicatus* (Meigen).   417
11. *P. peltatus* (Meigen)   374
12. *P. scambus* (Staeger)   12
13. *P. scutatus* (Meigen) 4354

Pyrophaena.   Adults never seen in garden. Larvae feed on aphids.

14. *P. granditarsa* (Forster)   33

Paragus.   Adults never seen in garden. Larvae feed on aphids. Prior to 1982, Coe (1953) was used for identification, leading to designation of all specimens as *tibialis* (Fallén). Since then, the key in Speight (1978) has been used, leading to designation of all males as *haemorrhous*. It is assumed (a) that all females captured since 1981 are also *haemorrhous*, and (b) that all individuals taken prior to 1982 were *haemorrhous* – certainly the voucher specimen is.

15. *P. haemorrhous* Meigen   128

Chrysotoxum.   Closely resemble *Vespula* wasps. Rarely seen in garden. It is assumed that the larvae feed on aphids, possibly root aphids, although there is no proof of this.

16. *C. bicinctum* (L.)   7
17. *C. festivum* (L.)   22
18. *C. verralli* Collin   23

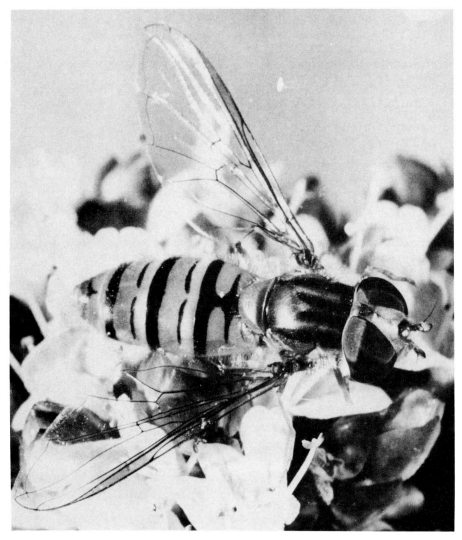

Photograph 7.1. Episyrphus balteatus *is well known as a migratory hoverfly. Influxes to the Leicester garden were particularly conspicuous in 1975, 1977 and 1985. Roger Whiteway.*

Dasysyrphus.   Never seen in garden. Larvae feed on aphids.

19. *D. albostriatus* (Fallén).   134
20. *D. lunulatus* (Meigen).   4
21. *D. tricinctus* (Fallén) 1 (1974)
22. *D. venustus* (Meigen)   3

Epistrophe.   *E. eligans* is frequently seen hovering motionless in sunlight above the lawn in early summer. Larvae feed on aphids.

23. *E. eligans* (Harris)   223

24. *E. grossulariae* (Meigen) 1 (1973)
25. *E. nitidicollis* (Meigen)   12

Episyrphus.   *E. balteatus* is common in the garden and breeds there, the larvae
feeding on *Brevicoryne brassicae* and other aphids. Large numbers regularly move
into the garden in July and August; immigration was particularly conspicuous in
1975, 1977 and 1985.

26. *E. balteatus* (Degeer) 5588

Leucozona.   Larvae feed on aphids.

27. *L. lucorum* (L.)   29

Melangyna.   Larvae feed on aphids.

28. *M. lasiophthalma* (Zett.).   7
29. *M. umbellatarum* (Fab.)   2
30. *M. cincta* (Fallén)   18
31. *M. triangulifera* (Zett.).   4

Meliscaeva.   Larvae feed on aphids; *M. auricollis* breeds in garden, the larvae
feeding on *Adelges cooleyi*.

32. *M. auricollis* (Meigen)   108
33. *M. cinctella* (Zett.).   3

Eupoedes.   *Eupoedes* is the *Metasyrphus* of Stubbs & Falk (1983); name change due to
Torp (1984). Large-scale immigration of *E. corollae* to the garden during August 1975
and July 1976. Larvae feed on aphids and *E. luniger* breeds in the garden, the larvae
feeding on *Macrosiphum euphorbiae* and other aphids.

34. *E. corollae* (Fab.) 6065
35. *E. latifasciatus* (Macquart)   123
36. *E. latilunulatus* (Collin)   9
37. *E. luniger* (Meigen)   773

Parasyrphus.   Larvae feed on aphids.

38. *P. malinellus* (Collin). 1 (1980)
39. *P. punctulatus* (Verrall)   2

Scaeva.   Larvae feed on aphids.

40. *S. pyrastri* (L.).   37
41. *S. selenitica* (Meigen).   3

Sphaerophoria.   Certain identification of most *Sphaerophoria* is possible for males
only. Larvae feed on aphids, and *S. scripta* breeds in the garden; it is occasionally
abundant and conspicuous on flowers, possibly as a result of immigration.

42. *S. menthastri* (L.).   69
43. *S. rueppellii* (Wiedemann).   18
44. *S. scripta* (L.). 1583

Syrphus. S. *ribesii* is one of the most conspicuous garden hoverflies. Larvae feed on aphids, and females of *S. ribesii* have several times been seen laying eggs among aggregations of aphids on garden plants. Larvae have been found feeding on *Brevicoryne brassicae, Aphis sambuci, Myzus cerasi, Betulaphis quadrituberculata* and other aphids.

45. *S. ribesii* (L.)   1426
46. *S. torvus* Osten-Sacken   2
47. *S. vitripennis* Meigen   300

Xanthogramma.   Larvae feed on aphids.

48. *X. pedissequum* (Harris) 1 (1972)

Subfamily: Milesiinae

Cheilosia.   Rarely seen in garden. Larvae feed on and in roots and stems of plants.

49. *C. albitarsis* Meigen   7
50. *C. bergenstammi* Becker   19
51. *C. pagana* (Meigen)   18
52. *C. proxima* (Zett.)   2
53. *C. vernalis* (Fallén)   140

Ferdinandea.   Larvae feed in sap runs on trees.

54. *F. cuprea* (Scopoli)   2
55. *F. ruficornis* (Fab.) 1 (1973)

Rhingia.   Often seen feeding at flowers, especially *Buddleia*. Larvae feed on and in cow dung.

56. *R. campestris* Meigen   243

Chrysogaster.   Aquatic larvae which feed on decaying organic material, and tap air spaces in underwater plants for oxygen with modified spiracles.

57. *C. hirtella* Loew 1 (1982)

Lejogaster.   Larvae feed on and in decaying organic material.

58. *L. metallina* (Fab.)   16

Neoascia.   Larvae feed on and in decaying organic material.

59. *N. podagrica* (Fab.)   75

Orthonevra.   Larvae feed on and in decaying organic material.

60. *O. splendens* (Meigen)   4

Eristalinus.   Larvae feed on decaying organic material. The aquatic larvae are 'rat-tailed', i.e. the posterior spiracles are extended into long, telescopic breathing tubes.

61. *E. sepulchralis* (L.)   6

Eristalis. Most, particularly the hairier species, bear some resemblance to bees; the drone-fly *E. tenax* is a honeybee mimic, and *E. intricarius*, polymorphic for colour and pattern, resembles various species of bumblebees. They are conspicuous at garden flowers in late summer and autumn. The aquatic, 'rat-tailed' larvae feed on decaying organic material. *E. tenax* and *pertinax* are under-represented in Malaise trap samples, perhaps because they are strong fliers. *E. interrupta* is called *E. nemorum* in Stubbs & Falk (1983); name change due to Torp (1984).

62. *E. abusivus* Collin . . . . 2 (both in 1973)
63. *E. arbustorum* (L.) 1541
64. *E. horticola* (Degeer) 24
65. *E. intricarius* (L.) 39
66. *E. interrupta* (L.) 56
67. *E. pertinax* (Scopoli) 215
68. *E. tenax* (L.) 286

Helophilus. *H. pendulus* is frequently seen basking in sunlight beside the garden pond. The aquatic, 'rat-tailed' larvae feed on decaying organic material.

69. *H. hybridus* Loew 14
70. *H. pendulus* (L.) 438
71. *H. trivittatus* (Fab.) 1 (1973)

Myathropa. The 'rat-tailed' larvae feed on decaying organic material in rot-holes and other water-filled cavities in trees. Under-represented in Malaise trap samples, perhaps because it is a strong flier.

72. *M. florea* (L.) 5

Eumerus. Frequently common in Malaise trap samples but rarely seen in garden. Known as lesser bulb flies because the larvae feed on and in bulbs, reputedly causing considerable losses.

73. *E. strigatus* (Fallén) 817
74. *E. tuberculatus* Rondani 933

Merodon. Hairy species occurring in many colour forms, most of which resemble different species of bumblebees. The inheritance of the colour patterns has been investigated by Conn (1972a & b). Conspicuous on flowers or hovering above them in June and early July when numbers of bumblebees are at a peak. Known as narcissus flies or large bulb flies and regarded as horticultural pests because the larvae feed on and in bulbs.

75. *M. equestris* (Fab.) 679

Heringia. Larvae feed on gall-making aphids.

76. *H. heringii* (Zett.) 4

Neocnemodon. Never seen in garden. Larvae feed on woolly aphids and on gall-making aphids.

77. *N. vitripennis* (Meigen) 104

Pipiza.   Never seen in garden. Larvae feed on aphids. The taxonomy of this genus is confused and confusing.

78. *P. austriaca* Meigen.   12
79. *P. bimaculata* Meigen   2
80. *P. fenestrata* Meigen   2
81. *P. luteitarsis* (Zett.)   4
82. *P. noctiluca* (L.).   113

Pipizella.   Larvae feed on root aphids. *P. viduata* is called *P. varipes* in Stubbs & Falk (1983); name change due to Thompson, Vockeroth & Speight (1982).

83. *P. viduata* (Meigen).   16

Triglyphus.   Larvae feed on aphids.

84. *T. primus* Loew. 1 (1985)

Volucella.   *V. bombylans* is hairy and occurs in three colour forms mimicking different species of bumblebees. Larvae feed as scavengers, occasionally as predators, in nests of social wasps and bumblebees.

85. *V. bombylans* (L.).   16
86. *V. pellucens* (L.).   21

Criorhina.   Robust, furry hoverflies resembling bumblebees. Larvae feed on dead wood. An additional individual of *C. berberina* was hand-netted in 1979.

87. *C. berberina* (Fab.). 1 (1976)
88. *C. floccosa* (Meigen). 1 (1972)

Syritta.   Common in the garden in summer and early autumn, the males defending territories and courting females over flowers. Larvae feed on decaying organic material.

89. *S. pipiens* (L.)   2473

Tropidia.   Larvae feed on decaying organic material.

90. *T. scita* (Harris). 1 (1973)

Xylota.   Larvae feed on dead wood and other decaying vegetable matter.

91. *X. segnis* (L.).   4

In a sample of 43 749 hoverflies collected over 15 years, the sex ratio might be expected to approximate to one, assuming equal production of males and females, but this is far from being the case. Overall, more females than males were caught, although there are differences from species to species and year to year. Of the 30 species trapped every year or represented by more than 50 individuals in any one year, 22 were represented by many more females than males, and in 13 of these the ratio of females to males exceeded two to one. Seventeen of the species in which females markedly outnumbered males have aphid-feeding larvae, and it may be that females abound in the garden because they are seeking egg-laying sites. Of the eight species in which males outnumbered females in the catch, only three have aphid-feeding larvae, and one of these, *Episyrphus balteatus*, is a migratory species, males being preponderant in influxes to the garden. Five of the species in which males outnumber females – *Eristalis arbustorum*, *E. tenax*, *Helophilus pendulus*, *Merodon equestris* and *Syritta pipiens* – more or less resemble bees or wasps, and are particularly associated with flowers. Their mimicry, however superficial, presumably confers most protection when they are in places frequented by bees and wasps. Males of *H. pendulus*, *M. equestris* and *S. pipiens* use flowers as courtship territories, and this may also be what brings males of the two species of *Eristalis* into the garden. *S. pipiens* is abundant in the garden in late summer, when every feathery head of goldenrod hums with activity as males 'buzz' every approaching hoverfly, chasing away male *pipiens* and other species, but zipping off in tandem behind female *pipiens*, and copulating just as soon as they alight.

The abundance and diversity of the garden hoverfly fauna is evident from the total 15-year sample of 43 749 individuals of 91 species taken in the Malaise trap, which collects only a proportion of the flying insects that spontaneously enter its air space. Differences in flight behaviour may mean that some species are more likely than others to be caught. For instance, large active species, such as *Eristalis tenax* and *Myathropa florea*, seem to be underrepresented in the catch, although this is a subjective view based on the numbers that I see in the garden. On balance, however, it is a reasonable assumption that the trap collects an unselected sample.

Analysis of annual changes in abundance and composition of the catch deals necessarily with common species, but rare species and occasional visitors also become involved in ecological relationships, however transient, and must be taken into account in any description of the fauna. Twelve species were taken in only one year in the Malaise trap, and 11 of these are represented by only one individual. A hoverfly species represented by a single individual in an unselected sample of 43 749 is an undisputed rarity; 1973 emerges as a year for rarities (see species list). Detailed recording over six years, in a $\frac{1}{2}$ square mile (1.3 square kilometres) area of rural Bedfordshire, yielded 83 species of which 19 were recorded once only (Laurence 1950). It seems that many hoverflies are genuinely rare. The rate of addition of new species to the garden list fell dramatically after 1973 (Fig. 7.1), suggesting that two years' trapping gives a fair indication of the number of species present in an area, but a further 13 years' trapping meant the addition of 20 species, or more than 20 per cent of the total. Despite the drop in rate of addition of species after 1973, every year brought some change in the composition of the catch, so that interest and anticipation persist year after year.

The frequency distribution of species in a sample can be expressed as an Index of Diversity, that used here being β, an estimate of the probability that an individual sampled at random will be different from the previous individual sampled. Table 7.1 shows the annual values of β for 1972 to 1986, and its overall value for the 15-year period. Diversity was highest in 1980,

Table 7.1 *Annual values of β-diversity for hoverflies caught in Malaise trap,*
1972–86

	Diversity (β)	Standard error	Number of individuals (N)	Number of species (S)
1972	0.903	±0.003750	1339	47
1973	0.910	±0.002466	3164	62
1974	0.862	±0.005029	2505	53
1975	0.802	±0.004344	6363	54
1976	0.803	±0.005718	3411	53
1977	0.762	±0.007357	2586	37
1978	0.890	±0.001921	6068	50
1979	0.896	±0.002653	3488	51
1980	0.921	±0.003437	1327	53
1981	0.866	±0.008796	885	44
1982	0.870	±0.003004	4256	52
1983	0.912	±0.003500	1283	47
1984	0.909	±0.002620	2064	48
1985	0.864	±0.003844	3553	51
1986	0.848	±0.006096	1457	43
1972–86	0.920	±0.000459	43 749	91

Note: β is derived from Simpson's Index of Diversity, λ; β=1−λ.

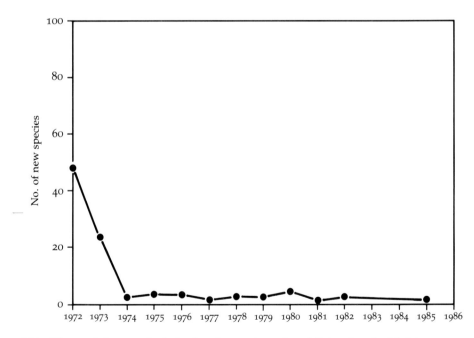

Note: The catch for 1972 includes hand-netted species additional to the Malaise trap total.

Fig. 7.1 *Rate of annual addition to hoverfly species list for Leicester garden, 1972–86.*

when N (number of individuals) was low but S (number of species) was high and the commonest species formed less than 19 per cent of the catch, and lowest in 1977, when S was very low but N about average and the commonest species formed nearly half the catch. The results of a test for heterogeneity on the 15-year series of computations of β is highly significant, so it is admissible to test for differences between pairs of years. Interestingly, β diversity was highly significantly different (p<0.01) between the years of peak abundance and the previous years (1974–75, and 1977–78), and between the year of minimum abundance and the previous year (1980–81), but not between these years of exceptionally high or low abundance and the subsequent years (1975–76, 1978–79 and 1981–82). This suggests that the composition of annual samples changed when numbers rose or fell dramatically, and that this change was sustained into the next year. The changes were not, however, consistent with any association of β with N; β-diversity fell in 1975 but rose in 1978, and fell in 1981. The diversity of the total 15-year sample is almost the same as the highest annual value, illustrating the insensitivity of Simpson's index to species richness, and the advisability of considering S as well as β when judging diversity.

ANNUAL FLUCTUATIONS IN NUMBERS AND IMMIGRATION OF HOVERFLIES

The catch varied considerably from year to year, as shown in Table 7.2 and Fig. 7.2. Two years, 1975 and 1978, stand out as being particularly good for hoverflies, with 6363 and 6068 individuals trapped, respectively. In 1975, vast numbers of hoverflies appeared in gardens in early August when it was unusually hot and sunny, and 'plagues' of them were reported on the east coast of Britain. But there was no conspicuous movement or influx in 1978, which was simply a 'good' year for hoverflies. The predominant species in August 1975 was *Eupoedes corollae*, which constituted over 40 per cent of the total catch. *E. corollae* formed almost as large a proportion of the catch in 1976, an unusually dry year. In 1977, however, when the total catch was just below average and the number of species was lower than in any other year, *Episyrphus balteatus*, a well-known migrant, was so abundant that it contributed 44.2 per cent of the catch. No one species formed as large a part of the catch in other years. 1981 was a particularly poor year for hoverflies, only 885 individuals being trapped. There are no clear associations between temperature or rainfall and the annual hoverfly catch. *Platycheirus cyaneus* was the commonest species in five years, *Episyrphus balteatus* in three, *Eupoedes corollae* and *Melanostoma mellinum* each in two, and *P. scutatus*, *Sphaerophoria scripta* and *Eumerus tuberculatus* each in one.

Forty-seven species, more than half the total known from the garden, were trapped in the first year, and new species were added in all but two years since. The annual total of species was highest in 1973 (62), a long, hot summer, and lowest in 1977 (37). The relationship between number of individuals (N) and number of species (S) was not constant, so that N/S ranged from 20.11 in 1981 to 121.36 in 1978. However, there is a strong correlation between N/S

Photograph 7.2. There was conspicuous and large-scale immigration of Eupoedes corollae *to the Leicester garden in August 1975 and July 1976. Roger Whiteway.*

and N (Fig. 7.2), showing that overall abundance is a result of increase in numbers of species already present, rather than addition of new species. This is supported by the relatively small number of species, eight, taken once only in 1975 and 1978, as opposed to a maximum of 16 in 1976. 1976 was a year of extreme drought conditions: by August, gardens were withered and the countryside parched and brown, and many nectar-feeding insects wandered far from their usual haunts in search of food.

Annual fluctuations in the total number of hoverflies caught may be the result of changes in the abundance of one or many species; several different patterns in annual variation are evident. Immigration was important in some years, particularly in 1975, 1976, 1977 and 1985, affecting numbers of *E. corollae* and *E. balteatus* (Fig. 7.3), as well as other species with aphid-feeding

Table 7.2 Size and composition of annual Malaise trap catch and 15-year catch of hoverflies in the Leicester garden, 1972–86

Year	Individuals (N)	Species (S)	N/S	Species taken once only (n_1)	n_1 as % S	Commonest as % N	Commonest species
1972	1339	47	28.49	10	21.28	18.67	Platycheirus cyaneus
1973	3164	62	51.03	14	22.58	19.53	P. scutatus
1974	2505	53	47.26	11	20.75	31.50	P. cyaneus
1975	6363	54	117.83	8	14.81	40.55	Eupoedes corollae
1976	3411	53	64.36	16	30.19	39.84	E. corollae
1977	2586	37	69.89	8	21.62	44.20	Episyrphus balteatus
1978	6068	50	121.36	8	16.00	20.40	P. cyaneus
1979	3488	51	68.39	14	27.45	23.25	Melanostoma mellinum
1980	1327	53	25.04	13	24.53	18.76	P. cyaneus
1981	885	44	20.11	9	20.45	32.09	P. cyaneus
1982	4256	52	81.85	10	19.23	27.23	M. mellinum
1983	1283	47	27.30	8	17.02	18.32	Eumerus tuberculatus
1984	2064	48	43.00	11	22.92	19.57	Sphaerophoria scripta
1985	3553	51	69.97	9	17.65	30.20	Episyrphus balteatus
1986	1457	43	33.88	9	20.93	29.86	E. balteatus
1972–86	43749	91	480.76	11	12.09	13.86	Eupoedes corollae

Note: Columns N and N/S are plotted in Fig. 7.2.

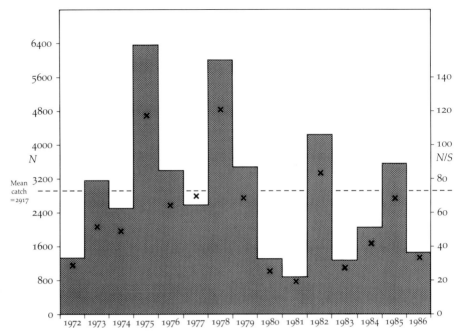

Fig. 7.2 Annual fluctuations in Malaise trap catch (N) of hoverflies, 1972–86, plotted as histogram, and in average number of individuals per species (N/S), plotted as crosses.

larvae. Both these species breed commonly on agricultural land, and in fine, warm springs and summers population explosions of their aphid food lead to a build-up in numbers. Depletion of their food supply caused by their predatory activities, or, perhaps in 1975, 1977 and 1985 by crop-harvesting, and in 1976 by withering of plants and death of aphids, led to mass movements away from the countryside bringing vast influxes to gardens with their abundant food resources. In 1975 and 1977, immigration occurred in the first two weeks of August, in 1985 during the second, third and fourth weeks of August, and in 1976 during the last two weeks of July. There was some evidence of movement into the garden during August in other years, but not on such a large scale as in 1975, 1976, 1977 and 1985.

In the first week of August 1975, hoverflies became conspicuous in gardens in central England, often entering houses where hundreds died on window-sills. Trap records show an increase in the total catch in the last week in July, followed by a phenomenal rise in numbers in the first and second weeks of August, after which the catch returned to normal. Seventy-one hoverflies were caught in the week ending 20 July, 139 in the week ending 27 July, 562 in the week ending 3 August, 2812 in the week ending 10 August, 1557 in the week ending 17 August, and 91 in the week ending 24 August. The increase was largely caused by the abundance of *E. corollae*, and to a lesser extent *E. balteatus* (Fig. 7.4), but the number and frequency of other species was also high. One hundred and sixty-two hoverflies were trapped on 2 August,

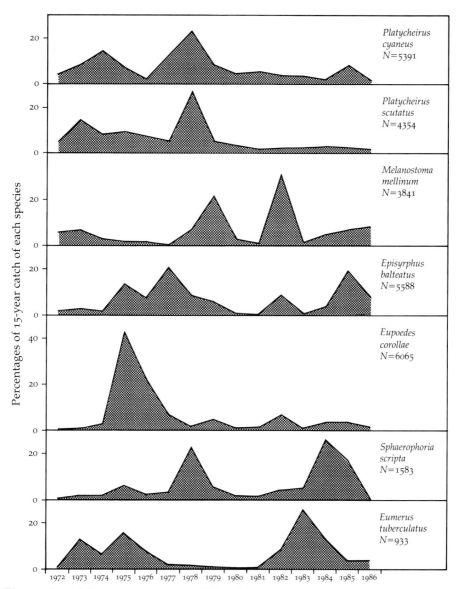

Fig. 7.3 *Annual fluctuations, 1972–86, of those hoverflies which were the commonest species in one or more years, expressed as percentages of 15-year catch of each species.*

three-quarters of the catch for the week ending 3 August. On 7 August, 672 hoverflies of 20 species were caught, including 376 *corollae* and 108 *balteatus*, and on 8 August, 954 of 24 species, including 551 *corollae* and 156 *balteatus*. By 12 August, the daily catch was less but still amounted to 636 of 19 species, including 401 *corollae* and 141 *balteatus*; on this day, hot and sunny throughout, the catch was examined at hourly intervals (Fig. 7.5). At 08.00 there were no hoverflies in the trap, but during the following hour they

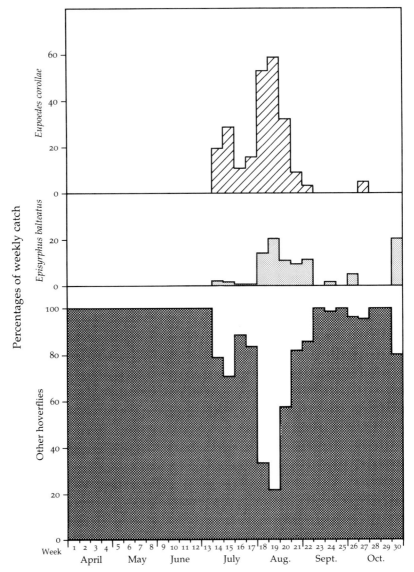

Fig. 7.4 *Relative frequencies of* Eupoedes corollae, Episyrphus balteatus *and all other hoverflies caught in Malaise trap weekly in 1975.*

entered at an increasing rate, totalling 29 by 09.00. Thereafter, the hourly number rose to 89 between 14.00 and 15.00, followed by a slight lull, and a rise to a maximum of 96 between 18.00 and 19.00. This hourly pattern was largely a consequence of the behaviour of E. *corollae*. A number of E. *balteatus* were caught in the early morning, particularly between 09.00 and 10.00; numbers declined in the afternoon and then, between 18.00 and 19.00, quadrupled on the previous hour. The peak in total number of hoverflies

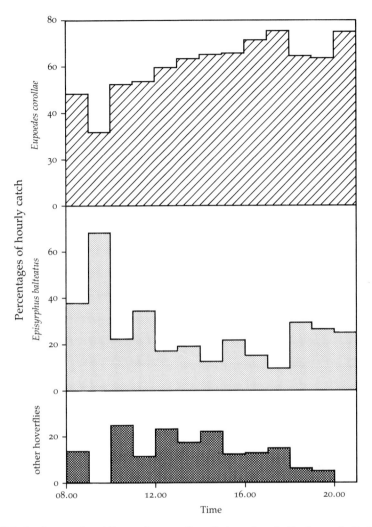

Fig. 7.5 *Relative frequencies of* Eupoedes corollae, Episyrphus balteatus *and all other hoverflies in hourly catches in Malaise trap on 12 August 1975.*

other than *E. corollae* and *E. balteatus* occurred between 14.00 and 15.00, when 20 individuals of ten species were caught. Captures of most species were spaced throughout the day, but all *Eumerus strigatus, E. tuberculatus* and *Syritta pipiens* were caught after 12.00.

Immigration is not, however, the only cause of annual variations in abundance of different species. Fig. 7.3 (p. 163) shows the relative numbers in each year of species which in one or more years were the commonest species; Fig. 7.6 gives the relative annual numbers of other species caught more than 500 times in the 15-year period; Fig. 7.7 shows relative numbers of species caught more than 200 but less than 500 times; and Table 7.3 gives actual

Table 7.3 Annual Malaise trap catches of species taken ten or more times in one year but absent in one or more years, 1972–86

	1972	1973	1974	1975	1976	1977	1978	1979	1980	1981	1982	1983	1984	1985	1986
Pyrophaena granditarsa	0	7	2	16	0	0	1	0	0	1	3	3	0	0	0
Paragus haemorrhous	1	19	5	19	27	1	13	5	1	0	13	11	8	5	0
Leucozona lucorum	0	6	1	1	10	3	0	0	3	0	4	0	0	1	0
Meliscaeva auricollis	0	1	3	5	1	55	16	10	4	0	4	4	0	4	1
Eupeodes latifasciatus	3	3	1	10	0	0	8	42	3	19	13	13	6	0	2
Scaeva pyrastri	2	0	2	2	0	7	1	2	0	0	0	1	2	15	3
Sphaerophoria menthastri	3	2	9	15	1	1	33	1	1	1	1	0	0	1	0
Cheilosia bergenstammi	10	8	1	0	0	0	0	0	1	0	0	0	0	0	0
Cheilosia vernalis	0	0	0	0	0	0	3	0	8	39	68	11	7	3	1
Neoascia podagrica	6	34	4	9	2	0	3	2	1	2	9	0	1	2	0
Eristalis interrupta	0	8	0	0	2	1	1	2	4	2	8	9	9	10	0
Neocnemodon vitripennis	8	24	9	8	3	0	30	9	1	0	0	1	3	4	3

Note: Platycheirus ambiguus (Fig. 7.7) and Melanostoma scalare (Fig. 7.6) were also each absent in one year, and Rhingia campestris (Fig. 7.7) in two years.

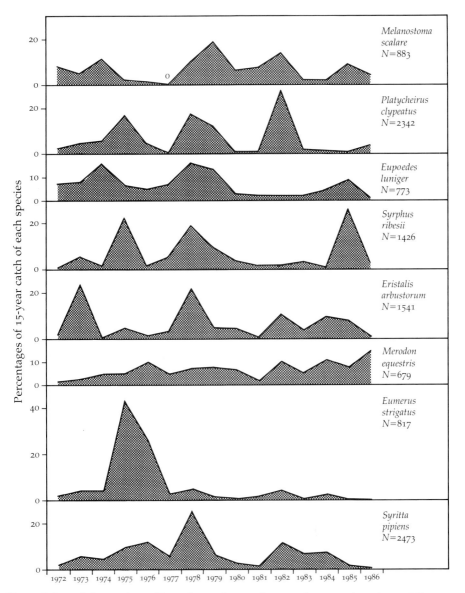

Fig. 7.6 Annual fluctuations of hoverfly species caught more than 500 times (1972–86), which in no year were the commonest, expressed as percentages of 15-year catch of each species.

numbers of species caught ten or more times in one year but not at all in one or more years.

E. corollae, in particular, and E. balteatus became more common in 1975, and both were abundant in 1976. Thereafter, E. corollae declined, whereas E. balteatus reached a peak in 1977, after which it too became less common; both,

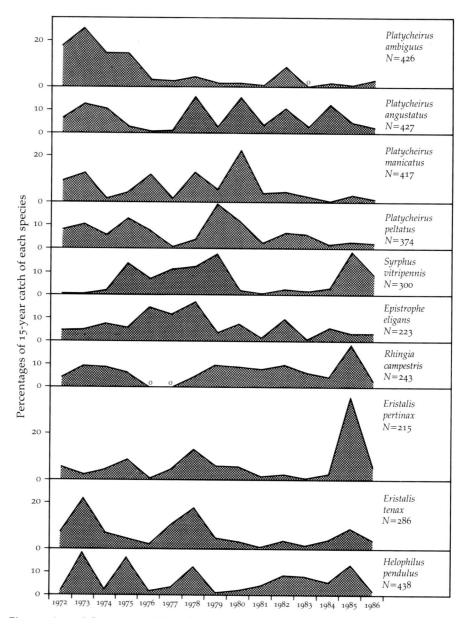

Fig. 7.7 Annual fluctuations of hoverfly species caught more than 200 but less than 500 times (1972–86), expressed as percentages of 15-year catch of each species.

however, increased again in 1982, and *E. balteatus* in 1985 (Fig. 7.3). *Platycheirus clypeatus* and *Syrphus ribesii* increased in 1975 and again in 1978, *P. clypeatus* in 1982, and *S. ribesii* in 1985 (Fig. 7.6). 1982 was a good year for a number of species, including *Melanostoma mellinum*, which was particularly abundant (Fig. 7.3), and *M. scalare* which also did well in 1979 and 1985 (Fig.

7.6). *Eristalis tenax* (Fig. 7.7), *E. arbustorum* (Fig. 7.6) and *Platycheirus scutatus* (Fig. 7.3) increased in 1973 and 1978, and *Helophilus pendulus* (Fig. 7.7) in 1973, 1975, 1978 and 1985. *Helophilus* and *Eristalis* continue flying in late summer and autumn and so might be expected to do well in the same years, such as 1973, when late summer and early autumn were fine and warm. Evidently 1975 was a good year for many types of hoverflies, for in addition to *H. pendulus* and several species of Syrphinae, *Eumerus strigatus* (Fig. 7.6) and *E. tuberculatus* (Fig. 7.3) were common. *E. tuberculatus* was also relatively common in 1973, as were *Rhingia campestris* (Fig. 7.7) and *Neoascia podagrica* (Table 7.3), and in 1983 *E. tuberculatus* was the commonest species. Several species additional to those already described were common in 1978, notably *Platycheirus cyaneus* and *Sphaerophoria scripta* (Fig. 7.3), and *Eupoedes luniger* and *Syritta pipiens* (Fig. 7.6). *P. cyaneus* and *E. luniger* were also common in 1974, *S. pipiens* in 1976, and *S. scripta* in 1984. *Platycheirus manicatus* increased in numbers in 1980 (Fig. 7.7). A number of species other than *E. balteatus*, *S. ribesii*, *M. scalare* and *H. pendulus* did well in 1985, notably *R. campestris* (Fig. 7.7), *P. cyaneus* (Fig. 7.3), *Eristalis pertinax* and *Syrphus vitripennis* (Fig. 7.7). In contrast, numbers of *Merodon equestris*, a large, early-flying species, have remained relatively constant apart from 1981, a bad year for hoverflies in general (Fig. 7.6).

Evidently species fluctuate markedly in numbers from year to year. Some years, 1975 in particular, were good for a variety of species, yet in some cases there are indications of an inverse relationship between species. The influx of *E. corollae* and *E. balteatus* was associated with a dramatic decline in numbers of *Melanostoma* spp. and to a lesser extent of *Platycheirus* spp. In 1978, when the numbers of *E. corollae* and *E. balteatus* fell, those of *M. mellinum* and, particularly, of *P. scutatus* and *P. cyaneus* reached unprecedented levels (Fig. 7.3). Numbers of *P. clypeatus* were high during the influx of 1975, but otherwise it followed the same pattern as the other *Platycheirus*, with high numbers in 1978 (Fig. 7.6). *Syritta pipiens*, on the other hand, whose larvae feed on decaying organic matter rather than on aphids, was largely unaffected either way by the influxes of 1975, 1976 and 1977, although, like many species, numbers were low in 1977 and high in 1978 (Fig. 7.6). However, despite this appearance of reciprocal fluctuations between different species whose larvae are predators of aphids, a detailed computer analysis of population dynamics and ecomorphological relationships has shown that hoverfly species respond independently to fluctuations in essential resources. There is little if any evidence of interactions between species and, indeed, the concept of a garden 'community' of hoverflies seems at best notional, the so-called community being merely a coincidence of species in space and time. Assessing the entire 15-year period, morphologically similar species that are presumably in competition, were not inversely correlated in population size (Gilbert & Owen, 1990).

The catches of many other species suggest that 1977 was a poor year for hoverflies, notwithstanding the large catch of *E. balteatus* and an increase in numbers of *Meliscaeva auricollis* (Table 7.3). *Melanostoma mellinum*, usually a conspicuous species, declined to two in 1977 (Fig. 7.3), and *M. scalare* was absent (Fig. 7.6), despite the fact that both breed in the garden. *Platycheirus clypeatus* fell to three, *P. manicatus* to six, and *P. ambiguus* to 11. *Rhingia campestris*, usually a regular garden visitor, and quite abundant in 1985, was absent in 1976 and 1977 (Fig. 7.7). Six other species which were absent in 1977 show otherwise anomalous annual patterns of frequency (Table 7.3): more than half the 15-year catch of *Neocnemodon vitripennis* was taken in two years, 1973 and 1978, and it was absent in two years; nearly half the total catch of *Neoascia podagrica* was taken in 1973 and it was absent in 1983 and 1986; *Pyrophaena granditarsa* occurred in only seven of the 15 years; *Cheilosia bergenstammi* has not been caught since 1974, whereas *C. vernalis* was first captured in 1978; and *Eupoedes latifasciatus* was absent in 1976 and 1985.

The variation from year to year of the total catch and its composition means that it is hard to generalize about the garden hoverfly fauna. Certainly it would be misleading to base assumptions on two or three years' investigation, which suggests that the results of student projects, spanning three years or less, have to be treated with some caution. Detailed computer analysis of 15 years' garden hoverfly catches has led to the somewhat unexpected conclusion that hoverfly populations are more stable than those of other invertebrate groups. There was no difference in the abundance and stability in general of those whose larvae are predators, phytophages and saprophages. Specialization of larval habitat leads to lower abundance, but has no effect on stability. Among predators, however, extreme specialists, such as *Platycheirus ambiguus*, are least and generalists, such as *Syrphus ribesii*, are most abundant, and both have less stable populations than moderate specialists, such as *Platycheirus scutatus*. This may be because the numbers of specialist predators are tied to the variability of their prey, whereas generalists tend to exploit irregular outbreaks of potential prey (Owen, J. and Gilbert 1989).

SEASONALITY OF HOVERFLIES

The hoverfly season in central England runs effectively from the beginning of April to the end of October, slightly earlier in some years, although most species fly for only part of the season. Those which hibernate as adults, such as *Eristalis tenax*, are occasionally seen basking in the sun on a mild winter's day, but this is exceptional.

August is usually the month of greatest activity, numbers trapped gradually building up from April onwards and declining rapidly in the autumn (Fig. 7.8). In 1972 and 1974, the May catch was higher than that of June or July, and in 1973, 1979 and 1986, there was a decline in July. Such annual differences in pattern of monthly catches can be attributed to the weather, cool and overcast or rainy conditions adversely affecting flight activity and hence likelihood of capture, even in years when the overall catch shows that hoverflies were abundant. The typical pattern of steadily increasing abundance to a summer peak is usually present even in years of overall small catch such as 1983. Up until 1979, I regarded the drought year, 1976, as an exception in that hoverfly

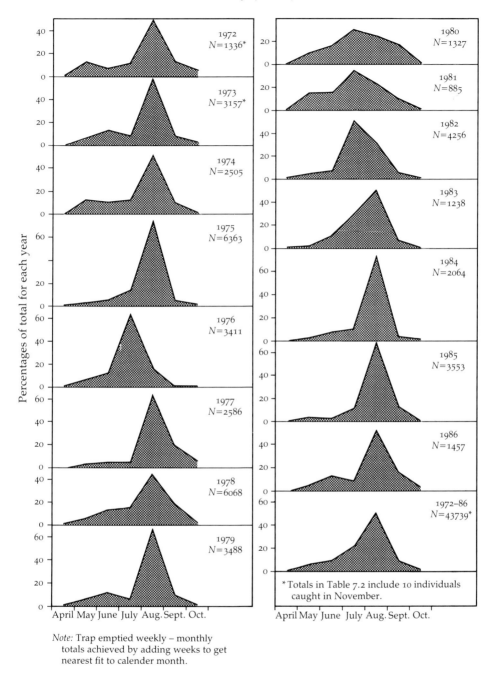

Fig. 7.8 Monthly fluctuations in numbers of hoverflies caught in Malaise trap (1972–86), expressed as percentages of annual total.

numbers reached a maximum in July and few were caught in September and October; in essence, the usual seven-month season was compressed into the period April–August. However, in 1980, 1981 and 1982 the July catch was larger than that for August, although September and October catches remained reasonably high. There is no obvious explanation for this in the temperature and rainfall figures. A slow start to the season does not necessarily augur a bad year: the overall 1983 catch was low, but 1977 picked up later in the year to produce the best September and October figures, and the 1984 August figures were high. In both 1975 and 1978, the total catch exceeded 6000, but the seasonal pattern differed fundamentally: more than two-thirds of the 1975 total was caught in August, when hoverflies entered the garden in large numbers, whereas in 1978 monthly catches were consistently high from May to October.

The number of species trapped reached a peak in August in every year except 1976, 1980 and 1982, when more were taken in July, and in 1986, when more were taken in May (Fig. 7.9). However, the total number of species taken in July and in August 1972–86 is almost the same (71 in July and 72 in August), whereas the total number of individuals caught in August during the 15-year period is more than double the number caught in July (Fig. 7.8). Diversity, in other words, was greater on average in July than in August when there was often an influx of one or a few species. There was great variation from year to year in which particular species occurred in any month, as is shown by the difference between the species totals (1972–86) for each month, and the number of species taken each month in any one year.

The number of species increased suddenly in May in all years, except 1983 when there was prolonged rain with hail and thunder in late April and May, and remained high throughout September and October, except in 1976 when drought shortened the season. The average number of individuals per species (N/S) also reached a maximum in August in most years, except in 1976, 1980, 1981 and 1982 when it peaked in July, as did the monthly catch (N). N/S for August was lowest in 1981 and highest in 1975, 7.7 and 113.1, respectively; it was relatively high in 1977, 1978 and 1979 (74.5, 67.0 and 67.2, respectively), and less than 20 in 1972, 1980 and 1983 (18.4, 10.4 and 18.5, respectively). The high values for 1975 and 1977 were due to influxes of *E. corollae* (1975) and *E. balteatus* (1977); in other years, they reflect increases in abundance of species already present as opposed to addition of new species. In 1976, and to a lesser extent in 1982, *E. corollae* and *E. balteatus* entered the garden in large numbers in July, when N/S reached a peak of 62.2 and 55.1, respectively. July values of N/S in 1980 and 1981 were low, 11.3 and 12.2, respectively, in line with them being poor years in general for hoverflies.

Monthly totals of individuals and species as presented in Figs. 7.8 and 7.9 obscure differences in seasonality between species. Some species are on the wing for several months, others for only restricted periods. Over the 15 years 1972–86, most of the common species were recorded in every month from May to October, and many from April to October. However, abundance varied greatly from month to month and year to year, and the 15-year records

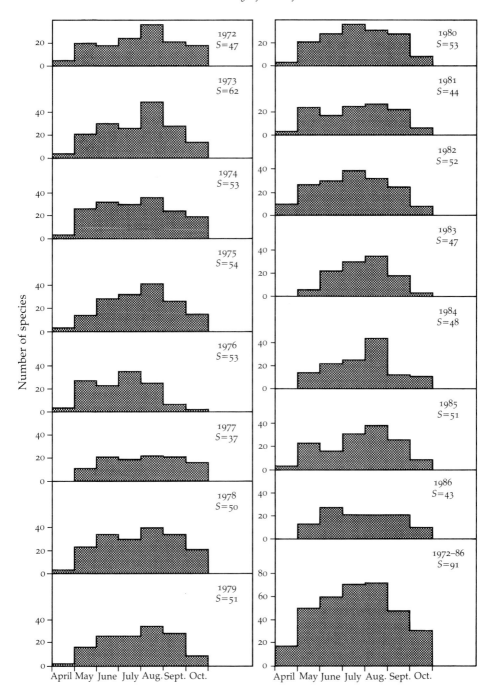

Fig. 7.9 Monthly totals of species of hoverflies caught in Malaise trap, 1972–86.

show seasonal patterns of occurrence that indicate the number of generations a year.

Analysis for seasonality was feasible for 37 species, which either occurred every year or were represented by more than 50 individuals in two or more years, or show marked seasonality. The species fall into six groups (Table 7.4): one generation a year and overwintering as larvae; one generation a year and overwintering as adults; two generations a year and overwintering as larvae; several overlapping generations a year and overwintering as larvae; two generations a year and overwintering as pupae; two generations a year and overwintering stage variable and facultative. I have estimated the number of generations a year on the basis of the pattern of weekly captures of males and females over the 15-year period. My assumptions do not accord in every case with the literature: according to Stubbs (Stubbs & Falk 1983), *Eristalis pertinax* may have several broods a year, *Melanostoma mellinum* and *M. scalare* may have more than two broods, and *Syrphus ribesii* is multiple-brooded. Since some species go through different numbers of generations in different years in my garden, I would also expect geographical differences, and I suggest that observations of seasonality in a particular species are accurate only for the location in which the observations are made. For instance, Schneider (1969), in a review paper, suggested that *Episyrphus balteatus* has several generations each year, whereas Pollard (1971), working at Monks Wood, established only one. Banks (1959) published trapping figures for suction traps operated in a garden at Rothamsted Lodge, which imply differences in seasonal occurrence over a small area. The confusion is compounded, in the case of many species, by differences of opinion as to how they overwinter. *Sphaerophoria scripta*, for instance, is reported as having two generations a year (Pollard 1971) and three or more generations (Bhatia 1939); it is said to hibernate as larvae (Scott 1939), as adults (Schneider 1969) or as either (Bhatia 1939). For many species given in Table 7.4, the overwintering stage is not known for sure, or may vary geographically, and I have deduced it from the trapping figures and the behaviour of similar species.

I sort hoverflies from trap samples on a weekly basis, providing at least 30 samples for the period 1 April to 31 October during which the Malaise trap is operated. Because the trap jar, for convenience, is always changed on a Sunday, a decision had to be made about fitting the 15 sets of weekly samples to a standard 30-week period. The end of week 1 therefore varies from 8 April to 14 April and the end of week 30 from 28 October to 3 November, depending on year. Biological events do not follow a man-made calendar, but the matching of weeks in the 15 years proves adequate, because a majority of the species that peak in numbers in the same week one year, peak together but in a different week in another year. Seven species listed in Table 7.4, one from each group and a migrant, *Episyrphus balteatus*, have been selected for detailed consideration of their seasonal patterns of occurrence. The weekly patterns of occurrence shown in Figs. 7.10 to 7.16 are summations of the 15 years, 1972–86.

Platycheirus ambiguus is an early spring species with a single short flight

Table 7.4 *Seasonal patterns of occurrence of 37 species of hoverflies in a Leicester garden*

Single flight period: single generation: overwintering larvae
*†*Platycheirus ambiguus*
   Leucozona lucorum
   Epistrophe eligans
   Eristalis intricarius
   Merodon equestris
   Volucella bombylans
   Volucella pellucens
Two flight periods: single generation: overwintering adults
*†*Episyrphus balteatus*
   Meliscaeva auricollis
 †*Eupoedes corollae*
 †*Syrphus vitripennis*
 †*Eristalis pertinax*
   Eristalis tenax
*†*Helophilus pendulus*
Two flight periods: two generations: overwintering larvae
 †*Baccha elongata*
 **Melanostoma mellinum*
   Melanostoma scalare
 †*Platycheirus angustatus*
   Platycheirus clypeatus
   Platycheirus manicatus
   Platycheirus peltatus
 †*Paragus haemorrhous*
   Eristalis arbustorum
 †*Emerus strigatus*
 †*Eumerus tuberculatus*
 †*Neocnemodon vitripennis*
 †*Pipiza noctiluca*
Long flight period: several overlapping generations: overwintering larvae
 **Platycheirus cyaneus*
   Platycheirus scutatus
 †*Syritta pipiens*
Two flight periods: two generations: overwintering pupae
 †*Cheilosia vernalis*
*†*Rhingia campestris*
 †*Neoascia podagrica*
Two flight periods: two generations: overwintering stage variable
   Dasysyrphus albostriatus
   Eupoedes luniger
   Sphaerophoria scripta
 **Syrphus ribesii*

Notes: *species for which weekly captures, 1972–86, are given in Figs. 7.10–7.16.
†overwintering stage unknown or in dispute, and so assumed on basis of trapping records and behaviour of related or ecologically similar species.

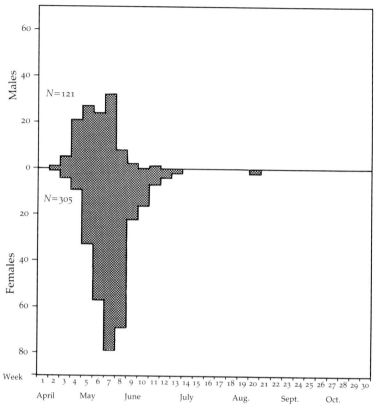

Note: End of week 1 varies from 8 to 14 April, and end of week 30 from 28 October to 3 November, depending on year.

Fig. 7.10 *Weekly captures of male and female* Platycheirus ambiguus, *1972–86.*

period (Fig. 7.10). The hoverflies probably hibernate as larvae, as do other species of *Platycheirus*, and the pattern of occurrence suggests a single generation. Earliest captures were in the weeks ending 16 April 1972 and 18 April 1982, and latest in the weeks ending 2 July 1972 and 7 July 1974 (apart from the anomalous capture of two females in the week ending 20 August 1972), with a peak in numbers in the third week of May. Annual captures ranged from one in 1981 to 107 in 1973, although none was caught in 1983: N = 426. Early captures were predominantly male, late captures exclusively female; females outnumbered males in every year, and overall by more than two to one.

Trapping records suggest that *Helophilus pendulus* overwinters as adults that produce a single generation in late summer to autumn, although overwintered adults were not trapped in every year (Fig. 7.11). None was caught in week 16 (the third or fourth week of July, depending on year), which presumably represents the interval between the flight of old adults and that of the new generation. The earliest capture was in the week ending 25 May 1980,

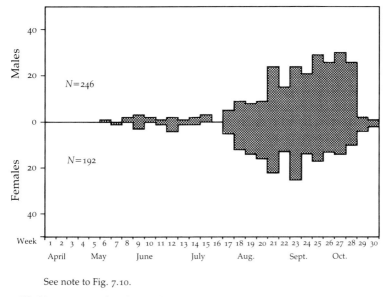

See note to Fig. 7.10.

Fig. 7.11 Weekly captures of male and female Helophilus pendulus, *1972 86.*

and the latest in the weeks ending 28 October 1973 and 2 November 1975, with maximum numbers in the period August to October. Annual captures ranged from 5 in 1979 to 79 in 1973: $N = 438$. Males outnumbered females in most years and in the total sample.

Melanostoma mellinum is usually abundant in the garden, and in 1979 and 1982 it was the commonest hoverfly. It has two well-marked generations, the second much larger than the first (Fig. 7.12), and has been reared from larvae found hibernating in the garden. The earliest capture was in the week ending 9 April 1972, and the latest in the week ending 28 October 1973. Peaks in numbers, corresponding to the two generations, occurred in the second half of May and from mid-July to the first week of September. Annual catches ranged from two in 1977 to 1159 in 1982, the decline in 1977 being associated with an influx of other species with aphid-feeding larvae: $N = 3839$. Early captures were predominantly male, late captures predominantly female, and females outnumbered males overall by three to two.

Platycheirus cyaneus is abundant in the garden, and in five years it was the commonest species in trap samples. There are several overlapping gener-ations a year with two peaks in numbers, the first from the second week in May to the first in June, the second from the third week in July to the first in September (Fig. 7.13). It has been reared from larvae found hibernating in the garden. The earliest capture was in the week ending 8 April 1973, and in six years it was caught in the last week of October. Annual catches ranged from 95 in 1986 to 1238 in 1978: $N = 5388$. Early captures were predominantly male, late captures predominantly female, and overall females outnumbered males by more than four to three.

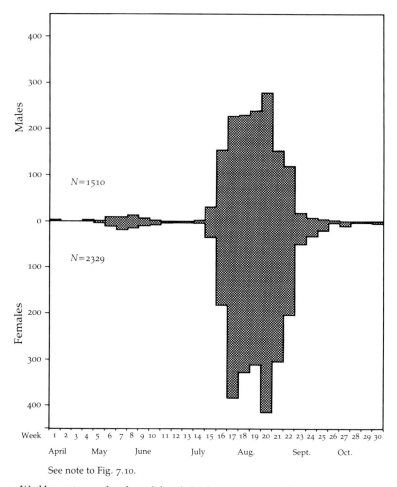

Fig. 7.12 *Weekly captures of male and female* Melanostoma mellinum, *1972–86.*

Rhingia campestris, whose larvae and pupae are found exclusively in cow dung, is unexpectedly numerous in the garden, presumably because adults visit to feed at flowers. It has two well-marked generations, separated by a three-week gap (Fig. 7.14). Only one individual was captured before the second week of May, and only one in October; consequently I believe that they overwinter as pupae in cow-pats or in the soil if the dung is dispersed. The earliest capture was in the week ending 5 May 1974, and the latest in the week ending 29 October 1978, with peaks in the second half of May to early June, and again in the last week of August and first week of September. Annual catches ranged from five in 1986 to 44 in 1985, although none was caught in 1976 and 1977: N = 243. More males than females were caught early in the year, but overall females outnumbered males by three to two.

Syrphus ribesii appears to have a small first and a larger second generation in most years (Fig. 7.15). They have been reared from larvae found hibernating

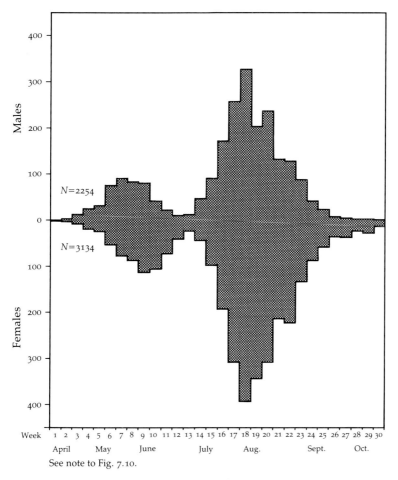

Fig. 7.13 Weekly captures of male and female Platycheirus cyaneus, *1972–86.*

in the garden, and are reported as overwintering as either larvae or as adults (Bhatia 1939, Pollard 1971, Schneider 1969). The earliest capture was in the week ending 12 May 1974, the latest in the weeks ending 22 October 1972, 26 October 1975 and 23 October 1977. In 1978, when the first generation was larger than usual, it peaked in numbers in the last week of May; the peak in numbers of the second generation varied from the first week of August to the first of September. Annual catches ranged from five in 1972 to 364 in 1985, when *S. ribesii* formed part of a hoverfly influx to the garden: N = 1426. Early captures were predominantly male, late captures exclusively female, and overall females outnumbered males by two to one.

Episyrphus balteatus is known to breed in the garden, and garden records suggest that it has one generation a year with overwintering adults, giving a small early flight and a larger second one (Fig. 7.16). Pollard (1971) found one generation with overwintering adults, although Bhatia (1939) found that the

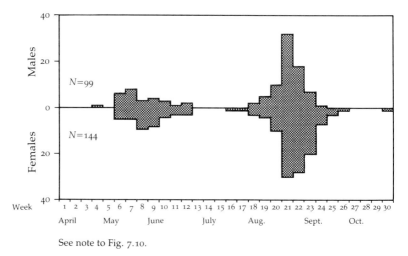

See note to Fig. 7.10.

Fig. 7.14 Weekly captures of male and female Rhingia campestris, 1972–86.

Photograph 7.3. In 1977, 1985 and 1986, Episyrphus balteatus was the commonest
hoverfly species in annual Malaise trap samples. More than 90 per cent of the adult food is
pollen, and they are particularly attracted to flowers of the alpine poppy Papaver alpinum
with their abundant, accessible stamens.

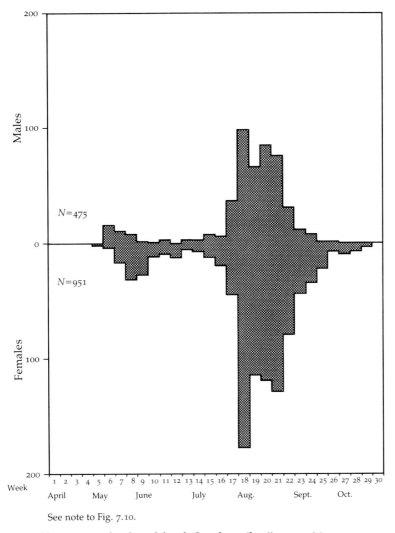

See note to Fig. 7.10.

Fig. 7.15 Weekly captures of male and female Syrphus ribesii, *1972–86.*

diapausing stage depends on development, and both Bhatia and Schneider (1969) found that it goes through three or more generations. The earliest capture was in the week ending 15 April 1979, although few were caught before mid-July, except in 1976, the drought year; in seven years it was caught in the last week of October. Peak numbers were trapped in August, except in 1976, when numbers peaked three weeks earlier. There were large influxes to the garden in 1975, 1976, 1978, 1982 and, particularly, in 1977 and 1985, consisting predominantly of males. In ten years, and in the total 15-year sample, males outnumbered females by approximately four to three. The preponderance of males may be related to the species' habit of undertaking long-distance migrations (see e.g. Pollard 1971, Stubbs & Chandler 1978). The

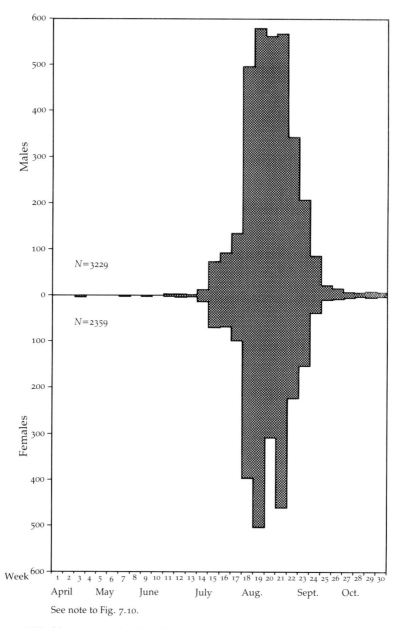

Fig. 7.16 Weekly captures of male and female Episyrphus balteatus, *1972–86.*

See note to Fig. 7.10.

particularly early record for the week ending 15 April 1979 was associated with a strong southerly airstream. Annual captures ranged from 12 in 1981 to 1143 in 1977: $N = 5588$.

Peaks in numbers of species in the same genus, which might be expected to be ecologically similar, tend either to be out of phase or, in some years, to be much more pronounced in one species than in the other(s). This suggests that peaks in numbers trapped are not simply a reflection of good flying weather, and that closely related species are responding differently to environmental conditions and hence exploiting the environment in different ways. This is the case with *Melanostoma mellinum* and *scalare*, *Platycheirus cyaneus* and *scutatus*, *Platycheirus clypeatus*, *manicatus* and *peltatus*, *Eumerus strigatus* and *tuberculatus*, and, particularly, with *Syrphus ribesii* and *vitripennis*, and *Eupoedes corollae* and *luniger*, where the congeneric species differ in the number of generations and overwintering stage.

Many research projects, particularly those funded by grants, have a duration of three years. In only ten of the 37 species listed in Table 7.4, e.g. in *Rhingia campestris*, would I have drawn the same conclusions about seasonality after three years' trapping as I have drawn after 15 years. Seven species would have been too scarce to yield any results, four would have given results difficult to interpret, six, including *Melanostoma mellinum* and *Platycheirus ambiguus*, would have given a misleading idea of seasonality or abundance, and ten, including *Platycheirus cyaneus*, *Episyrphus balteatus*, *Syrphus ribesii* and *Helophilus pendulus*, would have given results leading to misinterpretations of life histories. This emphasizes the value of long-term studies, and confirms that conclusions drawn from short-term projects should be treated with caution.

LARVAL HABITS OF HOVERFLIES

Adult hoverflies of different species are broadly similar in their feeding behaviour, but the food, feeding places, and hence life styles, of larvae are diverse. There are four categories: consumers of living roots, bulbs, stems, of tree sap, or fungi; predators of aphids and other Homoptera; dwellers in, and feeders on soft or liquid, decaying organic material, including rotting wood; and scavengers in the nests of ants, bees or wasps. Larvae in each of the four categories are thus operating at different trophic levels (Wiegert & Owen, D. F. 1971). Those that eat live plant material are primary consumers; predators of Homoptera are secondary consumers; those that eat decaying organic material are primary decomposers if they eat plant material, secondary or higher order decomposers if the dead material is of animal origin; and scavengers in nests of social Hymenoptera are secondary or higher order decomposers. Members of most groups are probably opportunist feeders: aphid-feeders occasionally attack other insects and each other; larvae in decaying organic material feed on faeces, on rotting plants, and on decomposing animal remains; and scavengers in Hymenoptera nests occasionally eat live larvae and pupae. As a family, hoverflies are unusual in exploiting the environment in such diverse ways. All four groups are represented by adults caught in the Malaise trap, and the numbers of individuals and species, and the trophic levels of the groups are shown in Table 7.5. The larvae of a number of garden species are undescribed, but they are assumed to have habits similar to those of closely related species.

Table 7.5 *Feeding sites and trophic levels of larvae of hoverflies trapped in the Leicester garden, 1972–86*

Trophic group	Feeding site	Trophic level	Representatives: species in each group in brackets	Species (S)	Individuals (N)	% of total N
A	Stems, roots, bulbs, tree sap	Primary consumers	*Cheilosia* (5), *Ferdinandea* (2) *Eumerus* (2), *Merodon* (1)	10	2618	5.98
B	Plants infested with aphids	Secondary consumers	Syrphinae (47), Pipizini (9)	56	35648	81.49
C	Decaying organic material, rotting wood	Primary to nth order decomposers	*Rhingia* (1), Eristalini (12) Chrysogastrini (4), Xylotini (5)	22	5446	12.45
D	Nests of wasps and bumblebees	Usually 2nd to nth order decomposers	*Volucella* (2)	2	37	0.08

Notes: Separation into trophic levels is based on the system suggested by Wiegert & Owen, D. F. (1971). This differs from the table of feeding sites in Owen, J. (1981a) in that the species are grouped in four not five categories: to minimize confusion the trophic groups have therefore been designated by letters not numbers (as in Owen, J. 1981a).

Four genera trapped in the garden, *Cheilosia, Ferdinandea, Merodon* and *Eumerus*, fall into group A. The foodplants of two of the five species of *Cheilosia* (see Stubbs & Falk 1983) grow in the garden: *C. bergenstammi* has not been caught since 1974, at about the time that greater efforts were made to control common ragwort, its foodplant, and *C. vernalis* was first caught in 1978, after its foodplant, yarrow, spread from the clipped confines of the lawn to a flowerbed. Both species of *Ferdinandea* are reported from sap runs associated with goat moth damage to trees, and only three individuals have been trapped. *Merodon equestris* and the two species of *Eumerus* feed by burrowing in living bulbs, and almost certainly breed in the garden. Group A is represented by 2618 individuals, almost 6 per cent of the total catch.

Most garden hoverflies, and the majority of the common species, have aphid-feeding larvae, which I often find among aggregations of aphids on such plants as cabbages, beans and elder. Nine species of Syrphinae have been reared from larvae collected in the garden: *Melanostoma mellinum, M. scalare, Platycheirus cyaneus, P. scutatus, Episyrphus balteatus, Meliscaeva auricollis, Eupoedes luniger, Sphaerophoria scripta* and *Syrphus ribesii*. As far as is known, the larvae of all species of Syrphinae and Pipizini feed on aphids or other herbivorous insects. Fifty-six species, and over 81 per cent of all individuals trapped, fall in group B.

The species in group C are taxonomically diverse but the larve of all are decomposers. Since, however, the nature of their food varies, they operate at several trophic levels. Larvae of *Eristalis* and *Helophilus* are usually found in water, feeding on minute particles of organic material, but *Eristalis tenax* also occurs in dung, in sewage and in wet carrion. *Myathropa florea* is particularly associated with wet rot-holes in trees. The larvae of *Syritta pipiens* occur in wet garden compost, but they have also been recorded from dung of cows, horses and humans; other Xylotini feed as larvae on and in rotting wood, and adults are rare in the garden. Of the Chrysogastrini, only *Neoascia podagrica* is regularly caught. Many of the species in this group undoubtedly breed in the garden or its immediate vicinity, either in compost, on dung or carrion, or in ponds and tree holes, and the proximity of farmland accounts for the regular capture of *Rhingia campestris*. Group C, represented by 22 species and 5446 individuals, over 12 per cent of the total catch, ranks second in importance to the aphid-feeders in group B.

Volucella bombylans and *V. pellucens* are the only representatives of group D. Their larvae scavenge in bee and wasp nests, not only eating dead Hymenoptera and detritus, but also stimulating the host larvae to produce excrement on which they feed. Occasionally, however, especially towards the end of the season, they eat live larvae, pupae, or weakened adults. Group D constitutes less than 0.1 per cent of the total sample.

Compared with the British list of hoverflies, group B, species with aphid-feeding larvae, are over-represented in the garden, and group A is under-represented; groups C and D form about the same proportions of the British and garden lists.

The trophic composition of the Malaise trap catch varied to some extent from year to year, as shown in Fig. 7.17. Although hoverflies with aphid-feeding larvae (group B) constituted the majority of the sample in every year, their relative frequency was considerably lower in 1973, 1976, 1983 and 1984. In 1973, this was associated with an increase in relative frequency of species with larvae that feed on decaying organic material (group C); the late summer of 1973 was warm and sunny, and species such as *Eristalis arbustorum* which continue flying late in the season were particularly abundant. In 1976, species with herbivorous larvae (group A) formed a greater proportion of the total catch than usual, and *Merodon equestris* and *Eumerus strigatus* were relatively common. In 1983, the relative frequency of group A was more than double that of any other year, largely because of the numbers of *Eumerus tuberculatus*, which was the commonest species; large catches of *Eristalis arbustorum* and *Syritta pipiens* relative to species with aphid-feeding larvae resulted in a high relative frequency of group C. Again in 1984, the relative frequency of groups A and C was greater than usual, largely because of the numbers of *E. tuberculatus* and *M. equestris*, and of *E. arbustorum* and *S. pipiens*.

Large numbers of hoverflies with aphid-feeding larvae, particularly *Eupoedes corollae* and *Episyrphus balteatus*, moved into the garden during the first two weeks of August in 1975 and 1977, and during the last two weeks of July in 1976; *E. corollae* formed about 40 per cent of the total catch in 1975 and in 1976, and *E. balteatus* over 44 per cent in 1977 (Table 7.6). This is not reflected in the relative frequency of group B hoverflies in these years (see Fig.

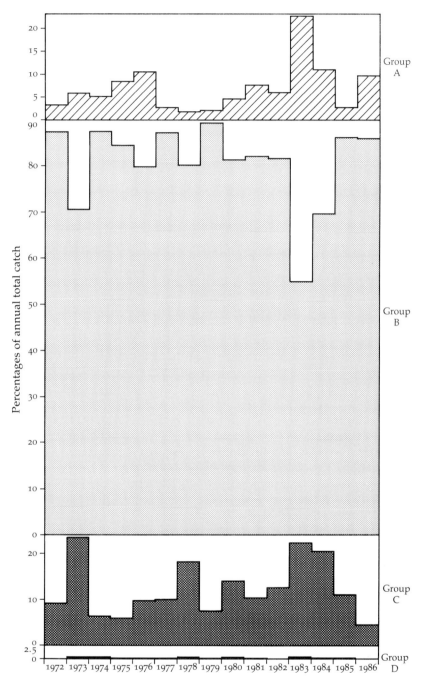

Note: Trophic groups slightly different from Owen, J. (1981a). Group 3 (1981) allocated to group 1 (Ferdinandea) or group 4 (Xyota and Criorhina). New groups designated as A, B, C and D: A=1 (1981); B=2 (1981); C=4 (1981); D=5 (1981).

Fig. 7.17 *Annual fluctuations (1972–86) in relative frequency of hoverflies with larvae in different trophic groups (see Table 7.5).*

Table 7.6 *Hoverflies with predatory larvae as percentages of monthly Malaise trap samples, and relative frequencies of* Eupoedes corollae. Episyrphus balteatus *and all hoverflies with predatory larvae, by year*

	April	May	June	July	Aug.	Sept.	Oct.	*Eupoedes corollae*	*Episyrphus balteatus*	Total predators
1972	100.0	99.4	82.4	80.1	88.7	91.5	61.8	1.3	6.1	87.2
1973	100.0	98.3	86.1	86.9	66.3	38.8	51.4	1.6	4.0	70.1
1974	83.3	90.7	62.2	87.8	92.1	90.9	64.6	6.6	4.1	87.3
1975	100.0	90.7	51.0	68.5	91.0	66.2	62.7	40.5	11.7	84.4
1976	100.0	93.1	45.2	85.6	76.9	71.4	66.7	39.8	11.8	79.7
1977	—	97.1	63.3	50.9	91.3	90.6	72.7	16.0	44.2	87.2
1978	100.0	96.9	78.8	55.1	85.5	85.7	62.4	1.9	7.6	80.0
1979	100.0	96.1	73.0	64.0	93.8	88.7	65.4	7.8	8.6	89.4
1980	100.0	91.5	76.5	87.7	89.5	56.9	75.0	5.0	2.0	81.3
1981	88.9	93.1	84.4	89.1	70.2	66.3	60.0	8.2	1.4	81.9
1982	95.2	85.7	48.7	84.6	87.7	54.7	47.8	9.7	11.1	81.4
1983	100.0	100.0	67.1	60.3	47.9	51.2	42.9	4.5	1.6	54.9
1984	—	81.5	53.0	58.3	71.6	86.6	66.7	10.7	9.2	69.4
1985	100.0	89.1	72.7	64.6	94.4	66.8	8.7	5.4	30.2	86.3
1986	—	97.0	50.5	57.1	97.2	95.0	62.5	4.8	29.9	85.9
Total	97.3	93.5	66.9	76.8	86.0	77.2	61.4	13.9	12.8	81.5

7.17), largely because the numbers of resident *Melanostoma* and *Platycheirus* in the catch were far lower than usual; *M. mellinum*, for instance, which breeds in the garden and is usually common, was represented by only two individuals in 1977, and *M. scalare*, which also breeds, was absent altogether. In 1978, when the numbers of *E. corollae* and *E. balteatus* fell, those of *M. mellinum*, and particularly of *Platycheirus cyaneus* and *P. scutatus* reached unprecedented levels. Thus the trophic composition of the garden hoverfly fauna was apparently unaffected by immigration of *E. corollae* and *E. balteatus* to the garden in 1975, 1976 and 1977.

Seasonal changes in the relative abundance of species with different larval habits is expressed in Table 7.6 as the percentage of each month's catch composed of aphid-feeders (group B). They were first to appear and increase in numbers in spring, and their frequency relative to those with other larval habits was high in April and May. It fell in June, when the catch increased in variety, but had usually risen again in August. By October, species whose larvae feed on decaying organic material, e.g. *Eristalis* and *Helophilus*, were common, and the relative frequency of aphid-feeders was lower, particularly in 1985.

The greatest departures from this pattern were in 1973 and 1983. In 1973, *Eristalis arbustorum* and other group C species became particularly abundant in August, such that, by September, the relative frequency of aphid-feeders had fallen to 38.8 per cent. In 1983, however, high numbers particularly of *Eumerus tuberculatus*, which has bulb-feeding larvae, and to a lesser extent of *E. arbustorum* and *Syritta pipiens*, depressed the relative frequency of aphid-feeders from July onwards. In 1975, the relative frequency of aphid-feeders fell to 51.0 per cent in June, when *Merodon equestris, S. pipiens* and *Eumerus* spp. were common, and rose to only 68.5 per cent in July with the appearance of a great variety of other species, especially *Eristalis* and *Helophilus*. In June 1976, abundance of the same species accounted for the low relative frequency of aphid-feeders, 45.2 per cent, but this rose dramatically to 85.6 per cent in July, when *Eristalis* and *Helophilus* became relatively scarce. Hoverflies with aphid-feeding larvae were relatively scarce in July 1977 and 1978, 50.9 and 55.1 per cent, respectively; in July 1977 the total catch was small and aphid-feeders unusually scarce, but in 1978, *E. arbustorum* and *S. pipiens* were relatively common. Abundance of *Merodon equestris* resulted in a low relative frequency of aphid-feeders in June 1982 and 1984, 48.7 and 53.0 per cent, respectively, and in June and July 1986, 50.5 and 57.1 per cent, respectively. Relative frequency of aphid-feeders was low in August 1981 and 1984, 70.2 and 71.6 per cent, respectively, but for different reasons. The total 1981 catch was low, and particularly that of aphid-feeders, but in 1984, *E. arbustorum, E. tuberculatus* and *S. pipiens* were more abundant than usual in August.

The usual situation in any month was for hoverflies with aphid-feeding larvae to outnumber those with other larval habits in the Malaise trap sample, although in the late summer, those whose larvae feed on decaying organic material became relatively common. Aphid-feeders were not only more abundant than the others, but also showed a more pronounced seasonal peak in numbers. Of the 35 648 hoverflies with predatory larvae, 19 726 (55.3 per cent) were caught in August, whereas of the 8101 with larvae of other habits, 3214 (39.7 per cent) were caught in August.

HOVERFLIES AS INDICATORS OF VARIETY OF HABITAT

Since publication of a Leicestershire list of hoverflies (Owen, J. 1979), corrections and additions have been made. *Pyrophaena rosarum* (Fab.) and *Cheilosia nasutula* Becker had been wrongly identified and should be deleted. *Meligramma triangulifera* and *Parasyrphus malinellus* have been trapped in the garden, and *Platycheirus tarsalis* (Schummel), *Cheilosia albipila* Meigen, *C. intonsa* Loew, *Portevinia maculata* (Fallén), *Chalcosyrphus nemorum* (Fab.) and *Xylota tarda* Meigen have been caught elsewhere in the county. This gives a county list of 133 species, but 27 of these have not been recorded since 1971, and many of them not since the nineteenth century. It is remarkable that 91 of the county species should have been trapped in my ordinary, medium-sized suburban garden. It is, however, noteworthy that in the five years prior to 1988, 131 species of hoverflies, including some unexpected finds, were recorded within an 8-km radius of the city centre of Coventry, only 36 km from the city of Leicester (Wright 1988). Moreover, Verlinden & Decleer (1987) report that of the ten 10 × 10 km squares in Belgium that are richest in hoverflies, six are mainly urban or suburban in character. Evidently, hoverflies find suitable habitats in and around towns and cities.

The size and diversity of the hoverfly sample from the Malaise trap indicates that gardens offer not only an abundance of resources but also a varied habitat, and lead to speculation as to what so many different hoverflies are doing in the garden. The dependence of adults on nectar and pollen for food accounts for large catches of resident species, and no doubt explains the presence of so many casual visitors. All the common, medium to large species are regularly seen at flowers such as shrubby cinquefoil *Potentilla fruticosa*, *Geranium cinereum*, *Buddleia davidii* and composites including marguerites *Chrysanthemum frutescens*, pot marigolds *Calendula officinalis*, goldenrod *Solidago canadensis* and various species of *Aster*. Some of the smaller hoverflies, such as *Melanostoma* spp., *Platycheirus* spp., *Sphaerophoria scripta* and *Syritta pipiens*, are also abundant on flowers, although others, such as *Paragus haemorrhous* and *Pipiza* spp., are only known from trap samples. There is abundant food for adults in gardens, which are especially important in late summer and autumn when hedgerows and meadows offer few flowers to compete with the colourful display of herbaceous borders. Females of species whose larvae feed on aphids find a multitude of egg-laying sites in gardens, which can also accommodate, although to a lesser extent, those whose larvae inhabit decaying organic material, bulbs or Hymenoptera nests.

The spatial complexity of gardens accommodates the habitat preferences of egg-laying females; for instance, *Syrphus ribesii* females are more often seen lapping honeydew and laying eggs on aphid-infested plants of rosebay willowherb *Epilobium angustifolium* in sun than in shade. In 1978, a second Malaise trap was operated in the front garden on an unmown lawn sheltered by trees, with few surrounding flowers. The two sites are not only different in vegetational and structural character, but also well-separated. Trap 2 caught only about a tenth as many individuals and fewer species than the original trap (trap 1); the average number of individuals per species was lower and the proportion of species taken once only, higher (Table 7.7). *Platycheirus cyaneus* was the commonest species in both samples, but the species composition of the catches differed. Eight species were relatively more abundant in trap 2: *Baccha obscuripennis*, *Melanostoma mellinum*, *M. scalare*, *Platycheirus angustatus*, *Eupoedes luniger*, *Sphaerophoria scripta*, *Syrphus ribesii* and *Rhingia campestris*. Six species were relatively less abundant: *Platycheirus cyaneus*, *P. scutatus*, *Epistrophe eligans*, *Eristalis arbustorum*, *Neocnemodon vitripennis* and *Syritta pipiens*. This almost certainly reflects behaviour in relation to sun, shelter and flowers. Eighteen species were taken in trap 1 but not in trap 2; one species, *Dasysyrphus lunulatus*, was taken only in trap 2, and was a new species for the garden, although it has since been caught in trap 1. The discrepancies between the catches of traps 1 and 2 demonstrate the patchiness of the garden environment and the consequent uneven distribution of certain species.

For six weeks, in the second half of July and August 1987, a Malaise trap was operated on the Scraptoft campus of Leicester Polytechnic, some 2 km from my garden. The trap was sited on an area of open grass, about 100 m from tall trees and undergrowth and from buildings, but nowhere near flowerbeds or other adult hoverfly food sources. The sizes of the six-week catches on the campus and in the garden were similar, but there were more

Table 7.7 *Comparison of catches from two Malaise traps operated in 1978*

	Individuals (N)	Species (S)	N/S	Species taken once only (n_1)	n_1 as % S	Commonest as % N	Commonest species
Trap 1	5661	49	115.5	8	16.3	20.6	*Platycheirus cyaneus*
Trap 2	594	32	18.6	7	21.9	17.7	*P. cyaneus*

Notes: Trap 1 is the original trap, operated every year. Trap 2 was taken down on 17 September — the catch from trap 1 is for the same period.

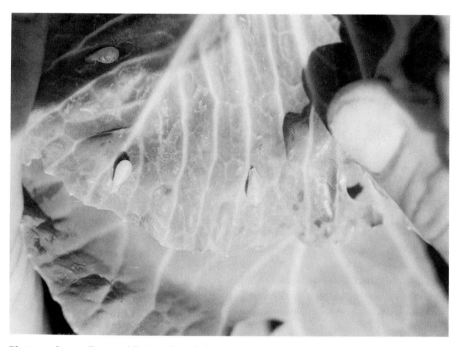

Photograph 7.4. *Pupae of* Episyrphus balteatus *on cabbage leaves. Larvae of this species of hoverfly are predators of aphids and feed particularly on cabbage aphids* Brevicoryne brassicae, *every year in the Leicester garden.*

species in the garden while β-diversity was higher on the campus. Twelve species occurred only in the garden, and five only on the campus, including two not known from the garden, *Xanthandrus comtus* (Harris) and *Cheilosia intonsa*. Weekly numbers reached a peak in the fourth week on the campus and then remained high, whereas garden numbers were low until the fifth week, when they rose dramatically to a higher level than ever attained on the campus.

The commonest garden species was *Episyrphus balteatus* but it ranked second on the campus with a six-week catch less than half that from the

Table 7.8 *Status of hoverflies recorded in a suburban garden, 1972–86*

Known or assumed to breed in garden (25 species)	Assumed to breed in surrounding area (26 species)
Baccha elongata	*Baccha obscuripennis*
Melanostoma mellinum	*Platycheirus immarginatus*
M. scalare	*P. scambus*
Platycheirus ambiguus	*Pyrophaena granditarsa*
P. angustatus	*Paragus haemorrhous*
P. cyaneus	*Chrysotoxum festivum*
P. clypeatus	*C. verralli*
P. manicatus	*Leucozona lucorum*
P. peltatus	*Melangyna cincta*
P. scutatus	*Eupoedes latifasciatus*
Dasysyrphus albostriatus	*Scaeva pyrastri*
Epistrophe eligans	*Sphaerophoria menthastri*
Episyrphus balteatus	*Cheilosia bergenstammi*
Meliscaeva auricollis	*C. paganus*
Eupoedes corollae	*Rhingia campestris*
E. luniger	*Neoascia podagrica*
Sphaerophoria scripta	*Eristalis arbustorum*
Syrphus ribesii	*E. intricarius*
S. vitripennis	*E. interrupta*
Cheilosia vernalis	*E. pertinax*
Eumerus strigatus	*E. tenax*
E. tuberculatus	*Helophilus pendulus*
Merodon equestris	*Myathropa florea*
Pipiza noctiluca	*Neocnemodon vitripennis*
Syritta pipiens	*Volucella bombylans*
	V. pellucens

Casual visitors (28 species)	Recorded in one year only (12 species)
Chrysotoxum bicinctum	*Dasysyrphus tricinctus*
Dasysyrphus lunulatus	*Epistrophe grossulariae*
D. venustus	*Parasyrphus malinellus*
Epistrophe nitidicollis	*Xanthogramma pedissequum*
Melangyna lasiophthalma	*Ferdinandea ruficornis*
M. triangulifera	*Chrysogaster hirtella*
M. umbellatarum	*Eristalis abusivus*
Meliscaeva cinctella	*Helophilus trivittatus*
Eupoedes latilunulatus	*Triglyphus primus*
Parasyrphus punctulatus	*Criorhina berberina*
Scaeva selenitica	*C. floccosa*
Sphaerophoria rueppellii	*Tropidia scita*
Syrphus torvus	
Cheilosia albitarsis	
C. proxima	
Ferdinandea cuprea	
Lejogaster metallina	

Table 7.8 Cont.

Casual visitors (28 species) Cont.

Orthonevra splendens
Eristalis sepulchralis
E. horticola
Helophilus hybridus
Heringia heringii
Pipiza austriaca
P. bimaculata
P. fenestrata
P. luteitarsis
Pipizella viduata
Xylota segnis

garden. *Platycheirus clypeatus* was the commonest species on the campus, but ranked only twelfth in the garden. *Platycheirus cyaneus* and *Sphaerophoria scripta* were considerably commoner in the garden, whereas *Eupoedes corollae*, *Melanostoma mellinum*, *M. scalare*, and *Pyrophaena granditarsa* were considerably commoner on the campus. *Syrphus ribesii* was a major component of the catch on both sites. A major qualitative difference between the catches was the almost total absence from the campus of hoverflies with larvae that feed on decaying organic material, except for *Rhingia campestris*. The inference is that the garden catch reflected the abundance of resident hoverflies and those seeking breeding sites, and also the way in which garden flowers attract migrant *E. balteatus*. However, the large campus catch confirms the extreme mobility of hoverflies, and is a sample of those that are continually on the move, circulating between feeding and breeding sites. It may be a more accurate, i.e. unbiased, sample of the suburban hoverfly community than is the garden catch.

Some garden hoverflies are resident, others visitors, sometimes casual, but probably seeking food or egg-laying sites. Of the 91 species, 30 (such as *Baccha*, *Ferdinandea*, *Criorhina* and *Xylota*) are usually associated with woodland or woodland edge, 11 (such as *Rhingia* and *Merodon*) with more open places, and 50 range over both types of habitat (compiled from Pollard 1971, Steele & Welch 1973, Stubbs & Chandler 1978, and Stubbs & Falk 1983). Gardens are a patchwork of open and shaded places, and so might be expected to accommodate species with no marked habitat preferences. On the other hand, species characteristic of open ground may be attracted to lawns and paths, and woodland species to sheltered and shady corners. On the whole, however, gardens are open habitats, and most of the woodland species are rare visitors. In six years' detailed recording on more than half a square mile (1.3 square kilometres) of woodland, pasture, hedgerow and pond in Bedfordshire, Laurence (1950) found 83 species of hoverfly, 30 of which were only in woodland, 12 only outside the woodland, and 41 in both.

On the basis of captures and breeding records, together with the statements about habitat given above, the garden hoverflies can be divided into four categories: 25 species are regular, common and are either known to breed or assumed to do so; a further 26 species are regularly present and are assumed to breed within surrounding gardens and suburban areas; 28 species have been recorded in more than one year, but only infrequently, and are probably casual visitors; and 12 species have been recorded in one year only, 11 of them represented by single individuals, and rank as chance visitors (Table 7.8). To summarize, 51 of the 91 species are regarded as part of the resident fauna of the garden and surrounding area, while 40, including most of the woodland species, are casual or chance visitors. The diversity of larval life styles among hoverflies resident in the garden or surrounding area demonstrates the complexity and variety of the garden habitat.

8

Ichneumonidae and other parasitic wasps

More than a quarter of a million Hymenoptera are known worldwide, about half of them parasitic, usually on other insects, as larvae. The checklist of British Hymenoptera (Fitton *et al.* 1978) includes 6641 species grouped in 66 families of which 40 are wholly or largely parasitic. The subdivision Parasitica of the suborder Apocrita includes 34 families in Britain; one of these, Cynipidae, however, consists largely of phytophagous species which cause gall formation. In addition, there are one family (containing one British species) of parasitic sawflies, suborder Symphyta, and six families of the subdivision Aculeata, of the suborder Apocrita, with parasitic members. (They are distinguished from other aculeate wasps as parasitic if they simply lay an egg on their stung and paralysed prey or host, without constructing a cell or nest structure around it.) These 40 families of parasitic British Hymenoptera include nearly 5500 species, of which more than 3000 belong to the superfamily Ichneumonoidea, which includes the families Braconidae, Aphidiidae and Ichneumonidae, the last-mentioned with 2028 recorded species, although the estimated fauna is 2900 (Owen, Townes & Townes 1981).

The interest of Henry Townes, a world authority on the family, and his readiness to identify specimens, led to an investigation of the Ichneumonidae captured in the Malaise trap in the Leicester garden. Serphidae and Heloridae (superfamily Serphoidea = Proctotrupoidea) inadvertently included with trap samples sent to Townes were also identified and are briefly described at the end of this chapter. It is important to recognize that these three families represent less than 40 per cent of the parasitic Hymenoptera on the British list and presumably, therefore, in my garden, the notable exclusions being Braconidae and the entire superfamily Chalcidoidea. However, because egg-laying ichneumonids tend to be niche-specific, rather than host-specific, laying on any broadly suitable host that they encounter in a particular sort of micro-habitat (Townes 1972b), the ichneumonid fauna of a garden is particularly interesting, being indicative of the garden's structural heterogeneity. Moreover, an astonishing number of species were identified in my garden – 533 in total, particularly bearing in mind that ichneumonids, like hoverflies, are a family.

NATURAL HISTORY OF ICHNEUMONIDAE

The Ichneumonidae comprises a vast and varied assemblage of parasitic wasps. They are slender, wasp-waisted insects, often with the abdomen rather compressed laterally. Some are quite large, up to 40 millimetres long, most are small, and many are tiny, down to 2 mm. They have a characteristic jerky walk over leaves, with the long, slender antennae in constant, quivering motion. The family is most easily distinguished from other small wasps by the venation of the front wing (the touching or fusion of costal and sub-costal veins and the presence of a second recurrent vein), which has a large stigma; some females, however, are wingless. The bodies of most are predominantly blackish or brown, sometimes spotted or banded with white, yellow or red.

The juvenile stages of ichneumonids are parasites of insect larvae or pupae, i.e. of the developmental stages of insects with complete metamorphosis, particularly Lepidoptera, sawflies and wood-boring beetles, or, less frequently, of spiders or spider eggs. The majority develop inside the host, i.e. as endoparasites, and, because the host is eventually killed, they are more akin to predators than parasites. Moreover, those that feed on spider eggs consume more than one individual, i.e. all the eggs in a spider egg cocoon. Consequently, they are often called parasitoids, to distinguish them from typical parasites, which do not kill their hosts. They are able to consume everything but the host's skin without risk of poisoning their environment by their own excreta, because no connection is established between their mid- and hind-guts until they have stopped feeding (Shaw & Askew 1976).

Females have an ovipositor, often conspicuous, which is used to lay one or more eggs on or actually inside the host. The ovipositor is longer in those species that habitually parasitize larvae or pupae concealed in burrows in wood or in soil. Eggs laid external to the host usually develop into ecto-parasites, although in some species small larvae enter the host, through an inter-segmental membrane, to complete their development. Before laying eggs which develop into external parasites, female Gelinae sting the host which is thus paralysed or killed (Shaw & Askew 1976). There may be several ectoparasitic larvae on one host, whereas endoparasites are usually solitary. In some cases this is the result of one parasitic larva attacking and killing others within the host, in others, because an egg-laying female detects chemi-cally, with her antennae or ovipositor, that a potential host is already parasitized. Eggs deposited in larvae may only hatch under the influence of the host's hormones when it pupates, and pupal parasites are often laid into the host's final larval instar, although others are laid directly into the pupa. Similarly, ichneumonid eggs laid into host eggs do not hatch and feed until the host is a larva, and only emerge from the host pupa or puparium. In *Diplazon* spp., which oviposit inside eggs or tiny larvae of aphid-eating hover-flies, the eggs are long and thin (Fitton & Rotheray 1982). Some, such as most Mesochorinae, are hyper-parasites, or secondary parasites, feeding on another, primary, parasite while it is still feeding on the host. Others, such as many Gelinae, are pseudo-hyper-parasites, feeding on a primary parasite after it is fully fed and has destroyed the host.

Many ichneumonids that parasitize insect species that diapause as larvae or pupae, come under the influence of the host's hormones, and themselves enter diapause. Conversely, substances injected by the egg-laying female ichneumonid may affect the development, behaviour, physiology and morphology of the host (Vinson & Iwantsch 1980). Parasitized hosts develop more slowly than unparasitized ones, and endoparasites consume fat and other replaceable tissues as they are formed, with the result that they may consume more than the final weight of the host (Shaw & Askew 1976). Females of an ichneumonid that parasitizes cocoons of burnet moths apparently assess the size, and hence food potential, of possible hosts visually and by antennal contact, before proceeding to lay eggs (Quicke 1986).

It was once assumed that adults do not feed, and only drink from dew, but many larger species are now known to feed on nectar, pollen, honeydew and plant sap. I have seen females of three species of Diplazontinae, which parasitize aphidophagous hoverfly larvae, feeding on the honeydew of *Brevicoryne brassicae* on cabbages in the garden. According to Fitton & Rotheray (1982), female Diplazontinae also feed on syrphid eggs and larvae. Furthermore, females of many species appear to feed on the body fluids oozing from the puncture made by the sting or ovipositor in the host's body (Askew 1971). It is likely that protein intake is required for egg maturation, and since individuals may be relatively long-lived, especially in those species which hibernate as adults, such as many Ichneumoninae, energy-rich food would seem to be essential.

Ichneumonids are active and wide-ranging, and adult females are very particular about where they seek hosts. Consequently, considerable amounts of energy may be expended, particularly on searching for egg-laying sites. They follow a hierarchy of cues, which finally lead them to lay eggs. For instance, a female of a given species may only search rotting vegetation, or growing tips of plants in open, sunny sites. The female inspects the selected micro-habitat very closely, and, particularly, any broadly suitable hosts that she encounters. As a response to colour, shape, texture, or chemical cues, she may lay eggs on the outside of the host, or insert her ovipositor and lay internally (Vinson 1976). In some cases, such as where the ichneumonid responds to the chemical composition of the host's haemolymph, this makes for a high degree of host specificity. In others, especially where eggs are laid externally to the potential host, there may be a wide range, so that the parasite may be said to be niche-specific, rather than host-specific.

For example, *Diplazon laetatorius*, one of the commonest ichneumonids of the British countryside, parasitizes the syrphids *Episyrphus balteatus*, *Sphaerophoria* spp., *Melanostoma* spp. and *Eupoedes (=Metasyrphus) corollae* (Fitton & Rotheray 1982). Rotheray (1981) has shown by experiment that female Diplazontinae locate aphid colonies by the odour of the aphids, then search for syrphid larvae, which they recognize chemically by touch, using the antennae. The initial response of a larva is to 'freeze', but if investigation by

Photograph 8.1. Dense and varied vegetation enhances the Leicester garden as a habitat for ichneumonids and other insects.

the ichneumonid continues, it moves its front end vigorously around, then directs regurgitated fluid at the attacking ovipositor. If this fails, it may roll over and over in an attempt to escape. *Enizemum ornatum* bites a prospective host, causing it to raise its front end, which the ichneumonid then stabs with its ovipositor. Lack of host specificity is confirmed by many ichneumonids having long flight periods, longer than any one potential host (Owen *et al.* 1981).

There are known to be more than 14 800 species of ichneumonid in the world, but allowing for a high percentage of unreported species, the world fauna is estimated at 60 000 species. There are certainly more species of Ichneumonidae than of vertebrates, and the only animal family that may exceed it in size is the Curculionidae (weevils). The estimated number of valid reported species of the western Palaearctic is 4023, thought to be 70 per cent of the actual total for this region (Townes 1969).

Although the group is large and worldwide in distribution, individuals are small and rarely abundant, and most people are unaware of them. Taxonomy and specific identification are difficult and, even among entomologists, few recognize any but the larger and more conspicuous species. Even the entomologist Frank Lutz, in his otherwise exhaustive account of the insects of his suburban lot in New York, identified only 22 species from the superfamilies Ichneumonoidea, Chalcidoidea and Proctotrupoidea, and admits 'Many of the others were here but I, like the "general public" did not notice them.'

197

Only 16 species are known from a survey of the garden of Buckingham Palace (Richards 1964), and in their record of Monks Wood nature reserve, Steele & Welch (1973) list only 46 species. Virtually nothing is known of the ichneumonid fauna of Leicestershire, and there is no county list. They remain a poorly known, little understood and largely neglected group of animals, and this is reflected in their classification. The geological age of the group is about the same as that of birds, and their structural and biological diversity is similar or greater. Despite this, the ichneumonids are ranked as a family, whereas the birds, with only a seventh as many species, are considered as a class comprising 160 families.

Ichneumonids in general, need a daily drink of water, usually in the form of dew clinging to foliage, and are restricted to areas with frequent dew or rain, where there is dense vegetation (Townes 1972b). This is one reason for their abundance in north temperate areas, particularly eastern North America, Europe excluding the Mediterranean region, and northern Japan. These areas have a rich ground flora, providing an array of micro-habitats and potential ecological niches. Moreover, the seasonality of north temperate areas tends to diversify food chains and hence potential hosts. Sawflies of the family Tenthredinidae and hoverflies (Syrphidae), both of which are extensively used as hosts, are much more abundant and diverse than in the tropics, and indeed the ichneumonid subfamily Diplazontinae, syrphid endoparasites with about 50 British species, are virtually absent in the lowland tropics.

DIVERSITY AND ABUNDANCE OF ICHNEUMONIDAE

The ichneumonid fauna of northern Europe as a whole is rich and varied, and the cool, damp climate of England is particularly favourable for them. Their abundance and diversity has become evident during the long-term study of the Leicester garden fauna. A Malaise trap is a relatively efficient means of collecting Ichneumonidae, although probably catching only 20 per cent of those that enter the trap air space (Townes 1972b). For three years, 1972–74, ichneumonids were sorted from the Malaise trap samples collected in the Leicester garden, and sent to Henry and Marjorie Townes at the American Entomological Institute for identification. In 1972 and 1973 the ichneumonid catches were tabulated in terms of individuals of each species. Even for a taxonomist of Henry Townes' calibre and dedication, this represented an enormous task, and only a list of additional species was prepared in 1974. For purposes of tabulation and analysis, it was necessary to sort the trap samples into species, but it was not necessary to identify them. Species were designated by generic name, and a number where there was more than one species of that genus. A total of 529 species were caught in the Malaise trap over the three-year period 1972–74 (Owen *et al.* 1981), and a further four species have been bred from parasitized larvae.

A striking feature of the ichneumonid sample from the garden compared with other groups, such as hoverflies (Syrphidae), is its extraordinary diversity. In 1972 and 1973, 6445 individuals of 455 species were caught. Of these, 141 species were represented by only one individual in the two years, and the commonest species, *Syrphoctonus pictus*, a parasite of hoverflies, formed only 5 per cent of the two-year catch. In other words, most of the garden ich-neumonids are rare, and none is particularly common. A further indication of the rarity of most species is that only 225 were collected in both years, 129 species being added in the second year of trapping. Furthermore, although the 1974 catch was not analysed, but only sorted for further species, it added 74, bringing the total number of species caught in three years to 529. The average number of individuals per species in 1972 and 1973 was 7.7 and 11.1, respectively. To emphasize by comparison how relatively rare ichneumonids are, the average number of individuals per species of hoverfly captured in the Malaise trap was 28.5 in 1972 and 51.0 in 1973. Evidently, ichneumonid populations are relatively small.

The Malaise trap catches of the two years differed considerably (Table 8.1). In 1973, the catch was much larger and consisted of more species, the average number of individuals per species was greater, as was the relative proportion of the commonest species, although the number of species taken once only was similar. A different species was commonest in each year: in 1972, a species of *Helictes*, parasites of flies that live in rotting vegetation, and in 1973, a species of *Charitopes*, parasites of brown lacewings (Hemerobiidae).

Of 70 species which were either represented by more than 50 individuals in the two-year sample, or by more than ten individuals in one year and by more than twice as many individuals in one year than in the other, 54 were more abundant in 1973 and only 16 in 1972. This could be a reflection of climatic differences between the years, the summer of 1973 being warmer, wetter and sunnier than that of 1972, or of availability of hosts. Among ichneumonids trapped in the garden, Lepidoptera and predators of aphids (Syrphidae, Chrysopidae and Hemerobiidae) are favoured hosts (see Table 8.3, p. 216). Nineteen of those species more abundant in 1973 use Lepidoptera as hosts, but only four of those more abundant in 1972. Fourteen of those more abundant in 1973 use predators of aphids, but only one of those more abundant in 1972. Butterflies and hoverflies (Syrphidae) were trapped more than twice as often in 1973 than in 1972, suggesting that two of the most important host groups were more abundant in that year.

However, weather conditions may have been more favourable in 1973 than in 1972 for hosts as well as for ichneumonids. The larger catch in 1973 was a consequence of better catches in May, June, July and August, suggesting that events in early to mid-summer affected the annual total of ichneumonids. May and June 1973 were both wetter and sunnier than in 1972, and June was much warmer (Figs. 8.4, 8.5 and 8.6). July 1973 was quite cool, but August was warmer than in 1972, although no sunnier and considerably drier. On the whole, 1973 was warmer and sunnier than 1972 and this probably accounts for a better Malaise trap catch, not only of ichneumonids, but of insects in general. Since a Malaise trap catches flying insects, a good catch is an indica-tion of high flight activity and suggests abundance, but a poor catch indicates only that flight activity was low and gives no indication of abundance. The low temperatures of July 1973 account for the similarity of the ichneumonid catch to that of the preceding month instead of the expected seasonal increase

Table 8.1 *Size and composition of annual Malaise trap catch and two-year catch of ichneumonids in Leicester garden, 1972–73*

	Number of individuals (N)	Number of species (S)	N/S	Species taken once only (n_1)	Commonest sp. as % N
1972	2495	326	7.7	118	3.2
1973	3950	355	11.1	117	6.6
Total	6445	455	14.2	141	5.0

in numbers. Ichneumonids are very sensitive to humidity, and conditions in May and June 1973 presumably produced more humid conditions, especially at ground level and around vegetation, later in the summer, even though relatively little rain fell in August. Warm, wet weather earlier in the year was especially favourable, and the warm, sunny conditions of August were ideal for flight (Owen *et al.* 1981). The summer of 1973 was better for ichneumonids, quite apart from the improved availability of hosts.

The number of species trapped in 1972–74 is more than a quarter the number of species known for the British Isles, an amazing proportion for a suburban garden. Moreover, it is known to include some rare species and others which, when identifications are completed, will prove to be additions to the British list. For instance, *Piogaster albina* had only once previously been reported for Britain, from King's Lynn in 1911, and the only previous record of *Lissonota stigmator* was from west Suffolk in 1916. *Apolophus borealis* had previously only been recorded from southern Germany and from North America, and seven species in the genera *Dichrogaster*, *Aclastus* and *Clistopyga* had not previously been recognized as distinct in museum collections. At least one of these (a *Clistopyga* sp.) is a hitherto undescribed species, and the others will be additions to the British list. *Mastrus longulus*, *M. mandibularis* and *Odontoneura punctiscutum* are all hitherto undescribed species. A further seven named species, four of which have been reared from parasitized host larvae, appear to be new to the British Isles. This indicates what surprises there are and what discoveries may be made in studying a poorly known group, such as Ichneumonidae, even in such an apparently mundane site as a suburban garden. A partial list of species of garden ichneumonids (see below) has resulted from rearing records and other identifications by Henry Townes. In the list, the species are grouped by subfamilies, and asterisks indicate species that do not appear in the British checklist (Fitton *et al.* 1978).

Ephialtinae

   Piogaster albina Perkins

Tryphoninae

   Netelia inedita Kokujev*
   Oedomopsis scabriculus (Gravenhorst)

Gelinae

  Lysibia nana (Gravenhorst)
  Dichrogaster aestivalis (Gravenhorst)
  Mastrus longulus Horst.*
  M. mandibularis Horst.*
  M. varicoxis Tasch.*
  Odontoneura punctiscutum Horst.*
  Megacara rusticellae Bridgman*
  M. vagans Gravenhorst*

Banchinae

  Lissonota stigmator Aubert
  Banchus volutatorius (L.)

Scolobatinae

  Rhinotorus atratus (Hgn.)

Porizontinae

  Dusona terebrator (Fab.)*
  Hyposoter tricolor (Ratzeburg)*

Metopiinae

  Apolophus borealis Townes*

Diplazontinae

  Syrphoctonus fissorius Gravenhorst
  S. pictus Gravenhorst
  S. tarsatorius Panzer
  Enizemum ornatum (Gravenhorst)
  Syrphophilus tricinctus (Ashmead)*
  Tymmophorus graculus (Gravenhorst)
  Diplazon laetatorius (Fab.)
  D. tetragonus (Thunberg)
  Promethes bridgmani Fitton
  P. sulcator (Gravenhorst)

Ichneumoninae

  Platylabus iridipennis (Gravenhorst)
  Hypomechus quadriannulatus (Gravenhorst)

The genera and number of species (in parentheses) of Ichneumonidae trapped in the Leicester garden 1972–73 are listed below and are followed by a list of additional species trapped in 1974. The data are from Owen *et al.* (1981).

Subfamily Ephialtinae: Scambus (4), Tromatobia (1), Zaglyptus (1), Clistopyga (1), Acrodactyla (1), Piogaster (1), Polysphincta (1), Sinarachna (1), Zatypota (2), Itoplectis (2), Coccygominus (3), Perithous (1).

Subfamily Tryphoninae: Phytodietus (2), Netelia (5), Neliopisthus (1), Oedomopsis (1), Erromenus (1), Dyspetes (1), Tryphon (2), Acrotomus (2), Cteniscus (1), Eridolius (2).

Subfamily Adelognathinae: Adelognathus (2).

Subfamily Gelinae: Encrateola (1), Eudelus (1), Acrolyta (3), Diaglyptidea (1), Lysibia (1), Arotrephes (1), Hemiteles (1), Aclastus (6), Xenolytus (1), Dichrogaster (4), Gelis (7), Mastrus (4), Stiboscopus (2), Lochetica (1), Ethelurgus (1), Charitopes (3), Endasys (6), Glyphicnemis (2), Bathythrix (4), Orthizema (2), Gnotus (1), Stibeutes (4), Theroscopus (4), Megacara (2), Phygadeuon (32), Stilpnus (6), Mesoleptus (2), Atractodes (5), Demopheles (1), Javra (1), Parmortha (1), Cubocephalus (3), Oresbius (3), Pleolophus (1), Aptesis (2), Agrothereutes (1), Gambrus (1), Aritranis (1), Pycnocryptus (1), Enclisis (1), Itamoplex (1).

Subfamily Stilbopinae: Stilbops (1).

Subfamily Banchinae: Apophua (1), Glypta (7), Lissonota (9), Exetastes (2), Banchus (1).

Subfamily Scolobatinae: Rhorus (3), Perilissus (1), Lathrolestes (1), Otlophorus (1), Campodorus (8), Mesoleius (1), Synomelix (1), Synodites (1), Mesoleptidea (1), Hypamblys (1), Euryproctus (1).

Subfamily Porizontinae: Sinophorus (2), Campoplex (8), Porizon (1), Casinaria (3), Bathyplectes (2), Leptocampoplex (1), Synetaeris (1), Campoletis (7), Dusona (3), Meloboris (2), Phobocampe (4), Tranosema (3), Enytus (2), Diadegma (23), Hyposoter (9), Olesicampe (6), Lathrostizus (1).

Subfamily Cremastinae: Pristomerus (1).

Subfamily Phrudinae: Phrudus (1).

Subfamily Tersilochinae: Probles (6), Tersilochus (4), Phradis (4), Diaparsis (2).

Subfamily Ophioninae: Ophion (3), Enicospilus (1).

Subfamily Mesochorinae: Astiphromma (1), Mesochorus (13), Stictopisthus (1).

Subfamily Metopiinae: Chorinaeus (3), Triclistus (1), Apolophus (1), Hypsicera (2), Exochus (2).

Subfamily Anomalinae: Trichionotus (2).

Subfamily *Microleptinae*: *Oxytorus* (1), *Pantisarthrus* (1), *Proclitus* (4), *Dialipsis* (1), *Plectiscidea* (3), *Aperileptus* (2), *Laepserus* (1), *Symplecis* (1), *Helictes* (1), *Megastylus* (2).

Subfamily *Orthopelmatinae*: *Orthopelma* (2).

Subfamily *Orthocentrinae*: *Orthocentrus* (8), *Picrostigeus* (2), *Stenomacrus* (12), *Leipaulus* (1), *Neurateles* (2).

Subfamily *Diplazontinae*: *Syrphoctonus* (14), *Enizemum* (1), *Woldstedtius* (2), *Phthorima* (2), *Syrphophilus* (3), *Tymmophorus* (2), *Diplazon* (5), *Promethes* (1), *Sussaba* (3).

Subfamily *Ichneumoninae*: *Ichneumon* (1), *Cratichneumon* (1), *Barichneumon* (2), *Stenobarichneumon* (1), *Homotherus* (1), *Pterocormus* (4), *Probolus* (1), *Triptognathus* (1), *Amblyteles* (1), *Exephanes* (1), *Hepiopelmus* (1), *Platylabus* (2), *Linycus* (1), *Apaeleticus* (1), *Heterischnus* (2), *Nematomicrus* (1), *Herpestomus* (1), *Dicaelotus* (2), *Mevesia* (1), *Diadromus* (3), *Centeterus* (1), *Oiorhinus* (1), *Aethecerus* (5), *Phaeogenes* (10), *Alomya* (1).

Additional species collected in 1974: *Dolichomitus* (1), *Grypocentrus* (1), *Erromenus* (1), *Eridolius* (1), *Hemiteles* (1), *Mastrus* (4), *Charitopes* (1), *Endasys* (1), *Bathythrix* (1), *Phygadeuon* (8), *Oresbius* (1), *Schenkia* (1), *Aptesis* (1), *Glypta* (2), *Lissonota* (2), *Alloplasta* (1), *Absyrtus* (1), *Lagarotis* (1), *Alexeter* (1), *Campodorus* (3), *Saotis* (1), *Perilissus* (1), *Anoncus* (1), *Synodites* (1), *Mesoleptidea* (1), *Phobetes* (1), *Euryproctus* (1), *Campoplex* (2), *Casinaria* (1), *Bathyplectes* (1), *Campoletis* (1), *Dusona* (1), *Cymodusa* (1), *Dolophron* (1), *Enytus* (1), *Diadegma* (1), *Hyposoter* (3), *Olesicampe* (1), *Temelucha* (1), *Probles* (1), *Tersilochus* (3), *Ophion* (1), *Mesochorus* (5), *Triclistus* (1), *Hypsicera* (1), *Megastylus* (1), *Syrphoctonus* (2), *Campocraspedon* (1), *Syspasis* (1), *Cratichneumon* (1).

Owen *et al.* (1981) calculated a numerical index of species diversity, H, for annual and monthly samples using the information theory of diversity (MacArthur 1965). Like Simpson's index, information theory indices take both evenness and species richness into account, and make no assumptions about the shape of the underlying species abundance distribution. H is moderately sensitive to sample size, and is affected by species richness rather than the abundance of the commonest species (Magurran 1988). The value of H for the 1973 sample was significantly smaller than that for 1972 (at the 1 per cent level), a reflection of the fact that in 1973 there were relatively fewer rare species (taken once only) and a higher average number of individuals per species. Diversity rose to a maximum in August, when ichneumonids were most abundant, suggesting that seasonal increase in abundance is a consequence of an increase in the number of species, rather than only an increase in abundance of species already present. During the summer peak of abundance (July–September), diversity was much the same in 1972 and 1973, but earlier and later in the season diversity differed in the two years. This was particularly marked in June, diversity in June 1973 being significantly lower (at the 1 per cent level) than that for June 1972. This is a reflection of a relatively large sample in June 1973 made up of fewer species than might have been expected, some of them relatively abundant.

The association of numbers of individuals and species with diversity is shown in Fig. 8.1. As a consequence of the average number of individuals per species rising to a peak in August in both years, the diversity curves flatten out at the time of maximum abundance. The value of H calculated for each total year is significantly greater than that for the month of maximum diversity in that year. This indicates that there is a succession of different associations of species during the year, the species present at different relative frequencies in different months. Evidently a whole year's sample is necessary to indicate the species diversity of a site or locality, rather than a sample taken at the time of maximum abundance.

DIVERSITY OF TEMPERATE AND TROPICAL ICHNEUMONIDS

It is generally recognized that species diversity is greatest in the tropics and declines towards the poles. Latitudinal gradients in diversity have been demonstrated using groups as diverse as trees, birds, bats and butterflies, and many theories, none of them entirely satisfactory, have been proposed to account for the phenomenon. Ichneumonidae, however, seem to be an exception to the rule.

Samples of ichneumonids were obtained using Malaise traps over 12-month periods from Kampala in Uganda, from Freetown in Sierra Leone and from Skåne in Sweden, and these were compared with the 1972 sample from Leicester (Owen & Owen 1974). The Kampala and Freetown sites were both gardens, although at opposite sides of tropical Africa. Like the Leicester garden, they were intensively cultivated, with a rich diversity of plants, many of them introduced from other parts of the world. The Skåne site was not a garden, but an area of disturbed land alongside a small stream. The African sites are tropical, the Leicester and Skåne sites temperate, the latter being somewhat further north. The information theory index of diversity, H (MacArthur 1965), was calculated for each sample (Table 8.2).

All the samples are large, but particularly those from Leicester and Skåne (the latter was obtained using four traps, giving an average catch per trap of 2748). More striking, however, are the large numbers of species and numbers of those taken once only, the small numbers of individuals per species and the small proportions formed by the commonest species. All samples, therefore, have high diversities. It is perhaps to be expected that in the Kampala sample, with the lowest diversity, the commonest species forms a greater proportion than in other samples, and that the average number of individuals per species is highest in the large Skåne sample despite its high diversity. The values of H for Leicester and Freetown are not significantly different, but all other pairs differ significantly from each other at the 1 per cent level. The four samples can therefore be ranked in terms of diversity: Kampala < Freetown = Leicester < Skåne.

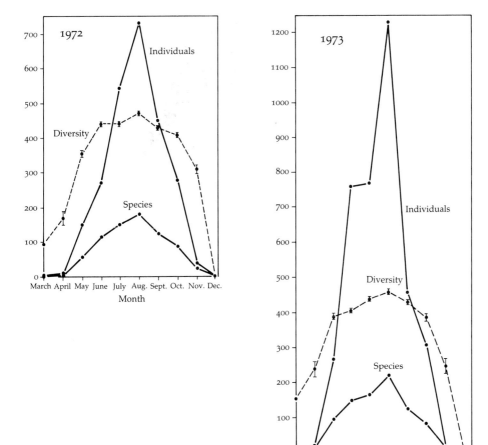

Fig. 8.1 Seasonal changes in numbers of individuals, species, and in diversity (H) of ichneumonids. (The values of H *have been scaled up* × 100: *the bars are standard errors.)*

Why should ichneumonids depart from the usual latitudinal gradient of diversity? One possibility, arising from the evidence that ichneumonids are niche-specific rather than host-specific, is that temperate sites are as heterogeneous, or even more so, than tropical sites and thus provide as many or more micro-habitats to be searched for hosts. Ichneumonids are certainly abundant and varied in temperate areas, finding conditions particularly favourable. It is unlikely to be a reflection of host availability, for Lepidoptera, much utilized as hosts, are more varied in the tropics; there were, for instance, more than 300 species of butterflies in the Freetown garden compared with 21 in the Leicester garden. Parasitized insects tend to be particularly prone to predation, and it may be that predation pressure is greater in

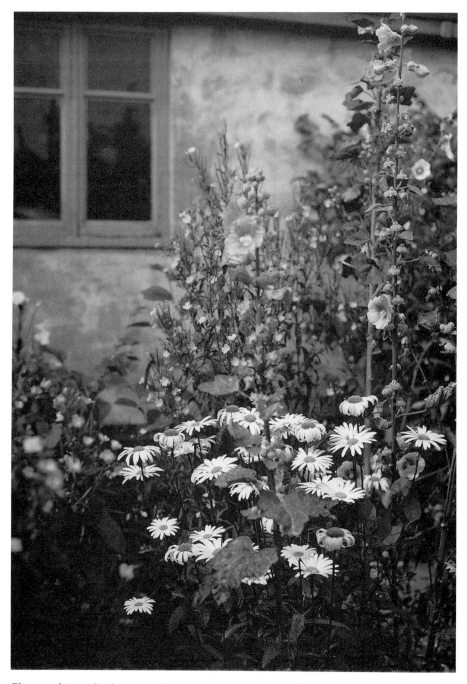

Photograph 8.2. Gardeners maximize plant and structural diversity thus making favourable habitats for ichneumonids. Here, cultivated marguerites Chrysanthemum frutescens *and hollyhocks* Althaea rosea *are mingled with great willowherb* Epilobium hirsutum *that has come into the garden of its own accord.*

Table 8.2 *Diversity of ichneumonids in tropical and temperate sites*

	Kampala	Freetown	Leicester	Skåne
Sample size (N)	2268	1979	2495	10 994
Species (S)	293	319	326	758
N/S	7.7	6.2	7.7	14.5
Species taken once only (n_1)	116	117	118*	203
Commonest species as % N	10.1	4.9	3.2	5.5
Diversity index, H,	4.524	4.934	4.937	5.481
and s.e.	±0.032	±0.029	±0.024	±0.014

*Differs from number given in Owen & Owen (1974) because of adjustments to species separation.

the tropics, thus selecting against parasitized hosts and their parasitoids (Rathcke & Price 1976), or perhaps other families of parasitoids, including Diptera, are particularly diverse in the tropics compared with temperate areas (Owen, D.F. 1983a).

It is possible that the cumulative number of ichneumonid species over several years is greater in the tropics, but it is difficult to compare temporal changes in ichneumonid numbers in tropical sites with those in temperate places because of the different nature of the seasons. In Kampala, for instance, ichneumonids fly in every month of the year with no obvious peaks of abundance or diversity; in Leicester, there is none in the winter, and numbers only gradually build up to a summer peak. In Kampala, although the rate of addition of new species fell off month by month, 12 per cent of those captured in the twelfth month were new, indicating that further trapping would have been required to fully describe the fauna. In Leicester, however, 36 per cent of the species taken in the second year were new, and yet more species were added in a third year. All that can be done is to compare diversity in samples taken over a similar period of time, and 12 months seems the logical period as it encompasses all seasons. Diversity of Ichneumonidae seems to be genuinely greater in temperate areas and it seems reasonable to suppose that this is because they can exploit more niches there.

SEASONALITY IN ICHNEUMONIDAE

Like other insects in Britain, ichneumonids are markedly seasonal in their flight activity. In 1972 and 1973, when the Malaise trap was operated throughout the year, only one was caught in December, none in January and February, and few in March. Thereafter, numbers of individuals and species trapped per month increased to an August peak then declined more rapidly (Figs. 8.2 and 8.3). The average number of individuals per species also reached a peak in August, indicating that the increase in numbers is a consequence of an increase in the abundance of species as well as the addition of species. From May to August 1973, the average number of individuals per

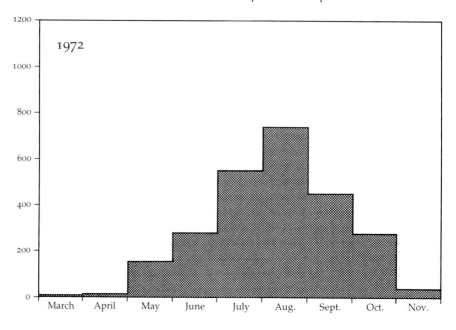

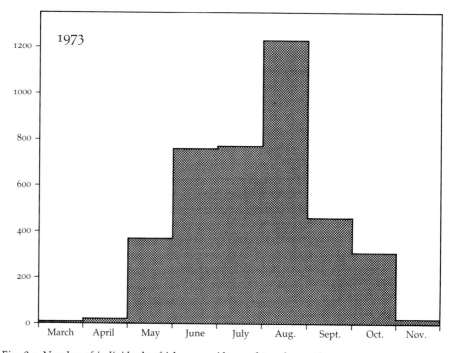

Fig. 8.2 Number of individuals of ichneumonids caught each month.

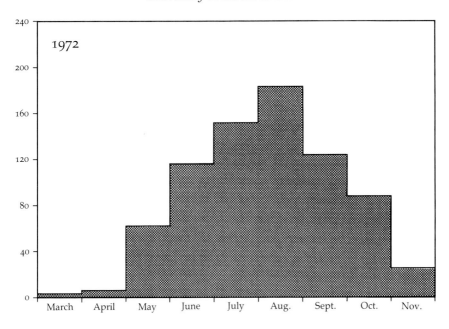

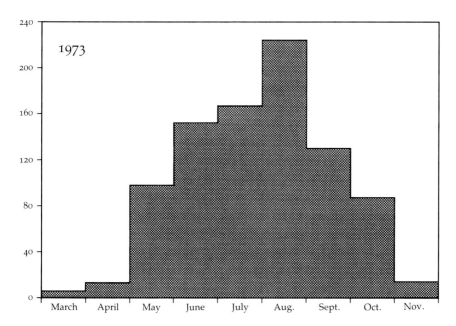

Fig. 8.3 Number of species of ichneumonids caught each month.

species was higher than in the equivalent months in 1972. The June 1973 catch was particularly large and diverse, making it similar to the July catch, which was poorer than might have been expected, with a lower average number of individuals per species than June, but the situation in 1972 was probably more

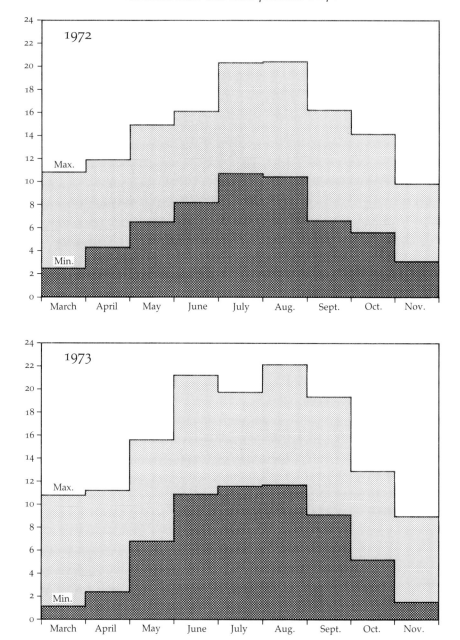

Fig. 8.4 Mean maximum and mean minimum temperatures (°C) for months in which ichneumonids were caught.

typical. As already indicated, June 1973 was wetter, warmer and sunnier than June 1972, as shown by Figs. 8.4, 8.5 and 8.6, whereas July 1973 was cooler and less sunny than July 1972. The higher temperatures and greater number of hours of sunshine for most summer months in 1973 account for the larger

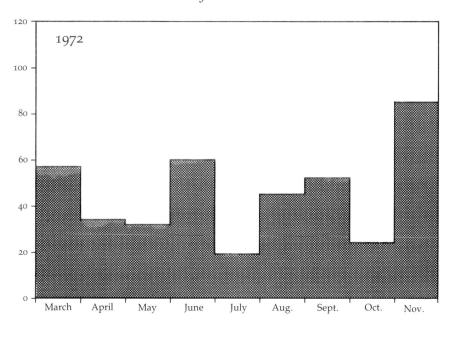

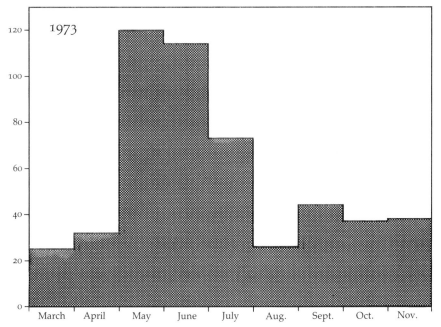

Fig. 8.5 Rainfall (mm) for months in which ichneumonids were caught.

catches than in 1972. Relatively little rain fell in August 1973, but the high rainfall in May and June had created conditions ideal for ichneumonids. The seasonal pattern of temperatures, rainfall and hours of sunshine are mirrored by the Malaise trap catches, and explain their seasonal nature.

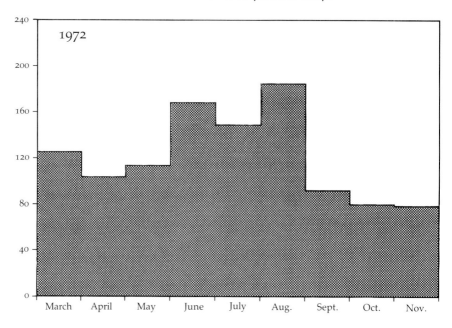

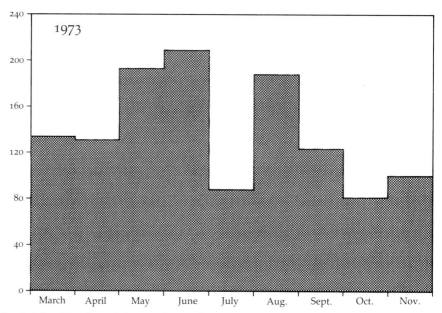

Fig. 8.6 Sunshine (hours) for months in which ichneumonids were caught.

In 1972, ichneumonids were sorted and tabulated as weekly catches. Numbers of individuals and species fluctuated markedly within the overall seasonal pattern. Comparison with weekly temperatures, rainfall and hours of sunshine show that these fluctuations reflected changes in weather and hence its favourability for insect flight (Fig. 8.7). 'For instance, the increases in

212

the third week of May (week 10), the third week of June (14), and the second week of July (18) coincided with increases in mean maximum temperature and decreases in rainfall, and week 18 was also sunny. The sudden decreases in the fourth week of June (15) and the first week of September (26) were correlated with lower temperatures; in week 26, but not in week 15, there were increased rainfall and less sunshine. The only climatic variable that is consistently mirrored in the size of the catch is temperature, and this is well illustrated by the last two weeks of September and the first week of October (weeks 28, 29 and 30), when the catch size fluctuated considerably ($N = 217$, 75 and 139, respectively). In weeks 28, 29 and 30 there was effectively no rain, the sunshine totals were 24.1, 26.2 and 38.8 hours, respectively, but the mean weekly maximum temperatures were 16.7, 15.0 and 16.7°C' (Owen *et al.* 1981). A single weekly sample may give a quite misleading impression of abundance of ichneumonids, and it is only by combining weeks into monthly samples that these fluctuations caused by weather are smoothed out into seasonal patterns. Consequently, in 1973, weekly trap samples were amalgamated into monthly samples before being sent to Henry and Marjorie Townes for sorting and tabulation of species.

Although no one species of ichneumonid is particularly common in the garden, certainly not by comparison with potential hosts, such as hoverflies, all the relatively abundant species were trapped over an amazingly long period of time. For instance, of the 33 species represented by more than 50 individuals in the two-year catch, six were trapped in eight consecutive months (i.e. March to October), nine in seven consecutive months, ten in six months, five in five months, and three in four months. Few potential hosts are present in a suitable stage of development for periods even of four months, let alone eight months. Such prolonged flight periods support the contention that ichneumonids are niche-specific rather than host-specific.

HOST GROUPS OF ICHNEUMONIDAE

The species of ichneumonid trapped in the garden in 1972–73 use a wide range of hosts. The dominant host groups were predators of aphids and other small insects, Lepidoptera, Diptera of humus and decaying plants and, to a lesser extent, spiders and spider eggs, and sawflies. Table 8.3 groups the trapped ichnuemonids by hosts and shows the relative frequencies of individuals and species in the groups, expressed as percentages of the total N (individuals) and S (species). A higher percentage of individuals than of species indicates unusual abundance of a group; a higher percentage of species than of individuals indicates annual scarcity of a group.

Parasites of aphid predators (Neuroptera and Syrphidae), of Lepidoptera, and of spiders and spider eggs were abundant, those of sawflies scarce; the relative frequencies of individuals and species of parasites of Diptera in decaying plant material were similar. Parasites of Lepidoptera are the largest group, and the hosts are very varied both taxonomically and ecologically. Parasites of micro-Lepidoptera were abundant, those of macro-Lepidoptera

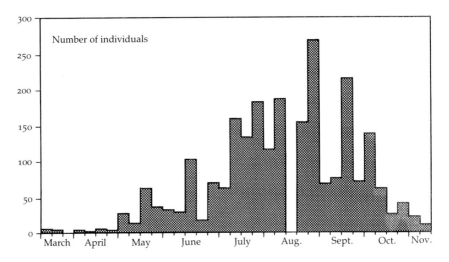

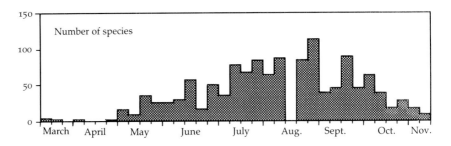

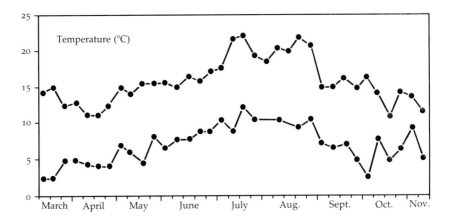

relatively scarce. Those, such as Ephialtini, that use exposed pupae were relatively abundant, those that use hidden pupae and hidden and exposed larvae, relatively scarce. Ichneumonids that use as hosts wood-borers, gall wasps (Cynipidae) and sphecid wasps were particularly scarce, being represented by two individuals of one species, three individuals of two spe-

214

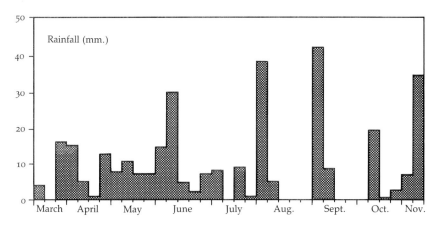

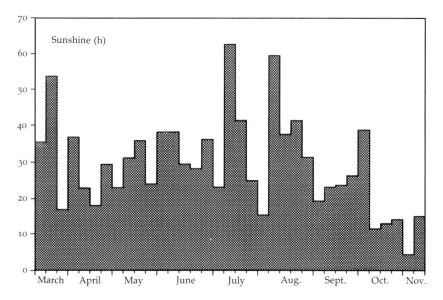

From Owen, Townes and Townes, 1981.

Fig. 8.7 Numbers of individuals and species of ichneumonids in weekly trap samples, 1972, together with mean weekly maximum and minimum temperatures, and weekly rainfall and sunshine totals. (The sample for the third week of August was lost in transit.)

cies, and eleven individuals of one species, respectively; however, a further species that parasitizes wood-borers was collected in 1974.

As far as is known, none of the ichneumonids listed in Table 8.3 as parasites of spiders or spider eggs is host-specific. The Polysphinctini (subfamily Ephialtinae) use nymphs and adults, but all the other genera parasitize eggs. The group includes four of the commoner garden species (i.e. represented by 50 or more individuals in the 1972–73 catch): *Hemiteles* sp., and three species of *Aclastus* (Gelinae), which are characteristic of moist places with

Table 8.3 *Numbers and relative frequencies of individuals and species of ichneumonids in the 1972–73 catch, grouped by host (from Owen et al. (1981))*

Host group	Ichneumonid parasites	Species (S)	% total S	Individuals (N)	% total N
Spiders and spider eggs	*Tromatobia, Zaglyptus, Clistopyga,* Polsphinctini (Ephialtinae), *Hemiteles, Aclastus*	16	3.5	428	6.6
Predators of aphids and other small insects	*Dichrogaster* (parasites of Chrysopidae), *Charitopes* (parasites of Hemerobiidae), *Ethelurgus* (parasites of Syrphidae), Diplazontinae (parasites of Syrphidae)	41	9.0	1777	27.6
Wood-borers	*Demopheles*	1	0.2	2	0.03
Phytophagous Coleoptera	*Bathyplectes* (parasites of Curculionidae), Phrudinae, Tersilochinae	19	4.2	95	1.5
Sawflies (Tenthredinidae)	Tryphonini, Cteniscini, (both Tryphoninae) Adelognathinae, *Endasys,* Echthrini (except *Demopheles*) (Gelinae), Scolobatinae, *Olesicampe, Lathrostizus*	54	11.9	412	6.4
Cynipidae	Orthopelmatinae	2	0.5	3	0.05
Sphecidae	*Perithous*	1	0.2	11	0.17
Lepidoptera	*Scambus,* Ephialtini (Ephialtinae), Phytodietini and Eclytini (Tryphoninae) *Encrateola, Eudelus, Acrolyta, Diaglyptidea, Xenolytus,* Mesostenini (except *Aritranis*) (Gelinae), Banchinae, Porizontinae (except *Bathyplectes, Olesicampe* and *Lathrostizus*), Cremastinae, Ophioninae, Metopiinae, Anomalinae, Ichneumoninae	183	40.2	1889	29.3
Diptera of humus and decaying plants	*Phygadeuon, Stilpnus, Mesoleptus, Atractodes,* Microleptinae, Orthocentrinae	87	19.1	1170	18.15
Other parasites	*Lysibia,* Mesochorinae	16	3.5	323	5.0

Table 8.3 Cont.

Host group	Ichneumonid parasites	Species (S)	% total S	Individuals (N)	% total N
Host unknown or variable	*Arotrephes, Gelis, Mastrus, Stiboscopus, Lochetica, Glyphicnemis, Bathythrix, Orthizema, Gnotus, Stibeutes, Theroscopus, Megacara, Aritranis*	35	7.7	335	5.2
Totals		455		6445	

damp vegetation, where they parasitize the eggs of small spiders among decaying plant material. Linyphiid spiders parasitized by ichneumonid larvae have been found in leaf litter.

A major group of ichneumonids in the garden parasitizes the predatory larvae of Syrphidae (Diptera), Chrysopidae or Hemerobiidae (Neuroptera), and made up more than 27 per cent of the two-year catch. Six parasites of syrphids, in the genera *Ethelurgus* (Gelinae), *Syrphoctonus*, *Diplazon* and *Sussaba* (Diplazontinae), and three parasites of Neuroptera, in the genera *Dichrogaster* and *Charitopes* (Gelinae), were among the commoner garden species. A *Charitopes* sp. was the commonest in the 1973 catch, and *Syrphoctonus pictus* was the commonest overall 1972–73 (see Table 8.1). Six species have been reared from predatory syrphid larvae, including one (*Promethes bridgmani*) not present in the Malaise trap catch (see p. 220). *Syrphoctonus pictus*, *Syrphophilus tricinctus* and *Diplazon laetatorius* were collected while feeding at honeydew excreted by the cabbage aphid *Brevicoryne brassicae*. The parasitization of predatory syrphid larvae by ichneumonids is evidently part of a major food chain in the garden, as more than 81 per cent of garden syrphids have predatory larvae (Owen, J. 1981a) (see Chapter 7).

Seven species that parasitize Lepidoptera were among the commoner species in the Malaise trap catch: species in the genera *Encrateola* (Gelinae), *Campoplex*, *Meloboris* and *Diadegma* (all Porizontinae) and *Netelia inedita* (Tryphoninae). Nine species have been reared from Lepidoptera larvae found in the garden, including one (*Hypomechus quadriannulatus*) not present in the Malaise trap catch (Table 8.5). The ichneumonid species that use Lepidoptera are clearly niche-specific, and can be characterized by the size, site and stage of development of their hosts. *Scambus* (Ephialtinae), some Phytodietini (Tryphoninae), several genera of Banchinae, and Cremastinae lay eggs in larvae hidden in leaf-rolls or stems; Eclytini (Tryphoninae) lay in small exposed larvae; *Banchus* and *Exetastes* (Banchini) and Ophioninae use large exposed larvae; and the species of Anomalinae captured lay eggs in medium-sized, exposed larvae. Ephialtini (Ephialtinae) lay eggs in exposed or semi-exposed pupae; *Encrateola*, *Eudelus*, *Acrolyta* and *Diaglyptidea* (all Gelinae) lay in small pupae hidden in cocoons; Mesostenini (Gelinae) use medium-sized

Table 8.4 *Numbers of individuals and species of commonest ichneumonids, grouped by host*

Host group	Number of individuals (N)	Number of species (S)
Syrphidae	819	6
Neuroptera	573	3
Smaller Lepidoptera	670	6
Larger Lepidoptera	61	1
Spiders and spider eggs	336	4
Cyclorrhapha in decaying vegetation	289	4
Other Diptera in decaying vegetation	275	3
Other parasites	228	2
Sawflies	104	2
Various	63	1
Unknown	78	1

to large pupae in cocoons or near the soil surface; and most Ichneumoninae lay eggs in medium-sized to large pupae hidden in soil or leaf litter.

Seven parasites of Diptera found in humus and decaying plant material are among the commoner garden species trapped in 1972–73: four are species of *Phygadeuon* (Gelinae), the others belonging to the genera *Helictes* (Microleptinae), *Orthocentrus* and *Stenomacrus* (Orthocentrinae). *Helictes* sp. was the most abundant species in the 1972 catch (see p. 200). *Phygadeuon*, *Stilpnus, Mesoleptus* and *Atractodes* (all Gelinae) occur in pupae of Cyclor-rhapha, and Microleptinae and Orthocentrinae parasitize Mycetophilidae, Sciaridae or Cecidomyiidae.

The other ichneumonid species represented by 50 or more individuals in the two-year catch use a variety of hosts. *Aptesis* (Gelinae) and *Olesicampe* (Porizontinae) parasitize sawflies (Tenthredinidae); an additional species of *Olesicampe* and *Rhinotorus atratus* (Scolobatinae) not caught in the Malaise trap, have been reared from *Croesus* sp. (probably *septentrionalis*) on birch (Table 8.5). *Lysibia* (Gelinae) and *Mesochorus* are secondary parasites: *Lysibia* occurs in cocoons of Microgasterinae (Braconidae) and Mesochorinae in Ich-neumonidae, Braconidae or Tachinidae (Diptera). *Gelis* spp. have many hosts, including Lepidoptera, spider eggs or cocoons of other parasites, but the hosts of *Megacara* spp. (Gelinae) are unknown.

The 33 species represented by 50 or more individuals in the two-year catch constitute more than 54 per cent of the individuals collected, although only seven per cent of the species, and may therefore be taken as representative of the garden fauna. They bear out the general statements made about the types of host used, as shown in Table 8.4.

Breeding records of Ichneumonidae

Twenty species of ichneumonid have been bred from host larvae found in the garden (Table 8.5). The preponderance of Lepidoptera and Syrphidae as hosts is simply a reflection of the custom of collecting these larvae for rearing so that they can be identified. It is nevertheless evident that plant – Lepidoptera larva – ichneumonid and plant – aphid – syrphid larva – ichneumonid are important food chains in the garden. It is astonishing that of the 20 species, four were not included in the 529 species collected in the Malaise trap in 1972–74. This suggests that the garden list of ichneumonids would have been increased considerably had sorting from the Malaise trap catch continued after 1974.

The garden ichneumonids in Table 8.5 are linked through food chains to 14 species of plants, seven alien and seven native. The equal importance of alien and native plants as bases for these food chains confirms that alien plants form a major component in biological interactions in a garden and promote the species richness of the garden habitat. For instance, when two individuals of *Hyposoter tricolor* were caught in the Malaise trap in 1972–74, it was a new record for Britain, although known from localities as far apart as Germany and Japan. Subsequently, two were bred from geometrid larvae feeding on *Ribes sanguineum* showing, not only that it is a part of the resident community, but also that its presence can be traced to cultivation of a North American shrub.

CHARACTERISTICS OF THE GARDEN FAUNA OF ICHNEUMONIDAE

The ichneumonid fauna of northern Europe as a whole is rich and varied, and the cool, damp climate of England is particularly favourable. More than a quarter of the total number of species recorded for the British Isles were trapped in the garden during the three-year period 1972–74; none was particularly common and many were rare. Most of the species collected are characteristic of damp habitats with dense vegetation, and species of drier habitats were almost entirely absent. The majority of species and individuals use as hosts Lepidoptera, predators of aphids (Syrphidae and Neuroptera), Diptera of humus and decaying plants, spiders and spider eggs, or sawflies, which tends to emphasize the association with damper habitats. Ichneumonid abundance and diversity in the garden are probably typical of partially disturbed areas of England.

Gardeners intentionally maximize the productivity and the plant and structural diversity of their plots, and these features of gardens make them favourable habitats for insects in general. High plant diversity generates a complex of food chains and, together with diversity in light and shade, in tall and low growth, and in living and dead vegetation, creates a detailed structural mosaic which accommodates a multitude of potential ecological niches (Owen, J. 1983b). This accounts for the high number of species of ichneumonids in gardens, more than in many natural areas. Gardens not only offer a wide range of ecological niches, but watering and application of fertilizers also promote plant growth, which further enhances gardens as habitats for ichneumonids.

However, the behaviour and activity of ichneumonids is such that the

Table 8.5 *Breeding records and associated food chains of Ichneumonidae from the garden*

Ichneumonid	Host
Tryphoninae	
Netelia (sp. 3) *inedita*	*Xanthorhoe fluctuata* (Geometridae) on *Iberis sempervirens*
Oedomopsis (sp. 1) *scabriculus*	Larva of Olethreutidae on *Rosmarinus officinalis*
Banchinae	
Banchus (sp. 1) *volutatorius*	*Lacanobia oleracea* (Noctuidae) on *Solanum dulcamara*
Scolobatinae	
*Rhinotorus atratus**	*Croesus* sp.† (Tenthredinidae) on *Betula pendula*
Porizontinae	
Campoletis sp. 2	Larva of Noctuidae on *Origanum majorana*
Dusona (sp. 1) *terebrator*	Larva of Noctuidae on *Epilobium hirsutum*
Diadegma sp. 5	Unidentified Lepidoptera larva on *Hesperis matronalis*
Hyposoter (sp. 9) *tricolor*	*Abraxas grossulariata* (Geometridae) on *Ribes sanguineum*
Hyposoter sp. 6	Larva of Geometridae on *Lonicera periclymenum*
Olesicampe sp. 8*	*Croesus* sp.† on *Betula pendula*
Diplazontinae	
Syrphoctonus (sp. 7) *fissorius*	*Episyrphus balteatus* on *Brevicoryne brassicae* on *Brassica oleracea*
	Unidentified syrphid larva on *Aphis sambuci* on *Sambucus nigra*
Syrphoctonus (sp. 12) *tarsatorius*	*Episyrphus balteatus* on *B. brassicae* on *B. oleracea*
Syrphophilus (sp. 1) *tricinctus*	*Syrphus ribesii* on *Betulaphis quadrituberculata* on *Betula pendula*
Tymmophorus (sp. 1) *graculus*	Hibernating syrphid larva
Diplazon (sp. 4) *laetatorius*	*Episyrphus balteatus* on *B. brassicae* on *B. oleracea*
Diplazon (sp. 2) *tetragonus*	*Platycheirus scutatus* on aphids on *Nepeta mussinnii*
*Promethes bridgmani**	Hibernating syrphid larva
Promethes (sp. 1) *sulcator*	*Episyrphus balteatus* on *B. brassicae* on *B. oleracea*
	Hibernating syrphid larva
Ichneumoninae	
Platylabus (sp. 1) *iridipennis*	*Gymnoscelis rufifasciata* (Geometridae) on *Buddleia davidii*
	Unidentified Lepidoptera larva on *Ribes sanguineum*
*Hypomechus quadriannulatus**	*Chloroclysta truncata* on *Rubus fruticosus*

Notes: Species numbers refer to tabulation in Owen *et al.* (1981). Ichneumonid larvae were also found on the linyphiid spiders *Stemonyphantes lineatus* and *Lethyphantes leprosus*.

* Ichneumonid species not taken in Malaise trap.

† The only *Croesus* identified from the garden is *C. septentrionalis*, larvae of which feed on birch tree in most years.

Photograph 8.3. Most species of ichneumonids that were collected in the Malaise trap are characteristic of damp habitats with dense vegetation as are provided by the lush plant growth of the Leicester garden.

Malaise trap in effect samples from an area much larger than that of the garden. Adults are highly mobile, and although they tend to frequent favourable clumps of plants, they make intermittent flights to other spots. Representatives of populations within a considerable radius undoubtedly visit the garden, and some may have travelled several kilometres. Any of these may find breeding sites in the garden, and if garden populations become locally extinct they are soon replaced by immigration. The sample of ichneumonids collected in my garden includes those breeding there and in neighbouring gardens, as well as visitors from meadows, woods and other habitats in the vicinity.

SERPHIDAE

Serphidae are small, black wasps, the front wing measuring 1.6–7.4 mm. They have weakly compressed bodies, and reduced wing venation with a large stigma in the forewing. The world list is 310 species, suggesting a world fauna of about 1200 species, more than half of them belonging to the genus *Exallonyx* (Townes & Townes 1981). They are internal parasites of beetles, particularly staphylinids and carabids, and of mycetophilid and sciarid flies, seeking their hosts in moist soil or rotting plant material; most species occur in damp habitats. They lay one or many eggs in the host larva with a quick

thrust. Mature serphid larvae emerge from the ventral side of the host, usually through inter-segmental membranes, and immediately moult to a pupa
which has the tail still embedded in the host remains. A revision of the
Serphidae by Nixon (1938) lists 28 British species, and 36 species are included
in the checklist of British Hymenoptera (Fitton *et al.* 1978). The garden list
(recorded between 1972 and 1974) is an astonishing 22 species, five of them
(indicated by asterisks) new to Britain (Owen *et al.* 1981).

The garden list, (see below) compiled by Henry Townes, uses the nomenclature of Townes & Townes (1981). This nomenclature is different from that
of Fitton *et al.* (1978), and the garden list includes eight species that appear not
to be in the 1978 checklist even allowing for synonymy.

> *Brachyserphus parvulus* Nees
> *Codrus niger* Panzer
> *C. picicornis* Foerster
> *Cryptoserphus aculeator* Haliday
> *Disogmus areolator* Haliday
> *Phaenoserphus chittii* Morley
> *P. viator* Haliday
> *Phaneroserphus calcar* Haliday
> *Serphus gravidator* L.
> *Tretoserphus laricis* Haliday
> *Exallonyx ater* Gravenhorst
> *E. brevicornis* Haliday
> **E. brevimala* Townes
> *E. ligatus* Nees
> *E. longicornis* Nees
> *E. microcerus* Kieffer
> **E. nixoni* Townes
> *E. pallidistigma* Morley
> **E. quadriceps* Ashmead
> *E. subserratus* Kieffer
> **E. trichomus* Townes
> **E. trifoveatus* Kieffer

Host records for the garden species have been compiled from Townes &
Townes (1981). *Brachyserphus* species use the beetle families Eurotylidae,
Melandryidae, Nitidulidae and Phalacridae as hosts. *B. parvulus* has been
reared from various small beetle larvae in Fungi and from larvae of *Meligethes*
beetles (family Nitidulidae) infesting crucifer flowers; there are at least half a
dozen species of *Meligethes* in the garden, and *M. aeneus* is common. *Codrus*
spp. use beetles of the family Carabidae as hosts. *C. niger* has been reared
from *Nebria brevicollis* (recorded in the garden) and other carabids, and *C.
picicornis* from *Notiophilus biguttatus* (recorded in the garden) and *N. rufipes*.
Cryptoserphus spp., including *C. aculeator* use various flies of the families
Mycetophilidae and Sciaridae (fungus gnats) as hosts. There appear to be no
host records for the genus *Disogmus*. *Phaenoserphus* spp. use carabid beetles as
hosts, *P. chittii* having been reared from *Carabus* spp. and *P. viator* from
Pterostichus spp. *Pterostichus madidus* is the commonest ground beetle in the

Photograph 8.4. Structural diversity and variations in light and shade increase the range of ecological niches available in a garden. The trunk of the apple tree is on the left of the picture.

garden. *Phaneroserphus* spp. parasitize Lithobiidae (centipedes) and beetles of the family Staphylinidae, *P. calcar* having been reared from *Lithobius forficatus* (the commonest garden centipede) and two species of staphylinids. *Serphus* spp., including *S. gravidator*, parasitize various carabids. The hosts of *Tretoserphus* spp. are unknown. *Exallonyx* spp. use staphylinid beetles as hosts, and several of the garden species have been reared from genera that occur in the garden.

HELORIDAE

The Heloridae, in the superfamily Serphoidea, are a small group comprising one genus of seven species. They are parasites of Chrysopidae (lacewings), laying into the host larva, which is killed after it has spun a cocoon, within which the parasite pupates. Two species have been recorded from the garden: *Helorus nigripes* Foerster and *H. ruficornis* Foerster. There is one additional species on the British list (Fitton *et al.* 1978). The species of *Chrysopa* recorded as hosts of *H. nigripes* and *ruficornis* (Townes 1977) have not been recorded in the garden, although several other species do occur.

The three families of parasitic wasps that have been looked at in the Leicester garden are represented there by an extraordinary number of species, which form a considerable proportion of the British list. Yet these families include less than half the parasitic Hymenoptera of the British Isles. Evidently, there is still a great deal more to discover about the parasitic wasps of a suburban garden. The diversity, and different life styles and microhabitats of their hosts illustrate particularly well the richness and variety of a garden.

9

Bees and wasps

Think of a rich and productive garden and you think of one alive with the diligent activity of bees at flowers, the air vibrant with their busy humming. In my garden, the mauve, hooded flowers of spotted dead-nettle *Lamium maculatum* dip and sway with the attentions of visiting bumblebees from early in spring to late in the autumn, the tall, violet-blue stems of sage flowers shimmer with all their comings and goings, and the trumpet-like hollyhock flowers seem to amplify the buzz of the fat, furry visitors. From high summer on into autumn, *al fresco* meals in the garden are enlivened by the uninvited visits of wasps lapping sweet drinks, licking up jam and the juices of soft fruits, and chewing morsels from cold meats.

The bumblebees so conspicuously active at flowers and the yellow wasps that disrupt picnics are social species. Colonies are founded in the spring by overwintered queens, most of whose offspring are non-breeding female workers that remain associated with the nest, building new cells, tending the queen's eggs and larvae, defending the colony and foraging for food. These social species are the commonest species of bees and wasps in the garden, but Malaise trapping has revealed an unexpected diversity of less conspicuous solitary species. Females of solitary bees and wasps make small nests often in burrows in the soil or in hollow plant stems or old beetle borings in wood. They make a few cells, usually necessarily end to end, each of which is in turn stocked with larval food on which an egg is laid. The majority have no further contact with their offspring. Although several nests of a particular species may be close together, there is no contact or collaboration on the part of the adults. Solitary bees may be quite conspicuous at flowers, particularly in early summer, when mining bees are also actively digging burrows in the light soil of vegetable beds or worn patches of lawn. Solitary wasps, however, are less common and are rarely noticed. Were it not for trapping, their diversity in the garden would not have been appreciated. In total, 51 species of bees and 41 species of wasps have been recorded in the Leicester garden, but these belong to 11 different families and so their diversity is not easily comparable with that of, say, ichneumonids.

NATURAL HISTORY OF BEES AND WASPS

Bees and wasps, like the parasitic wasps described in Chapter 8, belong to the suborder Apocrita of the insect order Hymenoptera, but comprise the subdivision Aculeata. Translating their name, these are insects in which the female's ovipositor has evolved into a venomous sting, eggs instead reaching the outside through an aperture near its base. However, the distinction between Aculeata and Parasitica is not that clear-cut, and there are many parasitic species in which the ovipositor retains its egg-laying function, although on other anatomical grounds they are regarded as aculeates. To conform with usually accepted classification and recognize these parasitic forms as aculeates, gives a grand total of 549 British species of bees and wasps (Fitton *et al.* 1978). Of these, 25 species belonging to the family Apidae are social bees, including bumblebees and the honeybee, and seven, belonging to the family Vespidae, are social wasps. Of the rest, 227 are solitary bees of seven different families, and 290 are solitary wasps, of ten different families.

The crucial difference between bees and wasps is in what they eat, and this has led to differences in their structure and appearance. Bees are entirely vegetarian, or first order consumers, in that they eat pollen and nectar, and feed them to their larvae. Adult wasps, however, although themselves feeding on nectar, honeydew and other sweet substances, catch animal food for their larvae. They are predators of other invertebrates, particularly insects, and hence operate as secondary or higher order consumers; they sometimes also eat carrion (when they function as decomposers). Solitary wasp larvae are supplied with live but paralysed (i.e. usually stung, but occasionally bitten) prey, whereas social wasp larvae are fed masticated pellets of 'meat'.

As adaptations to collecting and carrying pollen, bees in general are covered with branched hairs giving them a furry appearance, and the majority have particular areas of the body modified as pollen baskets for transporting pollen. Some species have become cuckoo bees in that they appropriate the nests and food stores of other species for their own larvae, and such bees are much less hairy and lack pollen baskets. Wasps, on the other hand, tend to be smooth-surfaced, even shiny in appearance. In general, bees, as nectar-feeders, have a much longer tongue, formed from the labium and associated maxillae, than do wasps.

Since they can sting, bees and wasps are distasteful to predators, and have consequently tended to develop conspicuous warning coloration. Usually, this involves some form of yellow and black banding. An inexperienced predator may try to eat a brightly patterned bee or wasp, but learns from the unpleasant experience, and avoids them in future. Not only do many bees and wasps share similar warning coloration (Mullerian mimicry), but also many palatable insects, such as hoverflies, mimic their appearance and thus gain protection (Batesian mimicry). A wasp pattern gives the message 'Danger!', whoever is wearing it, and is so effective as a bluff (in Batesian mimicry) because it is so widespread as a genuine warning (Mullerian mimicry).

Bees and wasps attract particular interest because of the various stages they display in the development of social behaviour. Clues to the evolution of social behaviour are sought from a study of existing species. One of the British solitary wasps of the genus *Ammophila*, for instance, practises progressive provisioning of its larvae. Instead of supplying all the food a developing larva will need and sealing the cell after laying an egg (mass provisioning), it returns to the nest, inspects the growing larva and its food supply, and provides more food as it is needed, an essential step towards social behaviour. On the other hand, wandering wasps, such as *Tiphia* spp., use no nest, merely paralysing and laying eggs on beetle larvae where they find them in the soil; behaviourally, this is a link with the Parasitica, such as Ichneumonidae.

Some solitary wasps, such as *Chrysis* spp., are parasitic on solitary bees and wasps, invading the host's nest and laying an egg which hatches to a larva that consumes the rightful occupant. Others, such as *Sapyga* spp., are cleptoparasites in the nests of solitary bees, in that the invasive larva destroys the rightful occupant and then devours the food store; this blurs the distinction between wasps and bees, since the parasitic wasp larva consumes stored pollen and nectar. It is, however, possible to make general statements about the many typical bees and wasps.

There are 18 species of British bumblebees, *Bombus*, although some workers also include a continental species captured in Kent in the nineteenth century and a form of *B. lucorum* that is recognized as a species on the continent. These, together with the honeybee *Apis mellifera* and six species of cuckoo bumblebee, make up the family Apidae, characterized by the development of social life. Queen bumblebees that have hibernated in various sites, such as in burrows in well-drained banks (*B. lapidarius*) or in holes in logs (*B. lucorum*), emerge in the spring and start building nests. *B. pratorum* is the earliest, often being on the wing in March, whereas queens of *B. lapidarius* rarely emerge before late May or early June. Species differ in their nest site preferences; *B. pascuorum*, for instance, nests beneath clumps of grass or moss on the soil surface, whereas *B. terrestris* takes over the abandoned nest of a small mammal at the end of a burrow a metre or more in length.

The queen hollows out a chamber, collects a mass of pollen, and around it constructs a cell from wax extruded between her abdominal segments. She lays several eggs on the pollen mass and also makes a wax honey-pot near the nest entrance as a bad-weather food store. As the larvae feed and grow, she supplies them with more pollen and honey and also incubates them by straddling the brood chamber with her warm, furry body. When fully grown, the larvae spin individual silken cocoons for pupation, and the queen recycles their wax brood chamber into new brood cells on the upper, outer edge of the cocoon cluster. The first new bees to emerge are all females and all workers, i.e. non-breeding and usually sterile. After a day or two, once the workers are ready to start foraging for nectar and pollen, the queen remains in the nest all the time, constructing new cells, laying eggs, and feeding and incubating larvae.

Female and worker bumblebees are highly adapted for provision of honey and pollen to the nest. Nectar is collected from flowers using the long tongue, which differs in length between species, such that the nectaries of different flowers are accessible to different species. Collected nectar is transported in a specialized length of the digestive tract, the honey stomach, where complex sugars are converted to simpler ones by enzymes. So capacious is the honey stomach that a bumblebee can transport a weight of honey almost equal to her own. Pollen adheres to the hairy body of a flower-visiting bumblebee; she uses her fore legs to brush this towards her mouth, moistens it with nectar, and crams it into pollen baskets, or corbiculae, on the tibiae of her hind legs. The tibia of each hind leg is concave and highly polished, and is surrounded by a fringe of long, inward-curving hairs; male *Bombus* lack these pollen baskets.

The colony steadily increases in size, those of *B. terrestris*, for instance, producing 300–400 workers from a comb 20–25 centimetres in diameter. As the colony grows, the ratio of workers to larvae, and hence the larval food supply, increases and it seems to be this that leads to development of new queens. Males are produced at the same time, sex determination in Hymenoptera being genetic and depending on fertilization; fertilized and therefore diploid eggs produce females, unfertilized or haploid eggs produce males. Males leave the nest, wait around the entrances of other nests of the same species, and mate with virgin queens. No more workers are produced, young queens take no part in nest activities, and the colony gradually disintegrates. Males, workers and the original, founding queen die, and only young, impregnated queens survive the winter.

Cuckoo bumblebees *Psithyrus* are different in that they develop no workers but instead exploit the social system of *Bombus* species. Female *Psithyrus* emerge from hibernation when nests of their hosts already contain workers. They infiltrate *Bombus* nests, often, although not always, killing the queen and start laying eggs, depending on the *Bombus* workers to feed and tend their young. Since they collect no pollen, *Psithyrus* females have no pollen baskets, and are also less generally hairy than *Bombus*. Each species resembles the *Bombus* species it habitually parasitizes, but it is still uncertain how they manage to gain entry to a host nest.

Solitary bees, like their social relatives, feed their young on nectar and pollen, although a few, such as *Hylaeus* spp., have no pollen baskets and transport pollen in the crop. Others, such as *Colletes* spp., have short blunt tongues, more like those of wasps. Many do not look particularly like bees, and are small, although others, such as *Anthophora* spp., are hairy and as large as the smaller bumblebees, for which they may be mistaken. Female solitary bees typically construct one or more cells in a burrow in soil or in a hollow stem; some just compact the soil, *Hylaeus* spp. use a secretion from the mouth, while *Megachile* spp. use discs cut from plant leaves. Each cell is mass-provisioned with pollen moistened with nectar, on which an egg is laid. Pollen may be carried on special brushes of hairs on the thorax, abdomen or hind legs, although the latter are never as highly modified as in social bees.

The resulting larva eats the stored food, pupates, and emerges in the following year, long after its mother, with which it has no contact, has died.

A few solitary bees, however, show the rudiments of social life. In some
species of Halictidae, the female remains near the nest and guards it. The bees
that emerge from the first cells are, in effect, worker females which build and
provision more cells in which the original female, perhaps better called the
queen, lays eggs. These produce males and normal females, which, after
mating, hibernate until the following year. In employing the activities of non-
breeding workers, females of these species are able to raise more offspring
than typical solitary bees, such as *Andrena* spp. There are also cuckoo bees or
cleptoparasites, such as *Nomada* spp., among the solitary bees. Less hairy,
since they never collect pollen, they lay their eggs in the nests of other solitary
bees, such as *Andrena* spp., and thus appropriate for their larvae the stored
food, usually destroying the host's egg.

The life cycle of social wasps is not unlike that of bumblebees; a nest is
founded by an overwintered queen, produces several batches of non-breeding workers, and is then abandoned after production of males and new
queens. Nests differ from those of bumblebees in several respects, however.
The nest is constructed of carton or paper, formed by masticating fragments
chewed from sound or slightly rotten wood, depending on species; it may
become very large, consisting of several separate horizontal combs of cells;
and it is usually enclosed in several overlapping, protective envelopes of
carton. *Vespula* spp. nest underground or in enclosed cavities, such as roof
spaces or hollow trees, whereas *Dolichovespula* spp. build aerial nests in
exposed or semi-exposed positions on trees. Nests of *Vespula* spp. may
become particularly large, those of *V. vulgaris* sometimes consisting of as
many as 12 combs composed of more than 15 000 cells. One species, *Vespula
austriaca* is a cuckoo wasp in that overwintered queens usurp *V. rufa* queens,
appropriate their nests and workers, and lay eggs that produce only males
and new queens.

Most species of wasps, however, are solitary. Each female constructs a
small group of individual cells in the soil, in hollow plant stems or in holes in
wood or masonry, stocks each with prey paralysed usually by stinging or
occasionally by biting, and lays an egg on the last item stored before sealing
the nest, in *Passaloecus* spp., for instance, with a plug of pine resin (Corbet &
Backhouse 1975). Prey is characteristic of the type of wasp. *Ectemnius* spp., for
instance, provision their nests with flies, including hoverflies, *Trypoxylon* spp.
with spiders, *Pemphredon* spp. with aphids, and *Ancistrocerus* spp. with
caterpillars. The majority of solitary wasps never return to the nest, and have
no contact with their offspring. As already indicated, however, some
Ammophila spp. practise progressive provisioning, and in *Ancistrocerus* spp.
there is a tenuous link between mother and offspring, as the egg is laid before
the cell is stocked. Non-British species in some genera, such as *Ectemnius* and
Ammophila, also show the rudiments of social behaviour in that two or more
females share and provision a common nest (Yeo & Corbet 1983). A few
chrysidids are cleptoparasites of solitary wasps and, as already indicated,

species of Sapygidae are cleptoparasites of solitary bees. Other members of the Chrysididae are ectoparasites of other sorts of insects, particularly solitary bees and wasps; they lay their eggs individually in occupied cells, the resulting larvae eating the occupants and occasionally the stored food as well.

DIVERSITY AND ABUNDANCE OF BEES AND WASPS

Social and solitary bees and wasps, representing many of the types described, occur in the study garden. Information on their numbers and diversity and on annual fluctuations and seasonality come from Malaise trapping. Social wasps were sorted from the trap sample during 1972–86, social bees in 1972–77 and in 1982–86, and solitary bees and wasps during 1975–86. The garden species are given here in four categories: social bees, social wasps, solitary bees and solitary wasps. Under each heading, some idea is given of what the insects are doing in the garden, and the frequency given for each species is the total Malaise trap catch. Social wasps and solitary bees and wasps were identified by M. E. Archer. Sequence within each category, classification and taxonomy follow Fitton *et al.* (1978).

A. Social bees

The social bees belong to the family Apidae. Eight species of bumblebees *Bombus* have been recorded, five of cuckoo bumblebees *Psithyrus*, and the honeybee *Apis mellifera*. Bumblebees are abundant in the garden, feeding and collecting pollen from flowers, and occasionally a cuckoo bumblebee has been seen. None has been found nesting in the garden, although *B. pascuorum*, the commonest, nests nearby. All *Bombus* except *ruderatus* have been recorded in every year that they were sorted from the Malaise trap sample, 1972–77 and 1982–86, a total of 11 years; *B. ruderatus* was recorded every year 1972–77, but was absent 1982–86. Captures of *Psithyrus* were sporadic. *Apis mellifera* is frequent, and is regularly captured, but no attempt has been made to monitor numbers. Bumblebees were identified using Alford (1973) and Free & Butler (1959).

1. *Bombus lucorum* (L.) 338
2. *B. terrestris* (L.) 611
3. *B. lapidarius* (L.) 225
4. *B. pratorum* (L.) 737
5. *B. hortorum* (L.) 787
6. *B. ruderatus* (Fab.) 86
7. *B. pascuorum* (Scopoli) 3208
8. *B. ruderarius* (Müller) 109
9. *Psithyrus barbutellus* (Kirby) 3
10. *P. campestris* (Panzer) 1
11. *P. rupestris* (Fab.) 1
12. *P. sylvestris* Lepeletier 4
13. *P. vestalis* (Geoffroy in Fourcroy) 2
14. *Apis mellifera* L.

Photograph 9.1. Bombus pascuorum *is by far the most abundant bumblebee in the* Leicester garden.

B. Social wasps

All seven British social wasp species are Vespidae; six of them have been caught in the Malaise trap. *Vespula germanica* and *vulgaris* are abundant every year in the garden, particularly in August and September, and overwintered queens are conspicuous in late spring and early summer in some years; both species were caught every year, 1972–86, in the trap. In 1973, *V. vulgaris* nested beneath a mat of saxifrage spreading over a path from a rockery. *V. rufa* has only been captured twice, in 1973 and 1985, and its cuckoo, *V. austriaca*, only once, in 1978. *Dolichovespula sylvestris* has been taken in six years, but *D. norvegica* only once, in 1980.

15. *Dolichovespula norvegica* (Fab.)........   1
16. *D. sylvestris* (Scopoli)   11
17. *Vespula austriaca* (Panzer)...........   1
18. *V. rufa* (L.)   2
19. *V. germanica* (Fab.)................ 1674
20. *V. vulgaris* (L.) 3365

C. Solitary bees

Thirty-seven species of solitary bees of five families were captured in the Malaise trap, 1975–86, some of them quite commonly, there being a total of 3796 individuals. However, it is important to appreciate that trap captures reflect flight activity rather than abundance *per se*. The commonest and most abundant species, caught more than 200 times, were *Colletes daviesanus*, *Andrena fulva*, *Hylaeus communis* and *Anthophora plumipes*. Eight species were taken once only in the 12-year period.

Females of *Andrena fulva* are regularly seen foraging from *Ribes uva-crispa*, females and males of *Colletes daviesanus* from *Tanacetum parthenium*, and megachilids take pollen from a variety of flowers. Females of *Hylaeus communis* feed from *Rubus fruticosus* flowers and *Anthophora plumipes* from *Lamium maculatum*, and in both, male–male aggressive interactions occur over flower patches (Archer 1990). Females of *Andrena fulva* nest in undisturbed soil, and a dozen neat leaf cells of *Megachile willughbiella* were found in 1985 beneath the compost in a seed-tray in the greenhouse. Bees consistently forage in the garden, although few species are known to nest. Seven species of solitary cuckoo bees have been recorded; *Nomada* spp. eat the stored food of *Andrena* spp., *Sphecodes* spp. that of *Halictus* spp., and *Coelioxys* spp. that of *Megachile* spp., each cuckoo species often being associated with one species of host.

Solitary bees are designated in the following list as common (C) if captured in 10–12 years, as frequent (F) if captured in 7–9 years, as occasional (O) if captured in 4–6 years, and as rare (R) if captured in 1–3 years.

Colletidae

21. *Colletes daviesanus* Smith, F. . . . . 1158 C
22. *Hyaleus communis* Nylander . . . .   457 C
23. *H. hyalinatus* Smith, F. . . . .   17 F

Andrenidae

24. *Andrena fucata* Smith, F. . . . .   5 O
25. *A. fulva* (Müller in Allioni). . . . .   742 C
26. *A. jacobi* Perkins, R.C.L. . . . .   176 C
27. *A. bicolor* Fab. . . . .   86 C
28. *A. nigroaenea* (Kirby). . . . .   94 C
29. *A. haemorrhoa* (Fab.) . . . .   71 C
30. *A. subopaca* Nylander . . . .   1 R

Halictidae

31. *Halictus rubicundus* (Christ). . . . .   4 R
32. *H. tumulorum* (L.). . . . .   1 R
33. *Lasioglossum albipes* (Fab.) . . . .   1 R
34. *L. fulvicorne* (Kirby). . . . .   1 R
35. *L. villosulum* (Kirby) . . . .   23 F
36. *L. cupromicans* (Pérez). . . . .   3 R
37. *L. leucopum* (Kirby). . . . .   25 F
38. *L. morio* (Fab.) . . . .   20 F
39. *L. smeathmanellum* (Kirby). . . . .   179 C
40. *Sphecodes fasciatus* von Hagens . . . .   8 O
41. *S. gibbus* (L.). . . . .   1 R

Megachilidae

42. *Anthidium manicatum* (L.) . . . .   13 O
43. *Osmia rufa* (L.) . . . .   129 C
44. *O. caerulescens* (L.) . . . .   57 C

45. *O. leaiana* (Kirby) . . . .   15 F
46. *Megachile centuncularis* (L.) . . . .   69 C
47. *M. ligniseca* (Kirby). . . . .   6 O
48. *M. versicolor* Smith, F. . . . .   1 R
49. *M. willughbiella* (Kirby). . . . .   70 C
50. *M. circumcincta* (Kirby) . . . .   1 R
51. *Coelioxys inermis* (Kirby) . . . .   2 R
52. *C. rufescens* Lepeletier & Serville . . . . . .   1 R

Anthophoridae

53. *Nomada flava* Panzer . . . .   2 R
54. *N. marshamella* (Kirby) . . . .   3 R
55. *N. panzeri* Lepeletier. . . . .   3 R
56. *Anthophora plumipes* (Pallas) . . . . 292 C
57. *A. furcata* (Panzer) . . . .   59 C

D. Solitary wasps

Thirty-four species of solitary wasps of four families were captured in the Malaise trap in 1975–86, and one *Mellinus arvensis* in 1973. As might be expected, as secondary or higher order consumers they are far less abundant than are solitary bees; a 12-year total of 1401 wasps was trapped compared with 3796 bees. The commonest and most abundant species, trapped more than 150 times, were *Passaloecus singularis*, *Pemphredon lugubris*, *P. lethifer* and *Omalus auratus*. Five species were taken once only in the 12-year period. Solitary wasps are rarely evident in the garden, although *Ancistrocerus gazellus* has been seen hunting for caterpillars beneath clover and lettuce leaves (Archer 1990).

Wasps of the family Chrysididae are parasitic on other insects or cleptoparasites, consuming the stored food of solitary wasps. *Omalus auratus*, for instance, feeds on aphids stored by *Pemphredon lethifer*, *P. inornatus* and *Passaloecus gracilis* for their larvae, and occasionally on spiders or flies stored by *Trypoxylon* sp. and *Corynopus coarctatus*, respectively (Danks 1971). *Cleptes semiauratus* is parasitic on sawflies, and *Chrysis* spp. parasitize solitary bees and wasps, eating host eggs or larvae and then, often, devouring the stored food. *Sapyga quinquepunctata* parasitizes *Osmia* and *Hylaeus* bees, the parasitic larva destroying the host larva then consuming the stored food.

Like solitary bees, wasps have been designated as common, frequent, occasional or rare on the basis of number of years in which they were trapped.

Chrysididae

58. *Omalus auratus* (L.). . . . . 151 C
59. *O. violaceus* (Scopoli) . . . .   31 C
60. *Chrysis ignita* (L.) . . . .   15 O
61. *C. impressa* Schenk . . . .   2 R
62. *Cleptes semiauratus* (L.) . . . .   4 R

Bees and wasps

Sapygidae

63. *Sapyga quinquepunctata* (Fab.) . . . . 2 R

Eumenidae

64. *Ancistrocerus gazella* (Panzer) . . . . 55 C
65. *A. trifasciatus* (Müller). . . . . 5 O
66. *Symmorphus mutinensis* (Baldini). . . . . . . 1 R

Sphecidae

67. *Trypoxylon attenuatum* Smith, F. . . . . 33 C
68. *T. clavicerum* Lepeletier. . . . . 15 F
69. *Crossocerus elongatulus* (Vander Linden) . 55 C
70. *C. annulipes* (Lepeletier & Brullé) . . . . . . 73 C
71. *C. capitosus* (Shuckard). . . . . 7 O
72. *C. megacephalus* (Rossius) . . . . 35 C
73. *C. podagricus* (Vander Linden) . . . . 3 R
74. *C. dimidiatus* (Fab.). . . . . 7 O
75. *Ectemnius cavifrons* (Thomson) . . . . 37 C
76. *E. ruficornis* (Zetterstedt) . . . . 1 R
77. *E. sexcinctus* (Fab.) . . . . 1 R
78. *E. continuus* (Fab.) 1R
79. *E. cephalotes* (Olivier) . . . . 7 R
80. *Rhopalum clavipes* (L.) . . . . 13 O
81. *Psenulus atratus* (Fab.). . . . . 54 C
82. *P. concolor* (Dahlbom). . . . . 21 F
83. *Stigmus solskyi* Morawitz, A. . . . . 5 R
84. *Pemphredon lugubris* (Fab.) . . . . 189 C
85. *P. inornatus* Say . . . . 20 F
86. *P. lethifer* (Shuckard) . . . . 166 C
87. *Diodontus tristis* (Vander Linden) . . . . . . 20 F
88. *Passaloecus corniger* Shuckard . . . . 13 F
89. *P. gracilis* (Curtis). . . . . 91 C
90. *P. insignis* (Vander Linden) . . . . 6 O
91. *P. singularis* Dahlbom. . . . . 262 C
92. *Mellinus arvensis* (L.). . . . . 1 R (in 1973)

All of the social bees recorded in the garden are widely distributed in the British Isles, the least common being *Bombus ruderatus*, which has a southerly distribution. Cuckoo bumblebees are never anything like as common as *Bombus* spp., but all species of *Psithyrus* recorded in the garden have wide distributions. Bumblebee ranges have contracted post-1960 compared with pre-1960 records, the Midlands being particularly affected, and Leicester is in the 'central impoverished region' where only 'mainland ubiquitous species' persist (Williams 1982), although the records analysed by Williams do not include *B. ruderatus*. Pre-1960 county records include *B. jonellus, humilis, sylvarum, distinguendus* and *subterraneus*, and *P. bohemicus*. Including the honeybee, the

234

garden list comprises 56 per cent of the national list of Apidae; possibly this is the maximum that can be expected, as the only other species recorded in the county post-1960 is *B. soroeensis* (Archer, M. E. pers. comm.).

Of the social wasps, *Vespula germanica, V. vulgaris, V. rufa, Dolichovespula sylvestris* and *D. norvegica* have wide distributions, the last named being the least common, particularly in Midland England. *V. austriaca*, the cuckoo wasp, however, is far less widespread, being mainly a northern and western species; the garden record is the first not only for the county of Leicestershire, but also for the Midlands. The only vespid not taken in the garden is the hornet *Vespa crabro*, which is confined to the southern half of England. They do occur sometimes in the county, and I saw one in 1979 elsewhere in the city of Leicester.

There are 517 species of solitary aculeates in the British Isles, and 176 species have been recorded in the county of Leicestershire; 41.6 per cent of the county's solitary bees and 38.6 per cent of the solitary wasps have been recorded in the garden. Of the bees, Megachilidae may be over-represented and Andrenidae and Anthophoridae under-represented in the garden; of the wasps, Chrysididae and Sphecidae may be over-represented (Archer 1990). *Megachile versicolor, Omalus violaceus, Ectemnius ruficornis* and *E. sexcinctus* are new records for the county, although not all records of other species have been published. About 93 per cent of the solitary species had been caught by the sixth year of trapping (1980), although new species continue to be added.

The percentage of garden solitary wasp species that are cleptoparasites, eating the host's food store, is similar to the percentage of cleptoparasitic bees but, whereas the percentage of cleptoparasitic wasps is nearly the same in the garden and in the county, the percentage of cleptoparasitic bees in the garden (18.9 per cent) is considerably lower than that in the county (28.1 per cent). More than 92 per cent of the solitary wasps (other than cleptoparasites) in the garden and more than 36 per cent of the solitary bees are aerial nesters, nesting in stems, walls or wood, as opposed to subterranean nesters, which nest in soil. The percentage of aerial nesters in the county is much lower for both bees and wasps. The garden is evidently providing a more satisfactory habitat for aerial than for subterranean nesters, almost certainly because the soil is not sandy and is frequently disturbed by cultivation. This would also explain why the percentage of cleptoparasitic bees is lower than in the county, because many are associated with subterranean nesters (Archer 1990).

In the 12 years 1975–86, β-diversity of solitary bees varied from 0.690 in 1985 to 0.885 in 1979 (Table 9.1). This is an estimate of the probability that an individual sampled at random will be different from the previous individual sampled. The series of 12 values is markedly heterogeneous (χ^2 = 205.75, df = 11, p<0.001) and so the values for pairs of years can legitimately be compared. For example, the β-diversity for 1976 (0.883) was highly significantly different from that for 1984 (0.831), another hot, dry, sunny summer. The fluctuations in annual diversity do not show any similarity to fluctuations in sample size (*N*) nor to fluctuations in numbers of species (*S*). In other

Table 9.1 *Annual fluctuations in β-diversity of solitary bees caught in Malaise trap, 1975–86*

	Individuals (N)	Species (S)	Diversity (β)	Standard error
1975	468	24	0.699	±0.017051
1976	477	27	0.883	±0.007693
1977	399	23	0.768	±0.019940
1978	530	22	0.840	±0.009072
1979	196	18	0.885	±0.010410
1980	173	18	0.871	±0.012410
1981	112	12	0.834	±0.016931
1982	359	19	0.786	±0.014952
1983	290	17	0.779	±0.016619
1984	308	20	0.831	±0.011271
1985	336	20	0.690	±0.023797
1986	148	16	0.821	±0.017397

Note: Diversity derived from Simpson's index, λ; $\beta = 1 - \lambda$.

Table 9.2 *Annual fluctuations in β-diversity of solitary wasps caught in Malaise trap, 1975–86*

	Individuals (N)	Species (S)	Diversity (β)	Standard error
1975	99	20	0.900	±0.013608
1976	192	23	0.908	±0.007449
1977	109	17	0.896	±0.012054
1978	75	20	0.926	±0.010655
1979	53	17	0.878	±0.031988
1980	80	18	0.859	±0.022090
1981	73	14	0.861	±0.021956
1982	166	20	0.882	±0.014505
1983	181	23	0.898	±0.009375
1984	186	24	0.874	±0.015007
1985	109	18	0.865	±0.021878
1986	78	21	0.918	±0.016640

Note: Diversity derived from Simpson's index, λ; $\beta = 1 - \lambda$.

words, the numerical composition of the catch seems to be independent of the size of catch and of the number of species it contains.

β-diversity of solitary wasps varied from 0.859 in 1980 to 0.926 in 1978 (Table 9.2), and was on average higher than that of solitary bees. As with bees, the pattern of variation in annual diversity was unlike that of variation in numbers of individuals or species. The 12-year series of β-diversity values is heterogeneous ($\chi^2 = 22.53$, df $= 11$, $p < 0.05 > 0.01$), although not as markedly so as that for bees. The difference between the values for 1976

Photograph 9.2. Lavatera trimestris *is a favourite bumblebee flower.*

(0.908) and 1984 (0.874) is significant ($p < 0.05 > 0.01$) showing that the numerical composition of the catch varied in good years.

Archer (1990) has shown that the rates of appearance of species in the garden and disappearance from it do not fit with the immigration and extinction curves that would be expected were the garden in effect an island. On the contrary, its solitary aculeate fauna is part of that of adjoining gardens, so that species move in and out by chance from the pool of species in the neighbourhood. This explains why there is no apparent pattern to the changes in diversity from year to year of either solitary bees or solitary wasps in the garden. Both are diverse, and reflect the heterogeneity of habitat of the neighbourhood of which the garden is a part. Nevertheless, the garden catch of solitary aculeates represents only about 40 per cent of the species recorded in the county as a whole and it seems likely that species will be added in future years. The situation for social bees and wasps is markedly different;

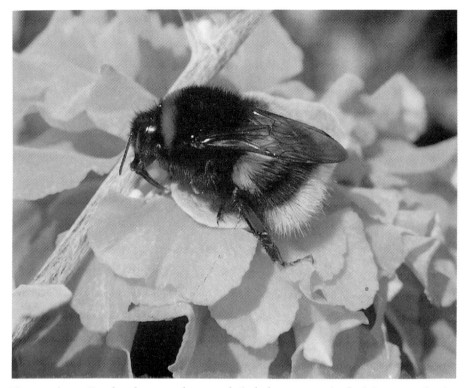

Photograph 9.3. Bombus lucorum *became relatively less common in the Leicester garden in the 1980s than it was in the 1970s.*

not only are some of the garden species much more numerous than any solitary species, but there are few additional species that could be caught.

ANNUAL FLUCTUATIONS IN NUMBERS OF BEES AND WASPS

Although it is possible over a time span of 15 years to generalize about absolute and relative numbers of species of bees and wasps, there are considerable differences from year to year in abundance in the Malaise trap sample. Often, although not always, these fluctuations can be attributed to weather differences from one year to the next, which affect not only the abundance but also the flight activity, and hence trappability, of insects.

Bumblebees were sorted from Malaise trap samples in 1972–77 and in 1982–86. Annual numbers trapped of all *Bombus* species and of *B. pascuorum*, the commonest species, are shown in Fig. 9.1. *Bombus* numbers were high in 1973, 1974, 1977 and 1982, and particularly low in 1972 and 1986. Numbers of *B. pascuorum* were particularly high in 1973 and 1982, and low in 1972, 1975 and 1986. Figures for 1977 and 1982, in particular, show that numbers of *pascuorum* do not reflect the situation for all *Bombus* species. In 1977, numbers of *hortorum*, *lucorum* and *pratorum* were relatively high. In 1982, numbers of

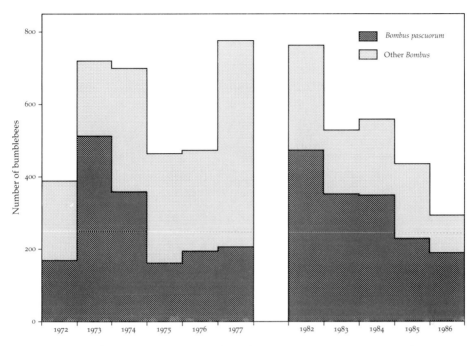

Fig. 9.1 Annual fluctuations in total numbers of Bombus *spp. and in numbers of* Bombus pascuorum *caught in Malaise trap, 1972–77 and 1982–86.*

lucorum were relatively high, but *B. ruderatus* was absent. In the period 1982–86, no *ruderatus* were taken, suggesting that the species had become locally extinct since 1977, lending support to the contention of Williams (1982) that it is a southern local species and hence only marginal in Midland England. Numbers of *lapidarius* were relatively high in 1982, and in 1983 numbers of *lucorum* were particularly low and *ruderarius* was absent.

As Fig. 9.1 shows, whatever bumblebee numbers were responding to, it was not quite the same factor(s) as affected numbers of butterflies, moths or hoverflies. 1973, one of the best bumblebee years, was also good for butterflies (Fig. 5.1) and moths (Fig. 6.1), and 1982 was also good for hoverflies (Fig. 7.2), but 1974 and 1977 were poor years for butterflies, moths and hoverflies. Of the poor bumblebee years, 1972 was also poor for butterflies, moths and hoverflies, and 1986 was also poor for butterflies and hoverflies. Of the good bumblebee years, 1974 and 1977 were cool, damp and only averagely sunny summers. 1973 and 1982 were on the warm side, rather damp and, while 1973 was sunny, 1982 was dull. The poor bumblebee years also differed, for 1972 was cool, dry and dull and 1986 average, if rather damp. In other words, bumblebees tended to do well in rather damp summers with average amounts of sunshine, and to be unaffected by temperature.

The relative status of different bumblebee species was different in 1982–86 from what it had been in 1972–77 (Table 9.3). *B. hortorum, lapidarius* and *pascuorum* became relatively more common, *B. lucorum, pratorum, ruderarius*

Table 9.3 *Malaise trap catches of* Bombus *spp. and percentage of total catch in 1972–77 and 1982–86*

	1972–77	% of 6-yr total	1982–86	% of 5-yr total
B. lucorum	247	7.0	91	3.5
B. terrestris	435	12.4	176	6.8
B. lapidarius	101	2.9	124	4.8
B. pratorum	500	14.2	237	9.2
B. hortorum	448	12.7	339	13.1
B. ruderatus	86	2.4	0	–
B. pascuorum	1610	45.8	1598	61.9
B. ruderarius	92	2.6	17	0.7
Total	3519		2582	

and *terrestris* became relatively less common, and *B. ruderatus* disappeared. The total bumble bee catch was less in 1982–86 than in 1972–77, even after adjustment for the difference in number of years in each period. More species of cuckoo bumblebees *Psithyrus* were caught in the period 1972–77 than in 1982–86, but only 11 individuals in total. *P. sylvestris* was caught in April 1975, April 1976, May 1982 and May 1983, *P. vestalis* in June 1974 and May 1976, *P. campestris* in August 1974, *P. rupestris* in May 1977 and *P. barbutellus* in May 1982 and twice in July 1982.

Numbers of the two common social wasps, *Vespula germanica* and *V. vulgaris*, caught in the Malaise trap fluctuated considerably from year to year (Fig. 9.2). Both species were abundant in 1974, 1977 and 1979, and *V. vulgaris* also in 1973 and 1981; 1983 and 1985 were poor years for *V. germanica*, and 1983 and 1984 for *V. vulgaris*. 1979 was a particularly good wasp year and 1983 a particularly poor year. In 1979, spring was wet and cool, and the summer dry with average temperatures, and it was neither particularly good nor particularly bad for other insects. It is, however, noteworthy that 1974 and 1977, when wasps were abundant, were good years for bumblebees; 1979 cannot be compared as there are no bumblebee figures for that year.

Archer (1985) has analysed the population dynamics of *V. vulgaris* and *V. germanica*, using in part the data from the Leicester garden. He identified several characteristics of yearly abundance: large variations in yearly abundance; a two- and possibly a seven-year cycle of yearly abundance; exceptional years of abundance and scarcity tend to occur in pairs; during exceptional years of summer abundance, spring queens are scarce and vice versa for exceptional years of summer scarcity. In the Leicester garden, the scarce year for *V. vulgaris* as measured by the number of workers captured (1978) preceded the most abundant year (1979), while for *V. germanica*, the most abundant year (1979) preceded the scarce year (1980). Furthermore, the year when spring queens were most abundant (1978 for *vulgaris* and 1980 for *germanica*) had the fewest workers, and the year when fewest spring queens were caught (1979 for both species) had the most workers. Archer has con-

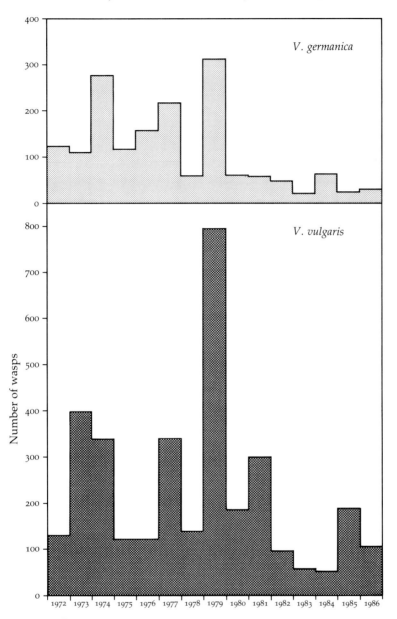

Annual fluctuations in numbers of bees and wasps

Fig. 9.2 *Annual fluctuations in numbers of* Vespula *wasps caught in Malaise trap, 1972–86.*

cluded that when spring queens are abundant, competition for nest sites and the risks of usurpation have an adverse effect leading to production of fewer workers in the population, but that when there are fewer queens, more nests succeed, leading to more workers. This endogenous factor generates the characteristics of population dynamics listed above, and is probably rein-

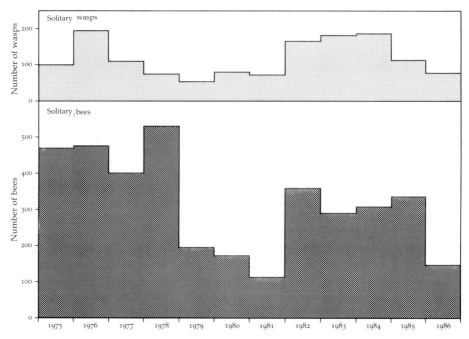

Fig. 9.3 Annual fluctuations in numbers of solitary wasps and bees caught in Malaise trap, 1975–86.

forced by differences in queen quality resulting from different rearing conditions, in that a queen reared in a nest full of workers tends to be better fed and vice versa. This endogenous mechanism is over-compensating because of the potentially large queen production of each colony, and tends to damp to equilibrium, taking about seven years to do so. It is then, however, as the endogenous factor weakens that the effects of an exogenous factor, probably a function of summer and autumn weather, become more evident. Hence the two- and seven-year cycles of abundance. To some extent, the figures plotted in Fig. 9.2 disguise the cycles, because queens, males and workers are all included, although it clearly shows the exceptional abundance of 1979, the key to the suggested two-year cycle.

The annual Malaise trap catch of solitary bees varied considerably between a high of 530 in 1978 and a low of 112 in 1981 (Fig. 9.3 and Table 9.4). The periods 1975–78 and 1982–85 were evidently good for bees, whereas 1979–81 and 1986 were poor. Thirty-seven species were caught in total, eight of them represented by only one individual; two of these were caught in 1975, one in 1976, one in 1977, two in 1980, one in 1984 and one in 1985. Numbers of species were high in 1975–78, 27 species being caught in 1976, the drought year. Fewer species were caught annually 1982–85, but the average number of individuals per species tended to be lower than in 1975–78 because the total catch was not so high. The average number of individuals per species (N/S)

Table 9.4 *Size and composition of annual and 12-year Malaise trap catches of
solitary bees in Leicester garden, 1975–86*

	Individuals (N)	Species (S)	N/S	Commonest as % N	Commonest species
1975	468	24	19.50	47.65	*Colletes daviesanus*
1976	477	27	17.67	24.74	*C. daviesanus*
1977	399	23	17.35	45.11	*C. daviesanus*
1978	530	22	24.09	26.42	*Andrena fulva*
1979	196	18	10.89	22.45	*A. fulva*
1980	173	18	9.61	24.86	*A. fulva*
1981	112	12	9.33	27.68	*C. daviesanus*
1982	359	19	18.89	38.16	*A. fulva*
1983	290	17	17.06	38.97	*C. daviesanus*
1984	308	20	15.40	25.65	*Hylaeus communis*
1985	336	20	16.80	51.79	*C. daviesanus*
1986	148	16	9.25	30.41	*Anthophora plumipes*
1975–86	3796	37	102.60	30.51	*C. daviesanus*

was lower in poor bee years (1979–81 and 1986), showing that the species that were present were less common.

Colletes daviesanus was the commonest species in six years, and the commonest overall; *Andrena fulva* was the commonest in four years, and *Hylaeus communis* and *Anthophora plumipes* each in one year. *C. daviesanus* tended to be the commonest in good bee years, and to form a higher percentage of the annual catch. Fig. 9.3 and Table 9.4 do not differentiate between spring and summer species, yet these may be fluctuating in different ways. Numbers of spring bees peaked in 1978 and 1982, because of large numbers of *Andrena fulva*, which was the commonest species in both years, whereas summer species peaked in 1976 and 1985, in both of which years *Colletes daviesanus* was the commonest species. In both 1978 and 1982, spring (March, April, May) was warm, dry and sunny; in 1976, summer (June, July, August) was hot, dry and sunny, but in 1985 it was cool, damp and only averagely sunny. However, 1979–81 and 1986, which were poor bee years, tended to be cool and dull, apart from spring 1980, although the rainfall varied from dry to wet. It seems that the weather affects numbers of solitary bees, so that there tend to be more when it is warm and sunny and, particularly, fewer when it is cool and dull, although there are exceptions, implying that other factors are important. For example, in three of the four years that the spring species, *Andrena fulva*, was the commonest species, spring was warm, dry and sunny. Apart from anything else, cool, dull weather may adversely affect flight activity and hence decrease numbers caught in the Malaise trap, whether or not population sizes are small.

The total 12-year catch of solitary wasps, 1401, was markedly smaller than that of solitary bees, 3796. Annual catches varied from 53 in 1979 to 192 in

Table 9.5 *Size and composition of annual and 12-year Malaise trap catches of solitary wasps in Leicester garden, 1975–86*

	Individuals (N)	Species (S)	N/S	Commonest as % N	Commonest species
1975	99	20	4.95	18.18	*Pemphredon lugubris*
1976	192	23	8.35	17.19	*Omalus auratus*
1977	109	17	6.41	19.27	*Passaloecus singularis*
1978	75	20	3.75	13.33	*Pemphredon lethifer*
1979	53	17	3.12	30.19	*P. lugubris*
1980	80	18	4.44	25.00	*P. lugubris*
1981	73	14	5.21	26.03	*Passaloecus singularis*
1982	166	20	8.3	25.90	*P. singularis*
1983	181	23	7.87	17.13	*Pemphredon lethifer*
1984	186	24	7.75	27.96	*Passaloecus singularis*
1985	109	18	6.06	30.28	*P. singularis*
1986	78	21	3.71	20.51	*P. singularis*
1975–86	1401	34	41.12	18.70	*P. singularis*

1976 (Table 9.5 and Fig. 9.3). The years 1976 and 1982–84 were good wasp years, and 1979 particularly poor, although the pattern of variation was not as marked as in solitary bees. The good wasp years were also years when the number of species was high; 34 were taken altogether, with a maximum of 24 in 1984, and a minimum of 14 in 1981. Four species were represented by only one individual in the 12 years, one each in 1975, 1979, 1980 and 1982. In three of the four good wasp years, summer weather was hot, dry and sunny, and in the poor year it was dry and dull, with only average temperatures. It is noteworthy that the poor year for solitary wasps was the year when most social *Vespula* were caught, emphasizing the independence of the latter from summer weather.

The average number of individuals per species was highest in the good wasp years (1976, 1982–84) indicating that abundance was a consequence of species already there being commoner as well as of addition of species. It is noteworthy that the average number of individuals per species was considerably lower than for solitary bees, in line with wasps being at a higher trophic level. *Passaloecus singularis* was the commonest species in six years and overall in the 12-year catch; *Pemphredon lugubris* was the commonest in three years, *Pemphredon lethifer* in two years, and *Omalus auratus* in one year. *O. auratus* is a cleptoparasite of nests of *Pemphredon lugubris*, *P. lethifer* and *Passaloecus gracilis* in bramble stems, and was the commonest species in 1976, the year in which the combined total catch of the three hosts was the second largest.

SEASONALITY OF BEES AND WASPS

The nature of the British climate means that, like other garden insects, the occurrence of bees and wasps is distinctly seasonal. The colonies of all but one

Photograph 9.4. The blossoms of the big, old apple tree in the Leicester garden hum with bees in May; in autumn, the fermenting fallen fruit attracts wasps.

social species are founded, expand, decline and disintegrate all in the time span between one winter and the next, the only exception to this being the honeybee. Each solitary species is on the wing for only as long as it takes to mate and to establish and stock a nest; for the greater part of the year they are only present in the nest as developing or hibernating pupae.

Seasonality of garden bumblebees is illustrated in Fig. 9.4 as separate graphs for the periods 1973–77 and 1982–86, and for the commonest species *Bombus pascuorum*, and for total *Bombus* spp. June is the peak month for capture of bumblebees in the Malaise trap, but the graphs of seasonal occurrence of all *Bombus* in 1973–77 and 1982–86 are different shapes. The graph for 1982–86 probably shows the typical seasonal pattern, with a steady decline in numbers after the June peak. The percentage occurring in August in the period 1973–77 was boosted by exceptional numbers of *B. pascuorum* workers and, to a lesser extent, males in August 1973 and 1974 when a total of 300 and 132 *pascuorum*, respectively, were captured.

Seasonal patterns of capture of species differ in line with times of emergence of their queens from hibernation and of production of new queens and males. For instance, only *B. pascuorum* was captured in April in all ten years; *terrestris* and *lucorum* were caught in seven years, *hortorum* in six, and *pratorum* in five. *B. ruderarius* was taken in April in three years, *ruderatus* in two years, and *lapidarius* in only one and, indeed, in May in only four years. *B. pascuorum* is not only abundant but also an early starter and one whose

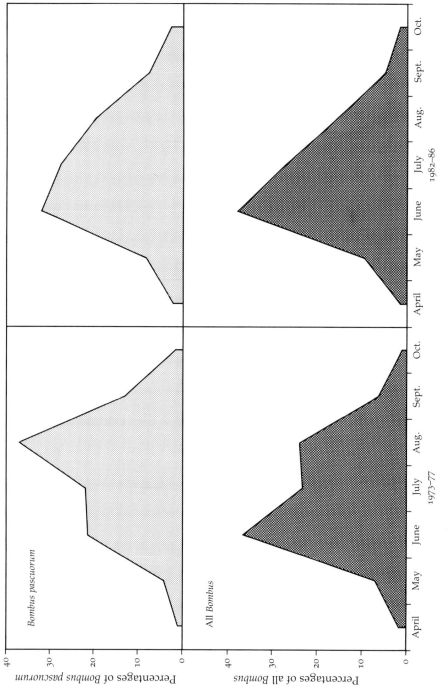

Fig. 9.4. Monthly fluctuations in numbers of *Bombus pascuorum* and of all *Bombus* spp. caught in Malaise trap, 1973–77 and 1982–86, expressed as percentages of five-year totals of each.

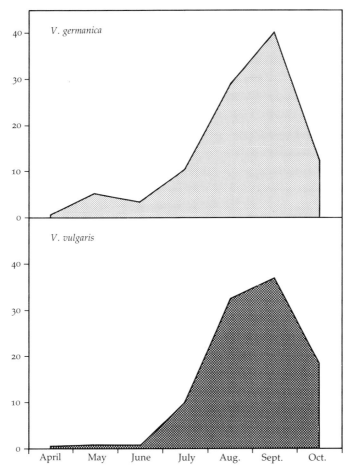

Fig. 9.5 Monthly fluctuations in numbers of Vespula *wasps caught in Malaise trap, 1972–86, expressed as percentages of totals for ten years.*

colonies persist into October, other than in the drought year. *B. lapidarius*, on the other hand, is a late starter, and colonies are active for a relatively short time, none having been caught in October and only one in September. Production of reproductives occurs when a colony has reached its peak, and so capture of males indicates the beginning of decline of a colony. *B. pratorum* males were captured as early as May in three of the ten years, confirming that this is an early species. Males of *pascuorum, terrestris* and *hortorum* were captured in June in several years, but males of *lapidarius* and *lucorum* not until July. No males of *ruderarius* or *ruderatus* were caught.

Numbers of social wasps caught in the Malaise trap reached a peak later, in September, as shown by graphs of the two abundant species, *Vespula germanica* and *vulgaris* (Fig. 9.5). The ten-year graphs (seasonal figures were not kept every year) mask differences in seasonality from year to year. In both species, worker numbers peaked in July in 1973 and in August in 1979, 1983 and 1984; workers of *V. vulgaris* also peaked in August in 1982. Overwintered queens were not captured in April or even in May in all years, but workers of both species were first caught in June. Males of *germanica* were first captured in September (in 1979), whereas a male *vulgaris* was taken in August 1973. The life cycles of the two species are very similar, although *germanica* queens tend to fly earlier in the spring and workers are produced earlier than in *vulgaris* (Edwards 1980).

The first solitary bees to be caught in April were mining bees *Andrena* spp., especially *A. fulva*, *Anthophora plumipes* and other Anthophoridae, and the megachilid, *Osmia rufa*. More Megachilidae, some Colletidae and female Halictidae appeared in the trap sample in May, and many solitary bees in June. There were no Andrenidae in June, other than the summer species, *Andrena bicolor*, which continued to be caught into August. Most male Halictidae appeared in July, but the numbers of most other solitary bees were declining, and by September only Halictidae were caught. Total numbers of individuals of solitary bees caught peaked in May and in July with a trough in June, whereas the number of species rose and declined gradually around a June peak. The spring peak in numbers was due to high numbers of the mining bees *Andrena fulva* and *jacobi*, and the summer peak to high numbers of *Colletes daviesanus* and *Hylaeus communis*. Slightly more female than male solitary bees were caught in total, with two distinct patterns of seasonal abundance of the sexes. In species such as *Colletes daviesanus* males and females were equally numerous and numbers of each peaked in July. In other species, females were more abundant than males, which were present during only the first part (*Andrena fulva*) or the second part (*Lasioglossum smeathmanellus*) of the females' flight period (Archer 1990).

The first solitary wasps, a few sphecids, were caught in May and most appeared in June; by the end of August most had gone leaving only a few sphecids in the September catch. Numbers of both individuals and species were high in June and August with a peak in July, particularly pronounced for individuals. Twice as many female as male solitary wasps were caught in total, female numbers reaching a peak in July, a month later than males. This reflects the shorter flight period of males, and probably also the greater activity of nesting females (Archer 1990).

Bees and wasps are abundant and conspicuous in the garden. Representation of social species that occur in Midland England is good, and few additional species can be expected. Less than half, however, of the species of solitary aculeates recorded in the county of Leicestershire have been caught, and additional species are likely. Relatively few species are known to breed in the garden, many evidently visiting, albeit regularly, to feed and to collect food for their larvae. The lack of pattern in changes from year to year of species diversity, β, of solitary bees and wasps reflects that species visiting the garden are part of the solitary aculeate fauna of a wider suburban area.

10

Beetles

Beetles are not conspicuous insects in gardens, but the gardener may be aware of several different types. Shiny, black ground beetles lurk beneath vegetation and under stones, and scurry away when disturbed; small flea beetles, yellowish-brown or black and yellow, are often abundant chewing holes in the leaves of many sorts of plants; tiny, shiny or metallic, dark beetles are common in flowers; bright ladybirds attract attention as they crawl from the shelter of vegetation in the first warm days of spring, and they and their splay-legged, prehistoric-looking larvae are often to be found among aggregations of aphids. There are more different sorts of beetles than of any other type of animal, and they are to be found just about everywhere, in all sorts of habitats, feeding on all sorts of foods. They are not, however, as amenable to trapping as some other sorts of insects, and the list of 251 species I have recorded in my garden probably represents only a fraction of the garden fauna. Furthermore, beetles form an order of insects, and so are a larger taxonomic group than other types of insects considered in Chapters 5–9.

NATURAL HISTORY OF BEETLES

There is usually little difficulty in recognizing a beetle. With their hard, often rounded, outer covering, they are one of the most distinctive types of insects. A detailed, authoritative account of their biology is given by Crowson (1981) and has been referred to for this section. Beetles constitute the insect order Coleoptera characterized, particularly, by the modification of the front pair of wings into hard, protective elytra, or wing-cases, which shield the functional yet typically flimsy hindwings. In their development, they undergo complete metamorphosis, from larval forms looking quite unlike beetles, through an immobile pupal stage to adult beetles. Typically, adults have biting mouth-parts, as do the active larvae characteristic of many families.

Elytra probably evolved as an adaptation for crawling into crevices or beneath soil, stones, leaves, bark or litter without damaging the functional wings. In evolutionary terms, this modification was evidently successful, for of the 1.82 million described species of animals and plants, half are insects and a quarter are beetles (Stork 1988). It is accepted, however, that there are many more undescribed species than are known, giving a total of perhaps 30 million species of animals and plants in the world (May 1987) or even more, for Stork estimates that there are up to 80 million arthropod species in tropical forest alone. On the higher estimate and extrapolating from known species, there would be in excess of 20 million species of beetles, and some justification for May's claim that, to a good approximation, all species are insects. The vast assemblage of beetles that are known, calls for detailed classification, and taxonomists are by no means agreed on how this should be achieved. In this chapter, I follow the Kloet & Hincks checklist of British insects (Pope 1977). The order is divided into three suborders, Adephaga (including six British families, such as Carabidae), Myxophaga (with one family), and Polyphaga (which incorporates 88 British families). The checklist recognizes 3729 species as British, with an additional 116 species either extinct or of doubtful status.

Adult beetles range in size from about 0.25 millimetres in the featherwing beetles (Ptilidae) to 10 cm in longhorn beetles (Cerambycidae), but considering the extraordinary diversity in how and where they live, their appearance is remarkably uniform. Most have a very toughened external skeleton, which in some is reputed to have the mechanical strength of mild steel, and the soft exoskeleton of cantharid beetles is exceptional. The body surface of terrestrial beetles is normally hydrofuge, and that of water beetles more hydrophilic, although areas where air is held for respiration while submerged, often beneath the elytra, are hydrofuge. The majority of beetles are black, but others are brightly coloured, as in ladybirds (Coccinellidae), where carotenoid pigments are responsible, or leaf beetles (Chrysomelidae), where the colours are structural, resulting from light interference in different layers of the cuticle. Many species, including black beetles, are iridescent because of micro-sculpturing of the cuticle. Patterning in cryptic colours, as in many weevils (Curculionidae) is often achieved by a dense covering of short, recumbent setae (hairs), which may be scale-like in structure.

Although elytra are characteristic of beetles, in some, particularly the rove beetles (Staphylinidae), they are abbreviated, exposing the abdomen. Furthermore, staphylinids tend to be elongate and sinuous, although they still generally have fully functional wings, which are folded in a complicated fashion beneath the short elytra. In some instances, such as in some ground beetles (Carabidae) and weevils (Curculionidae), the elytra are fused down the mid-line into a solid carapace, and the beetles cannot fly. In female glow-worms (Lampyridae), elytra and functional wings are lacking. Many species, such as ladybirds, hold the elytra raised and separated while in flight, but others, such as the rose chafer *Cetonia aurata* (Scarabaeidae), fly without extending the elytra.

Antennae are variable, being very short in whirligig beetles (Gyrinidae) and ladybirds, and sometimes very long, more than twice as long as the body, in Cerambycidae. In others, the individual segments are expanded, giving saw-like (some click beetles, Elateridae), comb-like (some Elateridae) or lamellate (cockchafers, Scarabaeidae) antennae, and in weevils the antennae are elbowed. Beetles' legs reflect the way of life of the insect: those of active, predatory ground beetles (Carabidae) are long and slender; the hind legs of flea beetles, such as *Phyllotreta* spp. (Chrysomelidae), have enlarged femora for jumping, those of dung beetles, such as *Geotrupes* spp. (Geotrupidae), are expanded and strengthened for digging; the tarsi of leaf beetles (Chrysomelidae) have adhesive lobes for clinging to vegetation; and the hind and sometimes the middle legs of many aquatic beetles are expanded and fringed with hairs, forming paddles for swimming.

Although the vast majority are recognizably beetles, their larvae are very different. Active predators on the ground, such as carabid larvae, on plants (ladybird larvae) or in water, such as diving beetle larvae (Dytiscidae), are more or less elongate with well-developed legs, and herbivores that live openly on plants, such as chrysomelid larvae, also have fully functional legs. In contrast, the many beetle larvae that live in their food, whether dung, decaying vegetation, wood, or some other substrate, have much reduced legs or even lack them altogether, so that they resemble maggots. Larval appearance reflects the needs for activity and mobility; newly hatched oil beetle (Meloidae) larvae, for instance, which climb vegetation and attach themselves to solitary bees, have well-developed legs ending in claws, but after being inadvertently carried by a bee back to its nest, where they feed on the stored pollen and honey, they become grub-like, with much reduced legs.

Life histories may be illustrated by ground beetles (Carabidae) and lady-birds (Coccinellidae). Some carabids breed in late summer or autumn and the resulting larvae hibernate; others breed in spring or early summer then die, and the new generation of adults, which appears in autumn, then hibernates. Others vary the time of breeding dependent on the weather, and some diapause as adults before reproduction. They usually lay eggs singly, either depositing them on the soil surface, or using the tip of the abdomen to dig a small hole. Most have three larval instars, and the active larvae are mainly carnivorous, whereas adults are often omnivorous or even phytophagous. The pupa is usually formed in a compacted cell in the soil, and lies on its back supported by specialized, long dorsal spines. The pupal cuticle is soft and white, but segments and head are recognizable, and all the appendages are visible and free although adpressed to the ventral side. Adults of *Nebria brevicollis* are active for only two or three weeks after emergence in early summer, feeding rapidly and building up fat reserves before entering diapause aggregated under stones or logs until the beginning of the breeding season, which lasts from mid-October until the following April (Forsythe 1987).

In contrast to ground beetles, most ladybirds lay their eggs in early summer in batches on vegetation, often among aggregations of aphids, the larval and adult prey of most species. Eggs hatch into active predatory larvae which range freely over plants, and, when food is short, turn cannibal. There are four larval instars, and the larvae pupate on leaf surfaces in a characteristic hunched position so that they appear rather featureless. The time taken from laying of an egg to emergence of an adult varies from 10 to 61 days, depend-ing on temperature in the 7-spot ladybird *Coccinella septempunctata* (Majerus &

Kearns 1989), and most adults emerge in mid- to late summer. There may occasionally, in a few species, such as 2-spot *Adalia bipunctata* and 14-spot *Propylea quattuordecimpunctata* ladybirds, be two generations a year. Ladybirds pass the winter as adults either, as in the 2-spot, hibernating fully, so that they are immobile all winter with a reduced metabolic rate, or merely slowing to immobility when it is cold but stirring on warm days, as in the 7-spot.

The mouthparts of most beetles are of a fairly typical biting type, but a few groups show extreme specialization. In some flower-feeding species, such as Rhipiphoridae for instance, each maxilla has developed one or more long, hairy lobes, and in Meloidae, the maxillae fit together to form an elongate suctorial proboscis; in some small species, such as Cerylonidae, which live under bark, in Fungi or in decomposing vegetable matter, mandibles, maxillae and labium are sharp and pointed for piercing and sucking. In many carnivorous beetle larvae, digestion is extra-oral, in that enzymes from the mid-gut are injected into the prey by the slender, grooved mandibles and the liquefied prey is then sucked into the narrow, muscular oesophagus.

Beetles are themselves subject to predation, and among the defences they use are chemical ones. Most Adephaga, for instance, have pygidial glands, opening dorsally in the region of the ninth abdominal tergite, which secrete an aqueous solution containing *inter alia* a toxin. The most extraordinary development of the pygidial glands is in certain carabids, such as the bombardier beetle *Brachinus crepitans*; a mixture of hydroquinones and hydrogen peroxide is exploded by enzymes in a 'combustion' chamber so that a hot, irritant liquid is forcibly ejected from the gland openings. As the abdomen is markedly flexible, this boiling jet can be discharged directly at the predator. Some *Stenus* spp. (Staphylinidae) use their pygidial glands to move on the surface film of water; the glandular secretion lowers the surface tension of the water behind the beetle so that it is rapidly propelled forwards by the pull of surface tension in front of it.

Aquatic adaptations of beetles are usually most evident in the locomotory and respiratory systems. Adult water beetles can be loosely divided into open-water swimmers, such as *Dytiscus* spp., and crawlers on vegetation, such as *Helophorus* spp. (Hydrophilidae). In most swimmers, the tibiae and tarsi of the hind legs are wide and hair-fringed, whereas in crawlers, the claws are enlarged. Many water beetles, such as *Dytiscus* spp., use for respiration a bubble of air trapped beneath the elytra; such beetles are good swimmers, making repeated trips to the surface to replenish the air bubble, and typically occur in water of 2 m depth or less. In addition, in many smaller dytiscids the air bubble behaves as a physical gill into which dissolved oxygen diffuses to replace that used by the beetle. Other water beetles, such as the non-swimming Elmidae, trap air bubbles in coatings of hydrofuge hairs or scales on the ventral side of the thorax and abdomen, and use this for respiration; although this plastron to some extent acts as a physical gill, they can also use the oxygen bubbles produced by photosynthesizing algae, and so remain submerged in deep water for weeks at a time.

Many beetles live on or in living or dead trees, often in association with

Fungi, which convert the nutrients in wood into a form the beetles can eat. Adult *Scolytus* beetles, for instance, when they feed on young shoots of elm, infect the trees with the sticky spores of Dutch elm disease *Ceratostomella ulmi*; as the fungus spreads, branches are killed, the beetles make their egg burrows under the bark of dying or recently dead fungus-infested branches, and the larvae tunnel and feed in the cambium, being apparently dependent for some elements of their nutrition on the presence of the fungus. Ambrosia beetles (Scolytidae) burrow more extensively into trees than *Scolytus* spp., and have a closer relationship with a fungus. Adults carry fungal spores in special cavities on the body, and these develop in the burrows to carpets of fungal growth on which the larvae feed. In other beetles, fungal spores are transported internally in the alimentary canal, infection of the tree being by defaecation. Not all tree-feeding beetles, however, have symbiotic relationships with Fungi, and they eat the actual tissues of the tree, including wood. In many cases, such beetles are known to have in the gut, symbiotic bacteria, which presumably aid digestion. Some, particularly Scolytidae, have a number of adaptations appropriate for burrowing: the body is markedly cylindrical, the prothorax is humped or hooded, the head is more or less bent down, the legs are short, and the elytra bend sharply down at the back, so that they block the end of a burrow, sometimes also being modified to scoop 'sawdust' backwards.

Beetles of many different families associate with social insects, particularly ants and termites, living in their nests, sometimes as scavengers, often as parasites or predators. In the majority of cases, those in ant nests, particularly Paussidae and many Staphylinidae, are tolerated because glands on the surface of their bodies produce sweet secretions that the hosts lick; many are tended, fed, and even carried about by the ants, which grasp the beetles' characteristically enlarged and thickened antennae. Their successful association with ants is dependent, in the evolutionary sense, on the beetles having broken into the ants' code of chemical and tactile signals; since they smell, taste and feel like ants, they are treated as such.

Beetles fill a great diversity of ecological niches, as predators, herbivores, scavengers, and even parasites, such as the louse-like *Platypsyllus castoris*, which parasitizes beavers. They abound on land and in aquatic habitats, and some have become adapted to the most physiologically stressful of habitats where water is almost totally lacking, such as hot deserts and stores of grain and dried foodstuffs. Adaptations to such difficult conditions demonstrate the diversity and evolutionary plasticity of beetles.

BEETLES FROM PITFALL TRAPS

Pitfall traps have been operated from 1 April to 31 October since 1979 in the Leicester garden. A total of eight are sited in four different areas of the garden, and since 1980 a further three have been positioned below the Malaise trap (see Chapter 3). In addition, the pitfalls were operated during winter 1985/86, i.e. from 31 October 1985 to 1 April 1986. The rationale for using pitfalls below the Malaise trap is that it is generally assumed that many beetles that fly into the open sides of a trap, drop to the ground on striking the central baffle and walk away. Beetles that have thus evaded capture in the Malaise trap, might be expected to fall into pitfalls beneath it.

The size of the catch of Carabidae in a pitfall depends on the density of vegetation and litter around the trap, more being caught where there is little to impede the movement of beetles near to the trap. Consequently, pitfall trapping cannot be used for assessing population sizes, other than by saturation trapping and removal from a confined area, or for comparing the beetle fauna in different habitats (Greenslade 1964). In my garden, however, pitfall traps have been used annually in identical sites, and can legitimately be used to compare the size of the ground beetle community in different years, as well as for compiling a species list. Formalin, which is used as a killing and preservative agent in pitfalls in the garden, appears to attract Staphylinidae, Curculionidae and Carabidae, particularly *Pterostichus madidus* (Luff 1968), militating against its use in population or community studies. Pitfalls with formalin are, however, very effective as traps, and the arguments against their use do not apply to pitfall trapping as conducted in the garden, nor compromise the way that the trapping data are used. Luff (1975) also found that plastic pitfalls were less effective than glass, and use of plastic drinking beakers in the garden may to some extent counter the effects of using formalin. The catch is removed from traps at regular intervals, at least every two weeks, and the formalin is changed when it becomes dirty or diluted.

Beetles have been identified as annual samples by D. Goddard, and nomenclature, sequence and classification follows Pope (1977). During the eight-year period 1979–86, 8218 individuals of 56 species of beetles were caught in pitfall traps, including 18 species of Carabidae, 20 species of Staphylinidae and eight species of Curculionidae. In the following list of species, total numbers caught in pitfalls (including the winter 1985/86) are given; MT indicates that the species was also caught in the Malaise trap, and MV that it was also caught in the mercury-vapour light trap.

Carabidae

1. *Carabus violaceus* L. 13
2. *Leistus fulvibarbis* Dejean. 1
3. *L. spinibarbis* (Fab.) 56
4. *Nebria brevicollis* (Fab.) . . . . 2405 and 138 larvae. MT
5. *Notiophilus biguttatus* (Fab.). . . . . 33 MT
6. *Trechus quadristriatus* (Schrank). . . . . 20 MT
7. *Bembidion aeneum* Germar. 1
8. *B. guttula* (Fab.) . . . . 2 MT
9. *Pterostichus madidus* (Fab.) . . . . 5001 and 16 larvae
10. *Calathus fuscipes* (Goeze). 1
11. *Synuchus nivalis* (Panzer) 7
12. *Amara familiaris* (Duftschmid). . . . . 2 MT
13. *A. plebeja* (Gyllenhal) 1

14. *A. similata* (Gyllenhal)   1
15. *Harpalus rufipes* (Degeer)   4
16. *H. rufibarbis* (Fab.) . . . .   1 MV
17. *H. affinis* (Schrank)   3
18. *Badister bipustulatus* (Fab.)   1

Dytiscidae

19. *Ilybius fuliginosus* (Fab.) . . . .   1 MV

Leiodidae

20. *Catops nigricans* (Spence) . . . .   3 MT

Staphylinidae

21. *Omalium excavatum* Stephens   2
22. *Anotylus inustus* (Gravenhorst) . . . .   1 MT
23. *A. sculpturatus* (Gravenhorst) . . . .   1 MT
24. *Lathrobium brunnipes* (Fab.)   1
25. *L. fulvipenne* (Gravenhorst)   1
26. *L. multipunctum* Gravenhorst   1
27. *Rugilus orbiculatus* (Paykull) . . . .   1 MT
28. *Gyrohypnus angustatus* Stephens   3
29. *Xantholinus linearis* (Olivier) . . . .   12 MT
30. *Philonthus cognatus* Stephens . . . .   2 MT
31. *P. marginatus* (Ström) . . . .   1 MT
32. *P. politus* (L.) . . . .   1 MT
33. *P. tenuicornis* Mulsant & Rey . . . .   2 MT
34. *P. varius* (Gyllenhal) . . . .   2 MT
35. *Staphylinus ater* Gravenhorst   109
36. *S. olens* Müller   4
37. *Quedius boops* (Gravenhorst) . . . .   1 MT
38. *Q. curtipennis* Bernhauer   2
39. *Q. tristis* (Gravenhorst) . . . .   6 MT
40. *Tachinus signatus* Gravenhorst . . . .   2 MT

Scarabaeidae

41. *Aphodius rufipes* (L.) . . . .   3 MV

Byrrhidae

42. *Simplocaria semistriata* (Fab.)   1

Elateridae

43. *Melanotus erythropus* (Gmelin in L.)   1
44. *Agriotes pallidulus* (Illiger) . . . .   1 MT

Photograph 10.1. *Continuous ground cover shelters numerous and diverse ground beetles (Carabidae).*

Nitidulidae

45. *Meligethes aeneus* (Fab.) . . . .   1 MT
46. *M. ?rotundicollis* Brisout   1

Lathridiidae

47. *Aridius nodifer* (Westwood) . . . .   1 MT

Apionidae

48. *Apion radiolus* (Marsham) . . . .   1 MT

Curculionidae

49. *Otiorhynchus porcatus* (Herbst)   47
50. *O. rugostriatus* (Goeze) . . . .   88 MT
51. *O. singularis* (L.)   141
52. *O. sulcatus* (Fab.) . . . .   18 MT
53. *Phyllobius maculicornis* Germar   1
54. *Barypeithes pellucidus* (Boheman) . . . .   142 MT
55. *Sciaphilus asperatus* (Bonsdorff)   3
56. *Barynotus obscurus* (Fab.) . . . .   56 MV

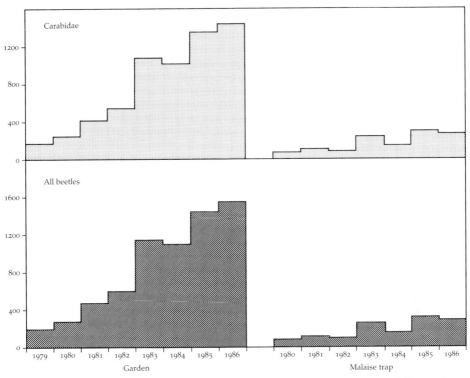

Fig. 10.1 Annual fluctuations in numbers of all beetles and of Carabidae in pitfall traps in
open garden and below Malaise trap, 1979–86.

The most striking aspect of the annual catches of beetles in pitfall traps in
the garden was the more or less constant rise in numbers from 1979 to 1986,
even in the small catch from pitfalls below the Malaise trap (Fig. 10.1, Table
10.1 and Table 10.2). As shown in Fig. 10.1, the pattern of rise in numbers is
accounted for by increases in catch of Carabidae, which formed 91.9 per cent
of the total catch. The commonest beetle in pitfalls was *Pterostichus madidus*,
which formed 60.89 per cent of the total catch and was the commonest in
every year in traps set in the open garden, i.e. away from the Malaise trap. It
formed a high percentage of the annual catch in 1979–83 (80.78 per cent in
1980), but a lower proportion in 1984–86 (53.05 per cent in 1985) when *Nebria
brevicollis* was relatively common. *N. brevicollis* and *P. madidus* together
account for 90 per cent of the total catch. Both the average catch per trap and
the average number of individuals per species (*N/S*) rose fairly steadily during
the eight-year period (Table 10.1).

In the hot, dry summer of 1983, followed by a warm autumn, numbers of
beetles trapped both in the open garden and beneath the Malaise trap,
increased markedly; the numbers trapped in 1984 were somewhat less than in
1983, thus interrupting the otherwise consistent annual increase. The reasons
for the increasing numbers of Carabidae and other ground-dwelling beetles

Table 10.1 *Size and composition of catch of beetles in eight pitfall traps set in open garden, 1979–86*

	Number of individuals (N)	Catch per trap N/8	Number of species (S)	N/S	Commonest as % N
1979	198	24.8	20	9.90	72.70
1980	281	35.1	16	17.56	80.78
1981	477	59.6	15	31.80	69.60
1982	598	74.7	18	33.22	77.93
1983	1146	143.2	17	67.41	80.19
1984	1104	138.0	19	58.11	54.35
1985	1442	180.3	16	90.13	53.05
1986	1550	193.8	18	86.11	56.97

Note: Pterostichus madidus was the commonest species every year.

Table 10.2 *Size and composition of catch of beetles in three pitfall traps set below Malaise trap, 1980–86*

	Number of individuals (N)	Catch per trap N/3	Number of species (S)	N/S	Commonest as % N	Commonest species
1980	86	28.7	11	7.82	70.93	*Pterostichus madidus*
1981	117	39.0	9	13.00	55.56	*P. madidus*
1982	97	32.3	13	7.46	59.79	*P. madidus*
1983	257	85.6	11	23.36	48.25	*Nebria brevicollis*
1984	158	52.7	11	14.36	43.04	*N. brevicollis*
1985	322	107.3	10	32.20	48.76	*N. brevicollis*
1986	293	97.7	9	32.56	49.49	*P. madidus*

are not entirely clear, even allowing for the apparent attractiveness for them of formalin. *Pterostichus madidus* is predatory as larvae and omnivorous as adults, and *Nebria brevicollis* preys on Collembola as larvae and as adults. Although impossible to quantify, it is probable that during the period 1979–86, the garden became lusher, with more dead and decomposing vegetation on the soil surface; this may have improved conditions for small soil animals, such as Collembola, on which both species of beetles feed.

The number of beetles caught on average in each of the three pitfalls below the Malaise trap was consistently smaller than the catch per trap in the open garden, and fewer species were caught. Moreover, numbers did not rise so consistently, falling, compared with the previous year, in 1982, 1984 and 1986 (Fig. 10.1 and Table 10.2). There is some evidence, although inconclusive, that these pitfalls were catching beetles 'lost' from the Malaise trap. For instance, seven species were trapped in pitfalls below the Malaise trap but not in pitfalls in the open garden, although 32 species were caught away from the Malaise trap but not beneath it. Of the seven only caught below the Malaise trap, four (*Anotylus sculpturatus*, *Rugilus orbiculatus*, *Agriotes pallidulus* and *Apion radiolus*) were also caught in the Malaise trap; the elaterid *Melanotus erythropus* was once seen and collected, and the carabids *Leistus fulvibarbis* and *Calathus fuscipes* are only known for the garden from single individuals caught in pitfalls below the Malaise trap.

There is, moreover, an interesting difference between the annual catches from the two categories of pitfall traps. In the open garden, *Pterostichus madidus* was the commonest species in every year, whereas below the Malaise trap, *Nebria brevicollis* was the commonest in 1983, 1984 and 1985, when it formed 48.25, 43.04 and 48.76 per cent, respectively, of the catch. *N. brevicollis* was only once taken in the Malaise trap, and it may be that flying individuals dropped to the ground on encountering the trap, and so were caught in the pitfalls.

Ninety-two beetles of 13 species were caught in the total of 11 traps in winter 1985/86, when, of course, the Malaise trap was not in position. Forty-six were *N. brevicollis*, and 23 *P. madidus*, all but one of the latter females; in addition, 102 larvae of *N. brevicollis* and eight of *P. madidus* were caught. *Lathrobium brunnipes* and *L. multipunctum* are known only from single individuals caught during the 1985/86 winter.

There is a pronounced similarity of the garden beetle fauna as recorded in pitfall traps to that of open ground and wasteland sites, and most of the recorded species are xerophilous or eurytopic. This accords with the widely held ecological view of gardens as open, disturbed sites, but interestingly there are also typically woodland species, such as *Leistus fulvibarbis*, and hydrophilic species, such as *Bembidion guttula*, the former caught beneath the Malaise trap, the latter in the open garden. *Pterostichus madidus* is common in open country, often on cultivated soil and in gardens, and *Nebria brevicollis* is extremely eurytopic. Of the other Carabidae that were most frequently captured, *Notiophilus biguttatus* is common in somewhat shady but dry places, and *Leistus spinibarbis* is widespread and fairly eurytopic (Lindroth 1974). The staphylinid *Staphylinus ater*, although common in Leicestershire, is often a coastal species. The 47 garden records of the curculionid *O. porcatus* are the first for the county, although, as all four species of *Otiorhynchus* captured are common garden pests, it may have been under-recorded. Of the other two abundant weevils, *Barypeithes pellucidus* is locally common, and *Barynotus obscurus* is widespread and common.

The species diversity of pitfall beetles was consistently higher below the Malaise trap than in the open garden (Table 10.3), in line with the smaller average number of individuals per species, N/S (Table 10.2). This difference was very highly significant ($p<0.001$) in 1981, 1982, 1983 and 1984, and significant ($p<0.05$) in 1986, although not significant in 1980 and 1985. The diversity index used was β, an estimate of the probability that a trapped individual will be different from the previous individual trapped. The striking difference in diversity of pitfall catches in the open garden between 1983 (0.341) and 1984 (0.570) is very highly significant ($p<0.001$), whereas there is no significant difference between the values for 1984 and 1985 (0.551); in 1983

Table 10.3 *Species diversity (β) of beetles caught in pitfall traps, 1979–86*

	Site	Number of individuals (N)	Number of species (S)	β-diversity	Standard error
1979	Open garden	198	20	0.465	±0.044214
1980	Open garden	281	16	0.344	±0.036952
	Below Malaise	86	11	0.487	±0.064629
	Total	367	19	0.380	±0.032609
1981	Open garden	477	15	0.488	±0.025558
	Below Malaise	117	9	0.613	±0.036722
	Total	594	17	0.519	±0.021999
1982	Open garden	598	18	0.381	±0.024660
	Below Malaise	97	13	0.617	±0.052241
	Total	695	21	0.419	±0.022906
1983	Open garden	1146	17	0.341	±0.016968
	Below Malaise	257	11	0.588	±0.016412
	Total	1403	20	0.426	±0.014188
1984	Open garden	1104	19	0.570	±0.009400
	Below Malaise	158	11	0.634	±0.022424
	Total	1262	22	0.580	±0.008589
1985	Open garden	1442	16	0.551	±0.006643
	Below Malaise	322	10	0.579	±0.013446
	Total	1764	19	0.559	±0.005800
1986	Open garden	1550	18	0.551	±0.008263
	Below Malaise	293	9	0.593	±0.015681
	Total	1843	21	0.559	±0.007392

Note: β-diversity is derived from Simpson's index, λ: $\beta = 1 - \lambda$.

the catch size increased markedly over the previous year, whereas it decreased slightly in 1984 (Fig. 10.1). In contrast, the diversity of catches in pitfalls below the Malaise trap was significantly less ($p < 0.05$) in 1985 (0.579) than in 1984 (0.634), but not significantly different in 1983 (0.588) and 1984. The largest annual total diversity of pitfall catches (all 11 traps), 0.580 in 1984, was very highly significantly greater ($p < 0.001$) than that of 1983 (0.426) and significantly greater ($p < 0.05$) than that of 1985 (0.559). It seems, therefore, that the frequency distribution of species in pitfall catches was not constant from year to year either below the Malaise trap or in the open garden or in total, and that it tended to differ between pitfalls in the open garden and below the Malaise trap. This suggests that the Malaise trap was having some effect on the frequency distribution of species caught in pitfalls beneath it, although it is not possible to single out particular species that were deflected from the Malaise trap and subsequently caught in pitfalls beneath it.

BEETLES, OTHER THAN LADYBIRDS, FROM THE MALAISE TRAP

A total of 2503 beetles, other than ladybirds, of at least 197 species in 30 families were caught in the Malaise trap during the period 1981–86. Ladybirds

Photograph 10.2. Mixed planting of cabbages Brassica oleracea, *black-eyed Susan* Rudbeckia hirta, *bergamot* Monarda didyma, *onion* Allium cepa *and flowering tobacco* Nicotiana alata *increases ground cover for carabids as well as diversifying feeding opportunities for other beetles.*

(Coccinellidae) are considered separately from other beetles caught in the Malaise trap for three reasons: ladybirds are conspicuous and familiar to most people; unlike other beetles, they have been sorted weekly from trap samples, identified and tabulated since 1972; and they differ from other beetles in that there are large-scale influxes to the garden of certain species in some years. In the following list, species also caught in pitfalls are given the same number (in parentheses) as in the previous section. Numbers refer to total Malaise trap catch, 1981–86. Species also caught in the mercury-vapour light trap are indicated by MV. Identification was by D. Goddard, and taxonomy and sequence follow Pope (1977).

Carabidae

(4.) *Nebria brevicollis* 1
(5.) *Notiophilus biguttatus* 1
57. *Loricera pilicornis* (Fab.) 4
(6.) *Trechus quadristriatus* 9
58. *Asaphidion flavipes* (L.) *sensu lato* 1
59. *Bembidion quadrimaculatum* (L.) 4
(8.) *B. guttula* 8

60. *B. lunulatum* (Fourcroy) 1
(12.) *Amara familiaris.* 1
61. *Bradycellus verbasci* (Duftschmid) . . . . . 4 MV
62. *Demetrias atricapillus* (L.). 2
63. *Dromius agilis* (Fab.) 1
64. *D. meridionalis* Dejean. 1

Hydrophilidae

65. *Helophorus brevipalpis* Bedel. 8
66. *Sphaeridium scarabaeoides* (L.). 4
67. *Cercyon haemorrhoidalis* (Fab.) 3
68. *C. lateralis* (Marsham). 1
69. *Cryptopleurum minutum* (Fab.). 2

Leiodidae

70. *Leiodes badia* (Sturm) 2
71. *Ptomaphagus medius* Rey 6
72. *P. subvillosus* (Goeze) 22
73. *Choleva angustata* (Fab.). 3
(20.) *Catops nigricans.* 2

Staphylinidae

74. *Megarthrus depressus* (Paykull). 1
75. *Proteinus brachypterus* (Fab.) 2
76. *P. ovalis* Stephens. 4
77. *Lesteva heeri* Fauvel 1
78. *Eusphalerum primulae* (Stephens). 1
79. *Phyllodrepa floralis* (Paykull) 2
80. *Omalium caesum* Gravenhorst 1
81. *O. rivulare* (Paykull) 1
82. *Platystethus arenarius* (Fourcroy) 3
83. *Anotylus complanatus* (Erichson) 16
(22.) *A. inustus* 3
84. *A. rugosus* (Fab.). 11
(23.) *A. sculpturatus* 15
85. *A. tetracarinatus* (Block). 1
86. *Oxytelus laqueatus* (Marsham) 6
87. *Stenus clavicornis* (Scopoli). 4
88. *S. nitidiusculus* Stephens. 1
89. *S. ossium* Stephens 1
90. *Achenium humile* (Nicolai) 1
(27.) *Rugilus orbiculatus.* 1
91. *R. rufipes* Germar 1
92. *Xantholinus glabratus* (Gravenhorst) . . . 2
(29.) *X. linearis* 1
(30.) *Philonthus cognatus* 24
93. *P. decorus* (Gravenhorst). 14

94.	*P. fimetarius* (Gravenhorst)	12 MV
95.	*P. laminatus* (Creutzer)	2
(31.)	*P. marginatus*	9
(32.)	*P. politus.*	2
96.	*P. sanguinolentus* (Gravenhorst)	1
97.	*P. sordidus* (Gravenhorst)	1
98.	*P. succicola* Thomson	2
(33.)	*P. tenuicornis*	3
99.	*P. varians* (Paykull)	203
(34.)	*P. varius*	26
(37.)	*Quedius boops*	12
100.	*Q. cruentus* (Olivier)	1
101.	*Q. nemoralis* Baudi	1
102.	*Q. picipes* (Mannerheim)	1
103.	*Q. semiaeneus* (Stephens)	1
104.	*Q. semiobscurus* (Marsham)	1
(39.)	*Q. tristis*	1
105.	*Mycetoporus clavicornis* (Stephens)	1
106.	*M. lepidus* (Gravenhorst)	10
107.	*M. nigricollis* Stephens	8
108.	*M. splendidus* (Gravenhorst)	2
109.	*Sepedophilus marshami* (Stephens)	1
110.	*Tachyporus chrysomelinus* (L.)	87
111.	*T. hypnorum* (Fab.)	211
112.	*T. nitidulus* (Fab.)	59
113.	*T. obtusus* (L.)	10
114.	*T. solutus* Erichson	5
115.	*Tachinus humeralis* Gravenhorst	5
116.	*T. marginellus* (Fab.)	1
(40.)	*T. signatus*	32
117.	*Cypha longicornis* (Paykull)	1
118.	*Falagria caesa* Erichson.	4
119.	*Tachyusa* sp.	2
120.	*Atheta fungi* (Gravenhorst)	3
121.	*Haploglossa pulla* (Gyllenhal)	1
122.	*Aleochara bilineata* Gyllenhal	1
123.	*A. bipustula* (L.).	9
124.	*A. curtula* (Goeze).	4
125.	*A. lanuginosa* Gravenhorst	12
126.	Undetermined Aleocharinae	93

Scarabaeidae

127.	*Aphodius contaminatus* (Herbst)	1

Scirtidae

128.	*Cyphon* sp.	1

Elateridae

129. *Athous haemorrhoidalis* (Fab.)   1
130. *Agriotes acuminatus* (Stephens)   1
(44.) *A. pallidulus*   1

Cantharidae

131. *Cantharis decipiens* Baudi   1
132. *C. lateralis* L.   1
133. *Rhagonycha femoralis* (Brullé)   1
134. *R. fulva* (Scopoli).   1
135. *Malthodes marginatus* (Latreille)   2

Dermestidae

136. *Anthrenus fuscus* Olivier   13
137. *A. verbasci* (L.).   20

Anobiidae

138. *Anobium punctatum* (Degeer).   1

Melyridae

139. *Dasytes aeratus* Stephens   1

Nitidulidae

140. *Brachypterus glaber* (Stephens).   12
141. *B. urticae* (Fab.).   4
142. *Pria dulcamarae* (Scopoli)   1
(45.) *Meligethes aeneus*.   154
143. *M. morosus* Erichson.   24
144. *M. nigrescens* Stephens   24
145. *M. ovatus* Sturm   2
146. *Meligethes* sp.   98
147. *Epurea pusilla* (Illiger)   1
148. *Soronia grisea* (L.)   1

Rhizophagidae

149. *Monotoma picipes* Herbst   2

Cryptophagidae

150. *Cryptophagus dentatus* (Herbst)   16
151. *Atomaria ?fuscata* (Schoenherr)   2
152. *Ephistemus globulus* (Paykull)   1

Byturidae

153. *Byturus tormentosus* (Degeer)   1

Phalacridae

154. *Olibrus aeneus* (Fab.)   3
155. *Stilbus testaceus* (Panzer)   66

Endomychidae

156. *Sphaerosoma piliferum* (Müller)   3

Lathridiidae

157. *Stephostethus lardarius* (Degeer)   22
158. *Aridius bifasciatus* (Reitter)   11
(47.) *A. nodifer*   10
159. *Enicmus transversus* (Olivier)   26
160. *Corticaria punctulata* Marsham   8
161. *Corticarina fuscula* (Gyllenhal)   38
162. *Cortinicara gibbosa* (Herbst)   26

Cisidae

163. *Cis boleti* (Scopoli)   3

Mycetophagidae

164. *Typhaea stercorea* (L.)   1

Tenebrionidae

165. *Lagria hirta* (L.)   1

Salpingidae

166. *Rhinosimus planirostris* (Fab.)   12

Scraptiidae

167. *Anaspis humeralis* (Fab.)   7
168. *A. maculata* Fourcroy   103
169. *A. regimbarti* Schilsky   54
170. *A. rufilabris* (Gyllenhal)   1
171. *A. thoracica* (L.)   2

Anthicidae

172. *Anthicus floralis* (L.)   2
173. *A. formicarius* (Goeze)   3

Aderidae

174. *Aderus populneus* (Creutzer in Panzer) .   1

Cerambycidae

175. *Grammoptera ruficornis* (Fab.).   14

Chrysomelidae

176. *Oulema melanopa* (L.).   21
177. *Plagiodera versicolora* (Laigharting).   1
178. *Lochmaea crataegi* (Forster).   1
179. *Phyllotreta ?cruciferae* (Goeze)   3
180. *P. nigripes* (Fab.).   23
181. *P. undulata* Kutschera.   23
182. *Longitarsus ganglbaurei* Heikertinger . . .   1
183. *L. ?jacobaeae* (Waterhouse)   3
184. *L. luridus* (Scopoli)   1
185. *Longitarsus* sp   2
186. *Alticao leracea* (L.)   1
187. *Chalcoides aurata* (Marsham)   1
188. *C. aurea* (Fourcroy)   1
189. *C. fulvicornis* (Fab.)   2
190. *Chaetocnema concinna* (Marsham).   50
191. *C. hortensis* (Fourcroy)   14
192. *Sphaeroderma testaceum* (Fab.)   1
193. *Psylliodes chrysocephala* (L.)   39

Apionidae

194. *Apion hydrolapathi* (Marsham).   1
195. *A. malvae* (Fab.)   3
196. *A. rufirostre* (Fab.).   1
(48.) *A. radiolus*.   46
197. *A. miniatum* Germar   1
198. *A. dichroum* Bedel   1
199. *A. nigritarse* Kirby.   1
200. *A. trifoliii* (L.)   2

Curculionidae

(50.) *Otiorhynchus rugostriatus*.   5
(52.) *O. sulcatus* . . . .   6 MV
(54.) *Barypeithes pellucidus*.   1
201. *Sitona cylindricollis* (Fahraeus).   1
202. *S. hispidulus* (Fab.)   4
203. *S. humeralis* Stephens   3
204. *S. lineatus* (L.).   38
205. *S. puncticollis* Stephens   25

206. *Magdalis ruficornis* (L.). 3
207. *Euophryum confine* (Broun) 2
208. *Cossonus parallelepipedus* (Herbst) 1
209. *Ceuthorhynchidius troglodytes* (Fab.) 1
210. *Ceutorhynchus assimilis* (Paykull) 199
211. *C. campestris* Gyllenhal 3
212. *C. erysimi* (Fab.) 3
213. *C. floralis* (Paykull) 2
214. *C. ?quadridens* (Panzer) 54
215. *Ceuthorhynchus* sp. 3
216. *Rhinoncus pericarpius* (L.). 1
217. *R. perpendicularis* (Reich). 2
218. *Amalorrhynchus melanarius* (Stephens). . 17
219. *Anthonomus rubi* (Herbst) 6
220. *Curculio pyrrhoceras* Marsham 3
221. *Miccotrogus picirostris* (Fab.) 1
222. *Rhynchaenus fagi* (L.) 1

Scolytidae

223. *Scolytus multistriatus* (Marsham) 3
224. *S. rugulosus* (Müller). 1
225. *S. scolytus* (Fab.) 2
226. *Hylesinus oleiperda* (Fab.). 3
227. *Leperisinus varius* (Fab.). 5
228. *Acrantus vittatus* (Fab.) 2

The annual catch of individuals and species of beetles, other than lady-birds, in the Malaise trap fluctuated markedly, with 1981 a particularly bad year, 1983 poor, and 1986 considerably worse than 1985 (Fig. 10.2). 1982, 1984 and 1985, however, were all good years. The average number of individuals per species followed much the same pattern, indicating that in 1981 the relatively few species present were each much less common than in other years. However, while the commonest species in 1981, *Anaspis maculata*, formed 18.3 per cent of the total catch, in 1983 and 1986, the commonest species were proportionately much more common, *Philonthus varians* forming 28.3 per cent of the 1983 catch and *Meligethes aeneus* 25.9 per cent in 1986 (Table 10.4). In 1984, one of the good years, when *Tachyporus chrysomelinus* and *Chaetocnema concinna* were equally common, however, their combined proportion of the total at 13.8 per cent was less than that of the commonest species in any other year. It seems that, on average, in poor years, one species particularly dominated the catch. The curious feature, however, of this annual pattern, was that a different species was involved every year, indicating that conditions that favoured one particular species also differed every year. This is in contrast to all other groups analysed in detail, although it is worth noting that beetles are taxonomically a larger group and also a more diverse group than any of the others. It may be this very fact that the difference from year to year in commonest species is confirming.

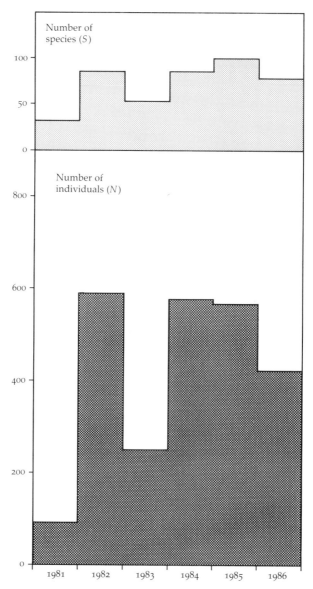

Fig. 10.2 *Annual fluctuations in numbers of individuals and species of beetles, other than
ladybirds, caught in Malaise trap, 1981–86.*

Of the commonest species (see Table 10.4), the scraptiid, *Anaspis maculata*,
is a flower-feeder, the larvae developing in rotting wood. The weevil
(Curculionidae), *Ceutorhynchus assimilis* is often a pest on cabbages and other
cruciferous plants and has become commoner with widespread cultivation of
oil-seed rape. The staphylinid, *Philonthus varians*, feeds on dung, carrion and
compost, but *Tachyporus* spp. (also staphylinids) are predators in leaf litter
and decaying vegetation on the ground and both genera occur in compost

Table 10.4 *Size and composition of catch of beetles, other than ladybirds, in Malaise trap,*
1981–86

	Number of individuals (N)	Number of species (S)	N/S	Commonest as % N	Commonest species
1981	93	33	2.8	18.3	*Anaspis maculata*
1982	589	86	6.8	18.0	*Ceutorhynchus assimilis*
1983	251	53	4.7	28.3	*Philonthus varians*
1984	578	86	6.7	6.9	*Tachyporus chrysomelinus*
				6.9	*Chaetocnema concinna*
1985	568	100	5.7	19.4	*T. hypnorum*
1986	424	78	5.4	25.9	*Meligethes aeneus*

Photograph 10.3. Beetles, such as the weevil Ceutorhynchus assimilis, *that feed on cruciferous plants have become commoner with widespread cultivation of oil-seed rape, seen here in flower in east Leicestershire.*

heaps. The chrysomelid *Chaetocnema concinna* is a herbivorous flea beetle, common on such plants as knotgrasses *Polygonum*, and sometimes a pest of mangolds. The flower-feeding *Meligethes aeneus* (Nitidulidae), a pollen beetle, sometimes so common as to be considered a pest, is non-specific as to flowers, but in some years is abundant on sweet peas in my garden.

In some years, additional species were particularly common relative to other years. For instance, 15 of the total catch of 24 *Philonthus cognatus*, another dung-, compost- and carrion-feeding staphylinid, were trapped in 1982; 18 of 23 *Phyllotreta nigripes*, a herbivorous flea beetle (Chrysomelidae), were caught in 1985; 30 of 46 *Apion radiolus*, a herbivorous weevil (Apionidae), were caught in 1985; and 30 of 38 mould beetles *Corticarina fuscula* (Lathridiidae), which feed on various Fungi, were trapped in 1984. It is noteworthy that 1982, 1984 and 1985 were years when beetle numbers in the Malaise trap were high (Fig. 10.2).

Ladybirds (Coccinellidae) are considered separately in the next section, but

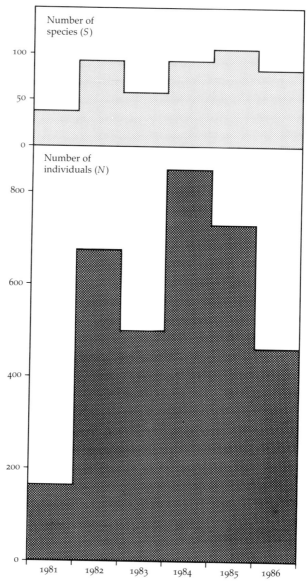

Fig. 10.3 Annual fluctuations in total numbers of individuals and species of beetles, including ladybirds, caught in Malaise trap, 1981–86.

in assessing annual fluctuations in numbers of beetles caught in the Malaise trap, it may be better to include them. When this is done (Fig. 10.3 and Table 10.5), although the overall annual pattern remains much the same, it becomes clear that 1984 was a very good year, and that 1983 was not as bad as it otherwise appeared. Moreover, in three years, ladybirds were the commonest species of beetle: *Adalia bipunctata* in 1981 and 1984, and *Propylea quattuorde-*

Table 10.5 *Size and composition of total catch of beetles in Malaise trap, 1981–86*

	Number of individuals (N)	Number of species (S)	N/S	Commonest as % N	Commonest species
1981	167	38	4.4	31.7	*Adalia bipunctata*
1982	676	92	7.4	15.7	*Ceutorhynchus assimilis*
1983	499	58	8.6	27.5	*Propylea quattuordecimpunctata*
1984	849	92	9.2	13.6	*A. bipunctata*
1985	729	106	6.9	15.1	*Tachyporus hypnorum*
1986	465	83	5.6	23.7	*Meligethes aeneus*

cimpunctata in 1983. 1981 remains the year when the average number of individuals per species (*N/S*) was the lowest at 4.4, but 1984 becomes the year when *N/S* was the highest, at 9.2, largely as a consequence of large numbers of *A. bipunctata*. On this basis, there were three years, 1981 as well as 1983 and 1986, when the commonest species formed a particularly high proportion of the total catch, again because of large numbers of *A. bipunctata*.

By and large, numbers of beetles each year seem to have followed the same trend as the other groups of insects, discussed so far, that were caught in the Malaise trap. Numbers of all groups were down in 1981, and all were up in 1982. Numbers of moths, hoverflies, beetles and, to some extent, solitary bees were also down in 1983, although those of social bees, solitary wasps and, particularly, butterflies were up. All groups, other than butterflies, were to some extent up in numbers in 1984, whereas in 1985 numbers of hoverflies, solitary bees and moths were up, and beetles had another good year, although social bees, solitary wasps and, particularly, butterflies were down. 1986 was another generally poor year, although moth and beetle numbers were quite good. In summary, 1982 and 1984 were generally good insect years, as monitored by the Malaise trap, 1981 and 1986 were generally poor years, and in 1983 and 1985 different groups responded in differing ways. That different insect groups, with different life styles and ecological requirements, respond differently in different years is well exemplified by fluctuations in abundance of lacewings and sawflies (see Chapter 11). Sawfly numbers were good in 1981 and 1983, followed by a decline to a low in 1986. Lacewing numbers, however, which were not high in either 1982 or 1984, were good in 1983, 1985, and to a lesser extent in 1986. The summer (June, July and August) of 1984 was hot, dry and sunny, after an average, if rather dry, spring (March, April and May); the summer of 1982 was mild, damp and dull after a warm, dry, bright spring. Spring of 1981 was wet and summer dry, but both were dull, with average temperatures, and spring and summer 1986 were damp and of only average brightness, spring being cool and summer of average temperature. The summers of 1983 and 1985 were very different, that of 1983 being hot, dry and sunny, but that of 1985 cool, damp, and of only average brightness; in both springs, temperatures were average, but 1983 was wet and dull, whereas in 1985 both rainfall and hours of sunshine

Table 10.6 *Species diversity (β) of beetles, other than ladybirds, caught in Malaise trap, 1981–86*

	Number of individuals (N)	Number of species (S)	β-diversity	Standard error
1981	93	33	0.941	±0.012979
1982	589	86	0.909	±0.006030
1983	251	53	0.878	±0.014729
1984	578	86	0.964	±0.002191
1985	568	100	0.946	±0.005847
1986	424	78	0.912	±0.010107
1981–86	2503	197	0.965	±0.001231

Note: β-diversity is derived from Simpson's index, λ: β=1−λ.

were average. Good insect years, particularly for the flower visitors, seem to have average or better temperatures in spring and summer, with a lot of sunshine in one or other season. The combination of cool to average temperatures, a fair amount of rainfall, and small to average amounts of sunshine seems to make for a poor insect year.

The annual pattern of species diversity (β) of beetles, other than ladybirds, caught in the Malaise trap (Table 10.6), does not follow that of catch size. Diversity was generally high, and highest in 1984 (0.964) but lowest in 1983 (0.878). β-diversity is an estimate that an individual trapped at random will be of a different species from the previous individual trapped. The series of diversity values is markedly heterogeneous (χ^2 = 121.733704, df = 5, p<0.001) and so any two can legitimately be compared. For instance, β-diversity for 1984 was very highly significantly greater than that for 1983 (p<0.001), and also, perhaps surprisingly, highly significantly greater than that for 1985 (p<0.01). It seems, therefore, that the chances of two consecutive captures being of different species differed from year to year, although always remaining good. β-diversity of the total 1981–86 catch did not differ significantly from diversity of the 1984 catch, suggesting that the 1984 catch was the most diverse possible in the garden during the six-year period. On comparing the diversity values for beetles with those of some other groups caught in the Malaise trap, the beetle values tend to be higher. This is to be expected, as the group is so diverse, so many species were caught, and the vast majority of species were captured in small numbers. The highest value for beetles (0.964 in 1984) is higher than the highest annual value for butter-flies, hoverflies, solitary bees and solitary wasps, and the lowest value (0.878 in 1983) is also higher than the lowest annual value for these other groups. Butterfly diversity was also highest in 1984, but diversity of hoverflies, soli-tary bees and wasps was not, although it was quite high in that year. More significantly, the annual values of β-diversity for total beetles caught in pitfall traps (Table 10.3), which ranged from 0.380 in 1980 to 0.580 in 1984, were considerably lower than the Malaise trap values, largely because of the dominance of pitfall catches by two species. It is noteworthy, however, that β-diversity of beetles caught in both pitfalls and Malaise trap reached a maximum in the same good insect year, 1984.

LADYBIRDS

Forty-two species of coccinellid are British, and 24 of these are generally considered as ladybirds (Majerus & Kearns 1989). This section deals with all coccinellids caught in the garden, although two of them, *Rhyzobius litura* and *Stethorus punctillum*, are not considered as ladybirds by Majerus & Kearns. In the 15-year period 1972–86, 5065 ladybirds of eight species were caught in the Malaise trap and one additional species in the mercury-vapour light trap. They were identified by D. F. Owen, who has published some of the results (Owen, D. F. 1982/83). The two other coccinellids were identified by D. Goddard. Taxonomy and classification follow Pope (1977) in the following list, which gives the numbers caught in the Malaise trap.

229. *Rhyzobius litura* (Fab.). 3
230. *Stethorus punctillum* Weise 1
231. Kidney-spot ladybird *Chilocorus renipustulatus* (Scriba). 1
232. 2-spot *Adalia bipunctata* (L.) 2589
233. 10-spot *A. decempunctata* (L.) 105
234. 7-spot *Coccinella septempunctata* L. 917
235. 11-spot *C. undecimpunctata* L. 225
236. 14-spot *Propylea quattuordecimpunctata* (L.) 1123
237. Eyed ladybird *Anatis ocellata* (L.) . . . . 1 in MV trap
238. Cream-spot ladybird *Calvia quattuordecimguttata* (L.) 1
239. 22-spot *Thea vigintiduopunctata* (L.) 104

It is frequently observed that numbers of ladybirds fluctuate markedly from year to year. Annual fluctuations in numbers of the six species regularly trapped are illustrated in Fig. 10.4. This shows that 1976 was a particularly good year for ladybirds when 1419 of the 15-year total of 5065 were trapped, and that 1986 was a particularly poor year when only 41 were trapped, but numbers of the six abundant species fluctuated in different ways.

The 2-spot ladybird *Adalia bipunctata* is a common, resident garden ladybird, whose larvae feed on aphids on woody shrubs and trees. It was the commonest ladybird in Malaise trap catches in every year except 1976, 1979 and 1983, and in 1981 and 1984 was the commonest beetle (Table 10.5). Numbers were generally high during 1972–78, but in 1979 the population collapsed and in succeeding years it never fully recovered. Numbers peaked in 1975, and it is possible that in that year and in 1976 (the drought year) some 2-spots moved into the garden from the surrounding countryside. The summers of 1975 and 1976 were hot, dry and sunny, and in the countryside at large there was a population explosion of aphids, followed by a rise in numbers of predatory ladybirds (and, in 1975, hoverflies). As their aphid food became depleted, either because there were so many predators, or when

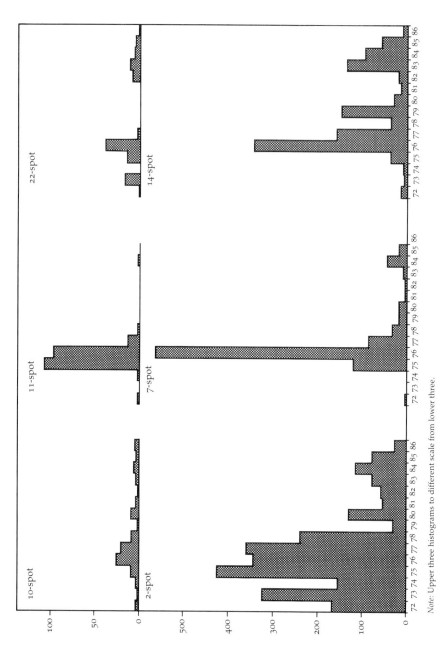

Fig. 10.4 Annual fluctuations in numbers of ladybirds caught in Malaise trap, 1972–86.

Note: Upper three histograms to different scale from lower three.

crops were harvested, or, in 1976, when the drought took hold and withered vegetation, large numbers of ladybirds (and in 1975, hoverflies) took to the wing and moved, often ending up in gardens.

The presence in the garden of 2-spot ladybirds and their capture in the Malaise trap changed from month to month during 1972–86, but the seasonal pattern was not the same in all years (Fig. 10.5). Adults emerge from hibernation in March, and in May and June mate and lay eggs, which produce a new generation of adults from July onwards. Adults captured in April–June were almost all hibernated individuals from the previous year, except in years such as 1976, when some freshly emerged adults appeared in late June. Adults taken in July–October were mainly the new generation, although a few hibernated individuals persisted until early July in some years. In 1973, 1978, 1980, 1984 and 1986 (although the catch in 1986 was only 26 individuals), numbers trapped peaked in May–June and then tailed off during the summer and autumn, i.e. there were more hibernated individuals early in the year than newly produced adults in the summer. However, in 1972, 1975, 1976, 1977, 1981, 1982 and 1983, numbers peaked in July–August, meaning that breeding in the garden was particularly successful (although relatively few were trapped in some years – 53 in 1981, 56 in 1982 and 88 in 1983). The situation was rather different in 1974, 1979 and 1985, when there was a double peak in numbers, with a drop in June (1985) or July (1974 and 1979), implying more or less equal hibernation and production of new individuals, although in both 1979 and 1985 numbers were small, 30 and 77, respectively. Evidently, breeding in the garden was more successful in some years, notably 1972, 1975, 1976 and 1977, than in others.

The 10-spot ladybird *Adalia decempunctata* was never common, although trapped in every year, numbers reaching a maximum of 25 in 1976. Its ecological requirements are similar to those of *A. bipunctata*, and it is suspected to have bred in the garden, although this is not known. Adults were often seen feeding on aggregations of aphids with *A. bipunctata*, and a few have hibernated in dense vegetation in the garden.

The 7-spot ladybird *Coccinella septempunctata*, the common large ladybird, is often reported as migratory, although more probably it periodically irrupts and undertakes mass movements. It was not caught in the Malaise trap in 1973, 1974 or 1986, but in other years numbers trapped varied from one to a spectacular 564 in 1976 (Fig. 10.4). This, the drought year, has been called 'The year of the ladybirds', for large numbers appeared in many areas, especially in eastern counties of England and on the east coast. This species, like most ladybirds, is aphidivorous, and is a countryside ladybird associated with aphids feeding on low-growing vegetation, particularly on field crops. In 1976, and to a lesser extent in 1975, numbers built up on the aphid population explosion in the countryside, and then moved away in vast numbers as the vegetation became parched and brown in the drought.

Four 7-spots were caught in 1972, but none in 1973 or 1974, although one was seen in the garden in March 1974. However, there was an influx to the garden in the first two weeks of August 1975, some individuals successfully

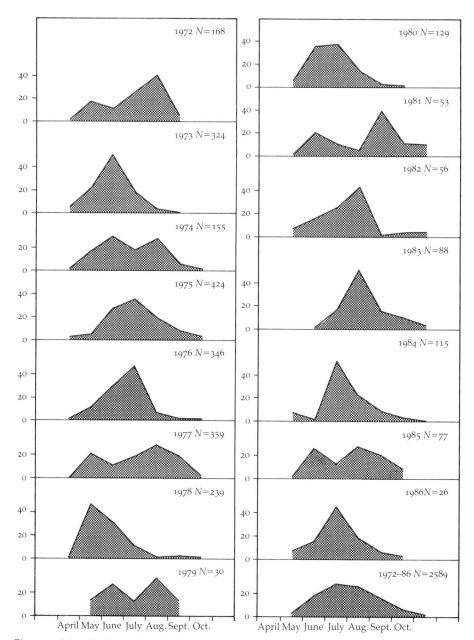

Fig. 10.5 Monthly fluctuations in numbers of 2-spot ladybirds caught in Malaise trap, expressed as percentages of annual total, 1972–86.

overwintering to be trapped in April–June 1976, and then another massive influx in July and August 1976 (Table 10.7). No larvae or pupae were found in 1975, strongly suggesting that the large numbers in August 1975 resulted from movement into the garden, but the individuals that successfully over-

Table 10.7 *Monthly fluctuations in numbers of 7-spot ladybirds* Coccinella septempunctata *caught in Malaise trap, 1975–85*

	April	May	June	July	Aug.	Sept.	Oct.
1975	1				95	20	5
1976	1	18	4	321	173	38	9
1977	1	49	11	4	2	14	5
1978	7	24	1		1		
1979						17	
1980	10	5	1	1	1		
1981		2					
1982						1	
1983		1		4		1	2
1984	7	2	3	2	30	1	1
1985	3	11			1	2	
Total	30	112	20	332	303	94	22

Note: None was caught in 1973, 1974 or 1986, and only 4 in 1972.

wintered, bred in the garden in 1976, when mating individuals, eggs, larvae and pupae were common. They again successfully overwintered, and bred again in 1977, but despite successfully overwintering to 1978, numbers tailed off. The 7-spot was never again as common as in 1976, despite small influxes in September 1979 and in August 1984, and none was caught in 1986.

The 11-spot ladybird *Coccinella undecimpunctata* is a salt marsh and coastal species, and is probably not a characteristic garden ladybird in central England (Owen, D. F. 1982/83), none having been trapped in 1973, 1979–83 or in 1985–86, although 108 were caught in 1975 and 98 in 1976 (Fig. 10.4). There was a movement into the garden in the first two weeks of August 1975, followed by successful overwintering, and another influx in July 1976. Although 11-spots have overwintered in the garden, there is no evidence of breeding, and they seem to be subject to irruptive movements similar to those of *C. septempunctata*, although on a smaller scale.

Although especially associated with deciduous woodland, the 14-spot ladybird *Propylea quattuordecimpunctata*, another aphid-feeder, was trapped every year in the garden in numbers varying from eight to 346 (Fig. 10.4). Fifteen were caught in 1972, eight each in 1973 and 1974, 38 in 1975, then 346 in 1976, when many invaded the garden in July and August (Table 10.8). The 14-spot has never been found breeding in the garden, but frequently overwinters in dense vegetation; the April–June captures were of overwintered individuals. There were evidently further invasions in 1977 and 1979, particularly in the third and first weeks of September, respectively, and again in July and August 1983 and August 1984. In 1983, the 14-spot was the commonest beetle caught in the Malaise trap (Table 10.5). The 1976 invasion may have been a response to drought, but the later ones have no particular association with weather, although 1983 and 1984 were hot, dry, sunny summers.

Table 10.8 *Monthly fluctuations in numbers of 14-spot ladybirds* Propylea quattuordecimpunctata *caught in Malaise trap, 1976–86*

	April	May	June	July	Aug.	Sept.	Oct.
1976		2	25	127	170	16	6
1977		3	14	26	26	85	6
1978		6	20	3	2	6	1
1979		1	1	1	9	137	
1980		8	9	4	7	1	
1981		1	3		8	2	
1982			1		4	15	
1983			2	43	87	5	
1984	1	1	12	8	72	2	
1985		12	8	15	12	10	
1986		4	2	2			
Total	1	38	97	229	397	279	13

Note: Annual totals before 1976 were small: 15 in 1972, 8 in 1973, 8 in 1974, and 38 in 1975.

The 22-spot ladybird *Thea vigintiduopunctata* is reputedly a mildew-feeder, and has been trapped in ten of the 15 years 1972–86, the best year being 1976 with 39 (Fig. 10.4). There is no evidence that it overwinters in the garden, but in August 1985 a pupa was found on a willowherb leaf and produced an adult in September.

Undoubtedly, the garden provides a good habitat for ladybirds. There is a resident population of 2-spots, which becomes large in some years, and periodic large influxes of 7-spots, 11-spots and 14-spots, the 7-spot occasionally breeding. The years 1975 and 1976 were evidently special for ladybirds, particularly for mass movements, and there has been no comparable repetition, despite the good summers of 1983 and 1984. Monitoring of numbers continues, and it will need another two summers comparable to 1975 and 1976 to know whether ladybird numbers have declined or not.

OTHER BEETLES

Although the vast majority of information about the garden's beetles comes from Malaise and pitfall trapping, a number of additional species have been captured at night in the mercury-vapour light trap (MV) or seen and collected (C). These have variously been identified by D. Goddard and K. Porter.

Carabidae

240. *Amara aenea* (Degeer) . . . .   2 MV
241. *A. apricaria* (Paykull). . . . .   1 MV

Hydrophilidae

242. *Helophorus* sp. (*minutus* (Fab.) group). .   1 C from pond

Silphidae

243. *Nicrophorus humator* (Gleditsch) . . . . . .   1 MV
244. *N. investigator* Zetterstedt. . . . .   2 MV
245. *Necrodes littoralis* (L.). . . . .   1 MV 1 C

Staphylinidae

246. *Tachinus subterraneus* (Fab.). . . . .   1 MV

Geotrupidae

247. *Geotrupes stercorarius* (L.) . . . .   2 MV

Scarabaeidae

248. *Aphodius rufus* (Moll) . . . .   9 MV

Cantharidae

249. *Cantharis livida* L. . . . .   1 C

Cleridae

250. *Thanasimus formicarius* (L.) . . . .   1 C

Chrysomelidae

251. *Phyllodecta vulgatissima* (L.). . . . .   1 C

Eight species, including all the Silphidae recorded and the only geotrupid, are thus known for the garden only from light-trapping at night, and four species, including the 'ant beetle', the only clerid, are known only from observation and collection. Evidently, pitfall and Malaise trapping, useful as they are for describing a beetle fauna, do not give the complete picture, and it is necessary to employ other methods, particularly to record species active at night.

BEETLES IN LEICESTERSHIRE

In all, a total of 251 species of beetles from the Leicester garden have been identified, at least to genus, with the possibility of additional species among undetermined Aleocharinae (Staphylinidae) and *Meligethes* (Nitidulidae). A few species, such as the carabid *Pterostichus madidus* and the coccinellid *Adalia*

Photograph 10.4. Dense mixed planting enhances the garden habitat for beetles and other insects.

bipunctata, were very common, but the majority were rare, and represented by only one or two individuals. The 251 species recorded in the garden form less than 7 per cent of the British list, and are, I assume, only a fraction of those really present or at least passing through, for A. A. Allen recorded more than 700 species in his southeast London garden between 1926 and 1973 (Davis 1978). Many beetles are fairly sedentary, others do not fly much, and only some walk about on the soil surface. The Malaise trap, pitfalls and light trap used in my garden can consequently catch only a proportion, perhaps only a small proportion, of those present.

The county of Leicestershire is poor for beetles, its central position making it too far north for southern species and too far south for northern species. Furthermore, destruction of ancient woodland in the last 50 years has meant the increasing rarity or disappearance from the county of many species associated with dead wood. The carabid and staphylinid fauna of the garden has a similarity to the fauna of open ground and of wasteland sites, which is perhaps what might be expected, as the majority are ground-dwellers and a garden is by definition a disturbed site. There is, however, an additional woodland element among the ground-frequenting beetles, reflecting, perhaps, the densely planted aspect of the garden. However, the entire beetle list, including the Malaise trap captures, is much more diverse, and includes many associated with dead wood, with dung or with compost, and also a good number of fungus- and mould-feeders.

Three of the 251 species captured in the garden seem not to have been previously recorded in the county: *Mycetoporus clavicornis* and M. *nigricollis* (Staphylinidae), small predatory rove beetles that live in moss and ground litter, and *Sphaerosoma piliferum* (Endomychidae), a tiny beetle which also lives in moss and ground litter, feeding on Fungi. The only previous county record of *Aderus populneus* (Aderidae) is from eastern Leicestershire in 1942; its life history seems to be a matter of some controversy, although other members of the family develop in decayed wood. The staphylinid *Eusphalerum primulae*, which feeds on pollen and other parts of flowers, has not been recorded in Leicestershire this century. *Stethorus punctillum*, a tiny, black coccinellid that feeds on red spider mites *Tetranychus* spp., was said by Joy (1932) to range in distribution from southern England north to Leicestershire. This implies a Leicestershire record, but the origin of this is unclear, and Leicestershire Museums know of no other record. Evidently, Malaise trapping in the garden is yielding beetle species either very rare in Leicestershire, or new to the county.

Worldwide, Coleoptera are a varied and large group, and it is perhaps to be expected that the list of garden species would be long. A few garden species are abundant, but most are rare, and species diversity, β, is higher than for any other group investigated.

11

Other insects

Other insect groups are less well known in the garden than those allocated separate chapters. Groups such as Odonata (dragonflies) are not typical of gardens and occur only occasionally, while others, such as true bugs (Hemiptera – Heteroptera), have only been sorted from trap samples for a short period. Groups such as Dermaptera (earwigs), are represented by only a single species, and many, such as most families of Diptera (flies), have not been systematically collected. Some of these groups, however, have proved to be abundant and diverse, and collectively they augment the garden list considerably. My knowledge of the insects dealt with in this chapter is patchy and the fact that there is more information, for instance, about Psocoptera, Heteroptera, Neuroptera – Planipennia (lacewings and their allies), and Symphyta (sawflies) is because specialists have been willing and able to identify them. Although there is much more information about some groups than others, 'other insects' are discussed here in taxonomic sequence, rather than separating off the better described groups, and named species are numbered in sequence.

WINGLESS INSECTS

Those insects that are generally assumed to be primitively wingless and to represent ancestral insects before the evolution of wings and the ability to fly, are included in the subclass Apterygota, although there is some question as to whether all are really insects. They are small to very small, many are difficult to identify, and few entomologists specialize in them or even collect them. I have no detailed knowledge of apterygotes in the garden, although Collembola (springtails) are common, many of the larger ones falling into pitfall traps. I occasionally see a shiny, sinuous silverfish (Thysanura) scuttling over the kitchen floor first thing in the morning, to disappear into a crack between floor tiles. The majority of apterygotes, however, live in leaf litter, beneath bark, in soil, or in similarly humid places.

Mayflies

Delicate, weak-flying, long-tailed mayflies, often called 'fishing flies', constitute the insect order Ephemeroptera. The nymphs are fully aquatic, breathing by means of plate-like abdominal gills, and the adult life is short, less than a day in many species. They are usually found close to water, and are not a constituent part of the insect fauna of suburban gardens, but they may get carried considerable distances when the wind catches their relatively large wings, and in most years occasional individuals are caught in the Malaise trap or at night in the mercury-vapour light trap, although none has been identified to species.

Dragonflies and damselflies

The long, slender body, prominent compound eyes, large wings, and rapid, agile flight of dragonflies make them one of the most easily recognized orders of insects. The nymphs are aquatic, and so dragonflies are particularly associated with water, although, being strong fliers, they are not infrequently encountered away from it. The adults are efficient predators of insects (including other dragonflies), which they catch on the wing, and the mouthparts include powerful toothed mandibles, hence the name of the order, Odonata, which means 'toothed'. Dragonflies, and their more fragile relatives, damselflies, fly too swiftly and strongly to be captured in the Malaise trap, but occasional individuals are seen in the garden. The common blue damselfly *Enallagma cyathigerum* was caught with a net in July 1983; large, powerful *Aeshna* sp. have occasionally been seen hawking over the garden and, in July 1974, one was caught and eaten by a starling; in July 1985 a blue *Libellula* dragonfly paused briefly in the garden; and in September in several years, solitary red *Sympetrum* sp. have been seen basking on flowers and vegetation.

1. *Enallagma cyathigerum* (Charpentier)
2. *Aeshna* sp.
3. *Libellula ?depressa* L.
4. *Sympetrum ?striolatum* (Charpentier)

Grasshoppers and crickets

Easily recognized by their robust body and head, long hind legs modified for jumping, and their propensity for chirping, grasshoppers and crickets, forming the order Orthoptera, are familiar to most people. They have biting mouthparts, and grasshoppers are vegetarians, feeding mainly on grass, although crickets are omnivorous, taking some animal food. The house cricket *Acheta domestica* is not native, but is well established, particularly in association with heated buildings and warm rubbish dumps. It was noisily conspicuous in the garden in the warm August of 1975. Grasshoppers are characteristic of meadowland, and cannot be expected in large numbers in a suburban garden where the lawn is kept closely mown. However, males of the common field grasshopper *Chorthippus brunneus* have occurred in many years, particularly in August, and this species has been captured in the Malaise trap. Curiously, a nymph of the meadow grasshopper *C. parallelus* was caught in the Malaise trap in June 1984, and another was seen on the lawn, suggesting that it may have bred in the garden.

5. *Acheta domestica* (L.)
6. *Chorthippus brunneus* (Thunberg)
7. *C. parallelus* (Zetterstedt)

EARWIGS

Despite the fearsome-looking pincers at the hind end of their bodies, earwigs (order Dermaptera) are harmless insects which subsist mainly as scavengers, although they often chew flower petals and occasionally prey on living insects. Their most endearing habit is that of maternal care, the female tending her young until they are well grown, and family groups are a common sight in spring. The only species recorded in the garden is ubiquitous and common in the British Isles.

8. *Forficula auricularia* L.

PSOCIDS

Psocids (order Psocoptera) are small, sometimes minute, insects, frequently wingless, when they are known as booklice, dustlice or barklice, although they are not parasitic. The antennae are long and thread-like, and in most, the large compound eyes are prominent on the sides of the broad head. At rest, the relatively large wings, which in some are marbled with pigment, are held sloping roof-like over the body. Many species are apterous or polymorphic, with winged and wingless forms. The biting mouthparts are unusual in that part of each maxilla forms a rod-like 'pick', believed to be used for scraping food from the substrate, and that upper and lower parts of the tongue-like hypopharynx, between the maxillae, are sclerotized and apparently grind together like a mortar and pestle (Imms 1964).

Psocids are found in most terrestrial communities, on the foliage of trees, on and beneath bark, in leaf litter, occasionally in old birds' nests or ants' nests, and also in buildings. Most outdoor species graze indiscriminately on organic debris, algae such as *Desmococcus* (= *Pleurococcus*), lichens, yeasts and farinaceous materials, and not infrequently eat their own and other insects' eggs and psocid corpses and moulted skins, although other species are more selective (New 1971). The indoor species feed on stored dried products, organic debris, moulds on damp books, paste on wallpaper or book-bindings, and on dried natural history specimens, such as pinned insects. The indoor species tend to be wingless, but those outdoors usually have wings at least at some stage in their life histories, and in many species flight is important, as, for instance, in species such as *Ectopsocus briggsi* that make spring flights from ground litter to the foliage of trees, where they spend the summer.

Most overwinter as eggs, although nymphs and adults of some species hibernate. The arboreal species, which are generally restricted either to bark or to foliage, have one, two, or occasionally three generations a year, but the domestic species breed continuously. Bark-frequenting Psocidae lay eggs in batches of up to 20 and cover them with a crust of chewed bark fragments and faecal pellets, whereas many foliage-dwellers cover their eggs with silk produced from glands in the labium (lower lip), and some lay eggs singly. Most overwintered eggs hatch in April and May, and after six nymphal instars, adults are produced in June. Parthenogenesis is widespread in the order, and in some species, males are unknown (New 1971).

About 90 species of Psocoptera are known in the British Isles, of which only about 50 live in natural surroundings out of doors (New 1974). The peak of flight activity appears to be in October, with a smaller peak in July, at any rate at Uppingham, some 24 kilometres from my garden (Clark 1979), and although I have sorted them as annual samples from the Malaise trap, most individuals were taken in the autumn. Obviously, the Malaise trap catches only winged psocids, and of those, there are many that fly little. Clark (1979) found that four species common on vegetation near to his Malaise trap were absent from trap samples. Nevertheless, 403 individuals of 14 species were caught in my garden Malaise trap in the period 1983–86, the commonest being *Ectopsocus briggsi* and *E. petersi*, which together comprised nearly 70 per cent of the four-year total. *E. petersi* was first described as a new species in 1978, from Ireland, but is very common in Leicestershire.

Fluctuations in numbers of individual species can be spectacular from year to year, as shown by the garden captures of *E. petersi*, the annual totals for 1983–86 being 17, 128, 2 and 3. The total annual catches also fluctuated, from 93 in 1983 to 245 in 1984, to 40 in 1985 and only 25 in 1986. Psocids thus followed the pattern of most other insect groups in the garden, as monitored by the Malaise trap, in having a good year in 1984, and a poor year in 1986. Six of the species captured in the Malaise trap in the garden were among the 11 species trapped at Uppingham School, 1972–78, where *Ectopsocus petersi* and *Graphopsocus cruciatus* were the most abundant (Clark 1979). Psocids from the garden were identified by J. Sellick, and total Malaise trap catches, 1983–86, are given in the following list. Taxonomy follows New (1974), except that *E. petersi* was first described as a new species after that publication.

Caecilidae

9. *Caecilius flavidus* (Stephens)   10
10. *C. burmeisteri* Brauer.   12

Stenopsocidae

11. *Graphopsocus cruciatus* (L.)   70
12. *Stenopsocus stigmaticus* (Imhoff & Labram)   2
13. *S. immaculatus* (Stephens).   1

BUGS

The order Hemiptera, bugs, is a large assemblage of insects with piercing, sucking mouthparts. Mandibles and maxillae are stylet-like and are held firmly together in a groove on the dorsal (or anterior) side of the segmented labium or rostrum. Within the rostrum, the mandibles lie peripheral to the maxillae which are W-shaped in cross-section, so that when held tightly together, they form two tubes, down one of which saliva is passed into a feeding puncture, and up the other of which liquid food is sucked. The mouthparts and feeding method are common to all bugs, but the order comprises two rather different groupings of insects, Heteroptera and Homoptera, which are sometimes given ordinal rank.

Heteroptera, the true bugs, are characterized by the forewings, the proximal halves of which are toughened and often pigmented, while the apical half is membranous. When the wings are closed, the forewings are held flat to the body and overlapping, covering the membranous hindwings. The head is normally held in line with the body, so that it is clearly visible from above. The beak-line rostrum arises from the front of the head, although, when not in use, it is held bent back to lie horizontally beneath the head and body, sometimes extending to the base of the hind legs. There are aquatic as well as terrestrial Heteroptera, and many are predatory, a considerable proportion of those that are primarily plant-feeders, taking animal food occasionally.

Homoptera, on the other hand, including cicadas, plant-hoppers of various sorts, aphids and scale insects, all have uniformly textured forewings, although they may be toughened or membranous; the hindwings are membranous. Moreover, at rest, the wings are closed roof-like over the body. The head is flexed downwards, and the rostrum originates from it posteriorly, so that in many the beak seems to arise from between the front legs. All Homoptera are plant-feeders, and none is aquatic. Many, particularly aphids, are injurious to plants not only because of feeding damage, but also because they transmit plant viruses, and hence they are economically important.

All Heteroptera were sorted from Malaise trap catches in 1984–86 as annual samples. In all, 354 individuals of at least 47 species (some females could only be identified to genus) were caught, 161 in 1984, 99 in 1985 and 94 in 1986. Thus, as for other insect groups, 1984 was evidently a good year, and it was also the year when most species were caught (33 as opposed to 29 in 1985 and 25 in 1986). The commonest species each year and overall was *Anthocoris nemorum*, which formed 47.6 per cent of the total catch. This is the common flower bug, one of the commonest Heteroptera in the British Isles. Adults overwinter in such places as under bark and in litter, and move in March or April to a variety of plants, where they feed on small animals, such as aphids, and occasionally on leaves. Eggs, which are inserted into leaf tissue, hatch in late May and early June, and the nymphs develop into adults in late June and early July. A second generation of adults is produced by early September and, in some years, a third generation occurs (Southwood & Leston 1959).

All but two of the Heteroptera recorded in the garden, 1984–86, are common and widespread in the British Isles. *Pilophorous perplexus*, of which one was caught in 1985 and one in 1986, is confined to the south of England and Wales, although it is common, at least in the southeast of its range. It is chiefly predatory, occurring on a variety of deciduous trees. *Elasmostethus tristriatus* was until recently confined to juniper *Juniperus communis*, feeding on the ripe berries, and therefore almost entirely restricted to southern English counties. Now, however, it is also found on exotic conifers, including *Chamaecyparis* spp., and is spreading in range on the new hosts. Three were captured in the Malaise trap in 1984, and it is noteworthy that *Chamaecyparis lawsoniana* grows in this and nearby gardens. Twelve species of the garden Heteroptera were not included by Clements & Evans (1973) in a list of Leicestershire bugs, but 11 of these have been recorded for the county by Leicestershire Museums or in other publications, and just one is a new record for the county: *Dicyphus errans* is partly predatory, occurring on a wide range of low-growing plants.

The shieldbugs (Acanthosomatidae) caught in the garden are herbivorous and *Acanthosoma haemorrhoidale* was abundant on the birch tree in 1981; the damsel bug (Nabidae) is predatory, the lace bug (Tingidae) feeds on creeping thistle *Cirsium arvense*, the three groundbugs (Lygaeidae) are respectively herbivorous, a moss-feeder, and a scavenger and fungivore, the flower bugs (Anthocoridae) are predatory, and the capsids (Miridae) are mostly herbivorous, although about a fifth of them are at least partly predatory.

The Malaise trap Heteroptera were identified by P. Kirby, who provided

much additional information, and occasional pitfall captures were identified
by D. Goddard. Taxonomy and sequence follow Kloet & Hincks (1964). Numbers refer to total Malaise trap captures, 1984–86, and P indicates that the
species was caught in pitfalls.

Acanthosomatidae

23. *Acanthosoma haemorrhoidale* (L.)
24. *Elasmostethus interstinctus* (L.)   3
25. *E. tristriatus* (Fab.)   3

Lygaeidae

26. *Kleidocerys resedae* (Panzer)   1
27. *Drymus sylvaticus* (Fab.) . . . .   5 P
28. *Scolopostethus affinis* (Schilling)   P

Tingidae

29. *Tingis ampliata* (Herrich-Schäffer)   2

Nabidae

30. *Nabis ferus* (L.)   1

Anthocoridae

31. *Anthocoris confusus* Reuter   2
32. *A. nemoralis* (Fab.)   18
33. *A. nemorum* (L.)   168
34. *Orius vicinus* (Ribaut)   1
35. *Orius* sp.   10
36. *Lyctocoris campestris* (Fab.)   1

Miridae

37. *Deraeocoris lutescens* (Schilling)   5
38. *D. ruber* (L.)   1
39. *Megalocoleus molliculus* (Fallén)   2
40. *Harpocera thoracica* (Fallén)   2
41. *Psallus ambiguus* (Fallén)   4
42. *P. lepidus* (Fieber)   1
43. *P. varians* (Herrich-Schäffer)   6
44. *Psallus* sp.   2
45. *Plagiognathus arbustorum* (Fab.)   10
46. *P. chrysanthemi* (Wolff)   3
47. *Asciodema obsoletum* (Fieber)   1
48. *Dicyphus epilobii* Reuter.   1
49. *D. errans* (Wolff).   2
50. *Pilophorus perplexus* Douglas & Scott . . .   2

51. *Dryophilocoris flavoquadrimaculatus* (Degeer)	1
52. *Heterocordylus tibialis* (Hahn)	1
53. *Heterotoma merioptera* (Scopoli)	6
54. *Blepharidopterus angulatus* (Fallén)	3
55. *Orthotylus marginalis* Reuter	1
56. *O. ochrotrichus* Fieber	4
57. *O. virescens* (Douglas & Scott)	5
58. *Orthotylus* sp.	1
59. *Lygus rugulipennis* Poppius	1
60. *Liocoris tripustulatus* (Fab.)	13
61. *Orthops cervinus* (Herrich-Schäffer)	16
62. *O. campestris* (L.)	6
63. *Orthops* sp.	9
64. *Lygocoris pabulinus* (L.)	7
65. *L. lucorum* (Meyer-Dür)	1
66. *L. spinolai* (Meyer-Dür)	7
67. *Plesiocoris rugicollis* (Fallén)	2
68. *Polymerus nigritus* (Fallén)	1
69. *Calocoris norvegicus* (Gmelin)	1
70. *Phytocoris longipennis* Flor	1
71. *P. tiliae* (Fab.)	2
72. *Stenodema calcaratum* (Fallén)	2
73. *S. laevigatum* (L.)	4
74. *Leptoterna dolobrata* (L.)	2

The Malaise trap captures vast numbers of the smaller Homoptera, such as aphids and cicadellids, but these have not been sorted systematically from the weekly catch. A number of spittlebugs (Cercopidae), and leaf-hoppers (Cicadellidae) were, however, sent to P. Kirby with the Heteropteran sample, and he identified these also. The 25 species of cercopid and cicadellid caught in the Malaise trap in 1984–86 form the basis of the garden list of Homoptera. A number of other cicadellids have been identified from Malaise trap catches by L. Jones-Walters and from pitfalls by D. Goddard. Only a few species of garden aphids have been identified: those familiar to most entomologists, and those whose feeding activities aroused interest or curiosity. It is not, therefore, possible to indicate any numbers for Homoptera, beyond saying that some species were very common. Furthermore, it is fully appreciated that the species list, a total of 42, is very incomplete.

Three species of spittlebugs (Cercopidae) have been recorded in the garden, of which two feed as nymphs and adults on a wide range of plants, and one of which, *Neophilaenus lineatus*, feeds on grasses. Their feeding method is unusual in that they insert their mouthparts into the xylem vessels of plants and extract the watery xylem sap. Since this contains only very low concentrations of amino acids, from which proteins are synthesized, feeding nymphs ingest vast quantities, and continually excrete the excess water. Air is pumped into the excreted fluid by extending the posterior segments of the abdomen out of the fluid so that air enters an 'air canal' beneath the tergites,

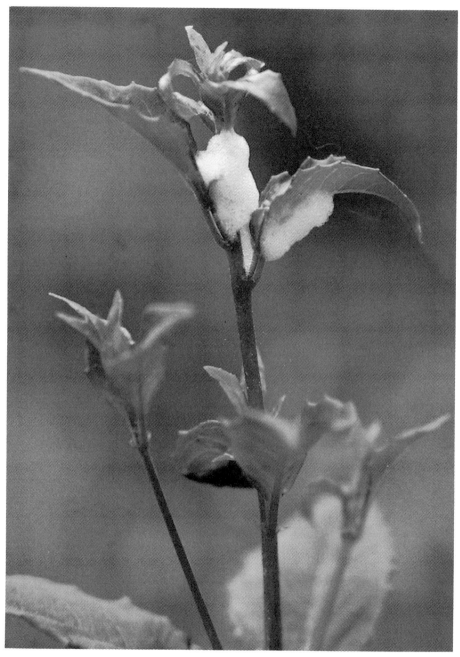

Photograph 11.1. Nymphs of the meadow spittlebug Philaenus spumarius *surrounded by cuckoo-spit were found on 142 species of plants in the Leicester garden.*

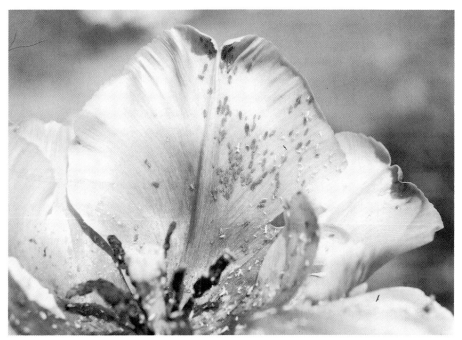

Photograph 11.2. Aphids tap sugary phloem sap and are found on all parts of plants, including petals.

and then contracting the body within the fluid so that a bubble of air is forced out (Weaver & King 1954). The resulting froth is the familiar cuckoo-spit that surrounds a spittlebug nymph.

The meadow spittlebug *Philaenus spumarius* is the commonest cercopid in the garden and one of the most conspicuous Homoptera. Weaver & King (1954) listed nearly 400 species of plants on which nymphs of *P. spumarius* have been recorded, and in my garden they have been found on 142 species of 46 plant families. The extreme polyphagy of spittlebugs is doubtless a consequence of feeding on xylem sap, which is unlikely to contain secondary plant substances and toxins, and indeed is little other than water. In contrast, most other Homoptera extract from plants sugary phloem sap, which originates in the photosynthetic tissues and is, therefore, likely to contain high concentrations of secondary plant substances, which give a plant its characteristic odour and taste. It is because they feed on a rich mixture of chemicals, that many of the phloem-feeders are restricted in which plants they can use as food. Aphids, for instance, are mostly monophagous, particularly in temperate regions, where there is a greater number of aphid species (Dixon 1985). Amino acids occur at only low concentrations in phloem sap, so phloem-feeders tend to ingest large amounts of the sugary sap and continually defaecate the excess sugar, familiar as the honeydew produced by aphids.

All the cicadellids, or leaf-hoppers, recorded from Malaise trap catches are common and widespread, particularly in southern parts of the British Isles.

Many are less than 5 mm in length, but they are often brightly coloured, and many are conspicuously marked. In many years, the whitefly, *Aleyrodes pro-letella* was abundant on brassicas, particularly curly kale. The aphids recorded in the garden belong to several different families. Five species were abundant in all or most years: *Betulaphis quadrituberculata* on birch, *Aphis sambuci* on elder, *Aphis fabae* on beans and *Philadelphus coronarius*, *Brevicoryne brassicae* on brassicas, and *Pemphigus spirothecae* on poplar. In 1976, *Macrosiphum euphorbiae* was recorded from *Buddleia davidii* and *Anoecia corni* from *Cornus sanguineum*, and in 1978 and other years, *Myzus cerasi* was abundant on suckers of *Prunus serratula*. In 1980, the green 'pineapple' galls of *Adelges viridis* were abundant on Norway spruce *Picea abies*, together with the wax-producing *A. cooleyi*, and the tree's growth was much affected. An unidentified unpigmented aphid was abundant in 1980 on roots of *Chrysanthemum frutescens* which were contained within a nest of the ant *Lasius niger*.

In the following list of Homoptera recorded from the garden, MT indicates identification from Malaise trap samples, and P identification from pitfalls. The taxonomy and sequence of spittlebugs and leaf-hoppers (Auchenorrhyncha) follows Le Quesne & Payne (1981) and that of aphids (Sternorrhyncha) follows Kloet & Hincks (1964).

Cercopidae

75. *Aphrophora alni* (Fallén) MT
76. *Philaenus spumarius* (L.) MT
77. *Neophilaenus lineatus* (L.) MT

Cicadellidae

78. *Cicadella viridis* (L.) MT
79. *Rhytidodus decimusquartus* (Schrank) MT
80. *Idiocerus nitidissimus* (Herrich-Schäffer) MT
81. *I. populi* (L.) MT
82. *I. vitreus* (Fab.) MT
83. *Oncopsis flavicollis* (L.) MT
84. *O. tristis* (Zetterstedt) MT
85. *Oncopsis* sp. MT
86. *Macropsis graminea* (Fab.) MT
87. *M. scotti* Edwards MT
88. *M. scutellata* (Boheman) MT
89. *Aphrodes bicinctus* (Schrank) P
90. *A. makarovi* Zakhvatkin MT
91. *Psammmotettrix* sp. MT
92. *Allygus mixtus* (Fab.) MT
93. *A. modestus* Scott MT
94. *Mocydia crocea* (Herrich-Schäffer) MT
95. *Empoasca vitis* (Gothe) MT
96. *Kybos betulicola* (Wagner) MT
97. *K. virgator* (Ribaut) MT

98. *Eupteryx aurata* (L.) MT
99. *E. florida* Ribaut MT
100. *E. urticae* (Fab.) MT
101. *Aguriahana stellulata* (Burmeister) MT
102. *Edwardsiana crataegi* (Douglas) MT
103. *Zygina flammigera* (Geoffroy in Fourcroy) MT
104. *Zygina* sp. MT
105. *Javesella* sp. MT

Aleyrodidae

106. *Aleyrodes proletella* (L.)

Callaphididae

107. *Betulaphis quadrituberculata* (Kaltenbach)

Aphididae

108. *Aphis sambuci* L.
109. *Aphis fabae* Scopoli
110. *Brevicoryne brassicae* (L.)
111. *Myzus cerasi* (Fab.)
112. *Macrosiphum euphorbiae* (Thomas, C. A.)

Thelaxidae

113. *Anoecia corni* (Fab.)

Pemphigidae

114. *Pemphigus spirothecae* Passerini

Adelgidae

115. *Adelges viridis* (Ratzeburg)
116. *A. cooleyi* (Gillette)

LACEWINGS AND THEIR ALLIES

The suborder Planipennia of the order Neuroptera includes the lacewings proper and a family (Coniopterygidae) of tiny insects, resembling psocids, which are covered with white, waxy powder, so that they look superficially like whiteflies. Some authors consider Planipennia as a separate order, Neuroptera, distinct from alderflies and snakeflies (Megaloptera). Lacewings are delicate, medium-sized insects characterized by their large, flimsy wings, densely netted with veins, which are held roof-like over the body at rest. Green lacewings (Chrysopidae) are often attracted to lighted windows at night, and some find their way into houses, particularly in winter before hibernating, when many assume a reddish tinge. Brown lacewings belong to the family Hemerobiidae.

Lacewings have rather prominent eyes on the sides of the head, and long, slender antennae. The mouthparts are of the biting type, with strong mandibles, and although some species are rarely seen to feed, others are predatory on aphids and other small animals. They quite often drink, and are attracted by sweet fluids, such as exposed nectar on flowers or the sugar solutions used as bait by moth collectors. They fly at dusk and during the night. Less is known about the tiny psociform Neuroptera (Coniopterygidae), but they too appear to be predatory. The wax covering the wings of conio-pterygids is produced from epidermal glands, mainly on the abdomen, and is spread over the wings using the hind legs (Killington 1936).

Eggs are laid singly or, more rarely, in twos or threes, and cemented to leaves, twigs or bark by Hemerobiidae and Coniopterygidae, while in spongeflies (Sisyridae), they are laid in batches covered with silk on vegetation overhanging water. Egg-laying in Chrysopidae is more curious, in that just before an egg is laid on a leaf, a drop of clear, viscous fluid is extruded from the female's abdomen, which is gently raised from the substrate so that the quick-drying fluid hardens into a stalk at the end of which an egg is attached (Killington 1936). These may be laid in groups, and, in some species, the stalks of several eggs coalesce. The eggs of Hemerobiidae, Coniopterygidae and Chrysopidae hatch into carnivorous larvae that prey on aphids, mites and other small animals, those of Sisyridae into larvae that drop into the water to feed on fresh-water sponges.

Larvae have quite conspicuous, thin, sharp, tubular 'jaws', that are used to suck fluids from their prey. Each of the pair of 'jaws' is formed from grooved maxilla and mandible, which are held firmly together so that each 'jaw' forms a tube. Fluid food is sucked up straight into the pharynx, often while the struggling prey is held aloft, impaled on the 'jaws'. They are extremely slender in Sisyridae, strongly inwardly curved in Chrysopidae and Hemerobiidae, and shorter in the tiny Coniopterygidae. The larvae of some species of Chrysopidae decorate their bodies with debris, including the drained skins of their prey. This covering of debris, which is attached to hooked setae on prominent tubercules on the thorax, serves as camouflage. The digestive tract of larval lacewings is blocked in the anterior region of the hind-gut, any waste products accumulating in the large mid-gut, to be voided after metamorphosis to the adult state. In third, and final, instar larvae, a number of the excretory Malphigian tubes that arise from the hind-gut become modified to secrete silk, which is stored in a silk reservoir, and later spun through the anus to form the cocoon in which pupation takes place. Cocoons and pupae are formed in crevices in bark, or other sheltered positions, including those of Sisyridae, whose larvae leave the water and travel considerable distances before selecting pupation sites.

There are 55 species of Planipennia in the British Isles, 18 of which have been caught in the Malaise trap, including five psociform Neuroptera (Coniopterygidae), one Sisyridae, and 12 lacewings proper – six Hemerobiidae and

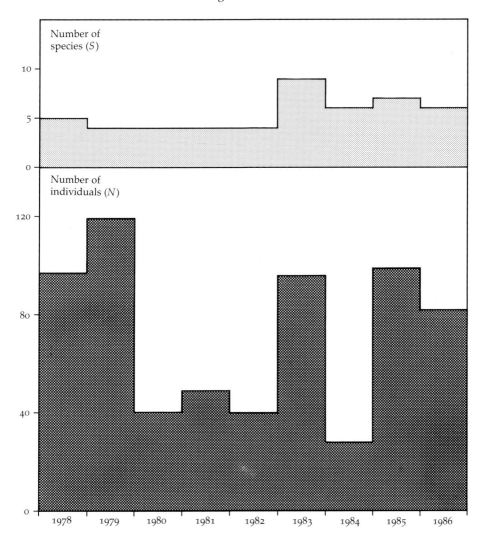

Fig. 11.1 Annual fluctuations in numbers of individuals and species of lacewings caught in Malaise trap, 1978–86.

six Chrysopidae. Lacewings were sorted from Malaise trap catches during 1978–86. Annual catches fluctuated considerably from year to year, the best year being 1979 and the worst, 1984 (Fig. 11.1). Lacewings, thus, did not follow the annual pattern of other garden insects caught in the Malaise trap, almost certainly because annual fluctuations in total numbers were largely a consequence of fluctuations in numbers of one species, *Wesmaelius subnebulosus*, whose population crashed in 1984. This is one of the commonest and most widespread hemerobiids, which is often found in gardens. It is often associated with birch, of which there is a large tree in my garden, and in

1981 many of those captured were teneral, suggesting that they might have been bred in the garden. The number of species of lacewing trapped was remarkably constant at four or five until 1983, when nine species were caught and the number of individuals also increased markedly over the previous year (Fig. 11.1). Two species of chrysopid were captured every year and in considerable numbers overall: *Chrysopa carnea* is the most common and widespread chrysopid, *C. albolineata* is generally more local in occurrence, but both are often found in gardens. The single individual of *Sisyra fuscata* caught in 1983 is the first adult spongefly to be recorded in the county of Leicestershire; larvae are common, but there are three species that cannot at present be separated.

Psociform Neuroptera (Coniopterygidae) were sorted from Malaise trap samples during 1983–86. Numbers were small, but the five species include *Coniopteryx borealis*, until recently only twice recorded in the British Isles, although recorded ten times since 1976, once in Leicestershire. These more recent occurrences may represent a spread from continental Europe, as although Coniopterygidae are easily overlooked, there are no mis-identified specimens of *C. borealis* in the British Museum (Natural History) collections (Barnard 1985). Coniopterygidae were identified by J. Sellick, other Planipennia by D. Goddard. In the following list, taxonomy and sequence follow Barnard (1978), and numbers refer to total Malaise trap captures (1983–86 for Coniopterygidae, and 1978–86 for the others).

Coniopterygidae

117. *Conwentzia psociformis* (Curtis) 5
118. *Coniopteryx tineiformis* Curtis. 8
119. *C. borealis* Tjeder. 2
120. *C. parthenia* Navás & Marcet 1
121. *Parasemiadalis fuscipennis* (Reuter) 2

Sisyridae

122. *Sisyra fuscata* (Fab.) 1

Hemerobiidae

123. *Micromus variegatus* (Fab.). 8
124. *Hemerobius humulinus* L. 2
125. *H. stigma* Stephens 1
126. *H. micans* Olivier 1
127. *H. lutescens* Fab. 13
128. *Wesmaelius subnebulosus* (Stephens). . . . 461

Chrysopidae

129. *Chrysopa flava* (Scopoli) 4
130. *C. ciliata* Wesmael. 3

131. *C. flavifrons* Brauer   7
132. *C. albolineata* Killington.   43
133. *C. carnea* Stephens   102
134. *C. perla* (L.).   4

CADDISFLIES

Caddisflies (Trichoptera) are closely associated with water and so not really garden insects, but larger species that fly more strongly are sometimes caught in light traps away from water, and the four species recorded in the garden have all been caught in the mercury-vapour light trap. The smaller ones may be mistaken for small moths, but caddisfly wings are covered with small hairs and they have biting mouthparts, although these are poorly developed and rarely used. To most people, the aquatic larvae of caddisflies are more familiar, especially to children who go pond-dipping, because the majority build protective, tubular cases of sand grains, plant material or other fragments on a foundation of sticky silk. Head and legs protrude from the case, which is open at both ends, and body movements maintain a through-current of water over the feathery gills along the sides of the abdomen. Most larvae are omnivorous, but some tend to be carnivorous, and a few species that live in running water make no cases, but trap food in silken nets spun among vegetation. Those caught in the garden were identified by K. Porter. Nomenclature follows Kloet & Hincks (1964).

135. *Phrygaena grandis* L.
136. *P. varia* Fab.
137. *Limnephilus rhombicus* (L.)
138. *Stenophylax* sp.

FLIES

Many true, two-winged flies (Diptera), other than hoverflies, have been recorded from the Leicester garden. The majority have been captured in the Malaise trap, but others have been seen and collected, or attracted by special baits. By no means all flies captured in the Malaise trap have been identified, however, inclusion in the garden list being dependent on the readiness of a specialist to identify them. Nomenclature and sequence of Diptera in this chapter follow Kloet & Hincks (1975).

Crane flies or daddy-long-legs (Tipulidae), with their fragile, gangling legs, are present in the Malaise trap during summer and autumn, and in some years are conspicuous in the garden in late summer. A female caught laying eggs in the lawn in September 1980 was identified by A. E. Stubbs as *Tipula paludosa*, which as a larva is the common garden leatherjacket. Each female *T. paludosa* lays several hundred eggs during the summer, inserting her ovipositor well into the ground. These hatch into grubs which, as in other crane flies, grow into tough-skinned, greyish leatherjackets, which feed with their biting mouthparts on the roots and lower stems of an enormous range of plants, and are the bane of farmers, gardeners and groundsmen. They feed throughout winter and spring, and pupate in early summer, still in the soil, the adult emerging about a fortnight later, after the pupa has worked its way to the soil surface using its downwardly directed spines (Colyer & Hammond 1968).

139. *Tipula paludosa* Meigen

Owl midges or moth flies (Psychodidae) are tiny flies whose body, legs and wings are clothed in long hairs, and which hold their wings roof-like over the body at rest. They may abound in damp, humid locations, and sometimes occur in sinks in houses or around drains, where they are recognizable by the jerky way they run and circle about. The larvae develop in dung, mud, rotting vegetation or any wet, decaying matter. The blood-sucking sand flies, which transmit several diseases, are psychodids, although absent from Britain.

From 1983 to 1986, Psychodidae were sorted from Malaise trap catches and sent to P. Withers, who identified 11 species, mostly in the genus *Psychoda*. The larvae of *Psychoda* spp. occur in a variety of decaying organic material, and *P. alternata, cinerea* and *severini* have colonized the bacteria beds of sewage purification works. A number of species occur on farmland, the larvae developing in cow dung, the most important numerically being *P. phalaenoides* (Satchell 1947). There are a number of suitable larval habitats for them in gardens, *P. albipennis*, for instance, breeding in kitchen waste and *P. minuta* in rotting lawn clippings, and as the adults fly weakly and do not disperse far, it can be assumed that those caught in the Malaise trap are garden residents.

140. *Pericoma pilularia* Tonnoir
141. *P. trivialis* Eaton
142. *Psychoda albipennis* Zetterstedt
143. *P. alternata* Say
144. *P. cinerea* Banks
145. *P. grisescens* Tonnoir
146. *P. minuta* Banks
147. *P. parthenogenetica* Tonnoir
148. *P. phalaenoides* (L.)
149. *P. setigera* Tonnoir
150. *P. surcoufi* Tonnoir

Soldier flies (Stratiomyidae and Xylomyiidae) are small to fairly large, somewhat flattened flies, often with a colourful, metallic lustre. They are rather sluggish in behaviour, and are often found resting on vegetation. They vary considerably in shape, *Microchrysa* spp., for instance, being short and broad, and *Sargus* spp. elongate. Adults are fluid-feeders and occasionally take exposed nectar from flowers, whereas larvae occur in a variety of sites, feeding on decomposing organic material. Some larvae are aquatic, but all the species recorded in the garden have terrestrial larvae that feed in and on dung, rotting wood or organic debris of one sort or another. Soldier flies were sorted from Malaise trap catches in the years 1980–86 and identified by J. Mousley.

Total annual catches for the seven-year period were 56, 66, 95, 264+, 99, 40 and 97, respectively. In other words, 1983 was far and away the best year and 1985 the worst. 1983 was a hot, dry, sunny summer, when many insect groups were abundant, although most, unlike soldier flies, fared even better in 1984; the summer of 1985 was cool and damp, with only average hours of sunshine, and insect numbers in general were down. The most abundant soldier fly overall was *Microchrysa polita*, a small shining-green fly, which is generally common. Numbers in the following list refer to Malaise trap captures 1980–86.

Stratiomyidae

151. *Beris chalybeata* (Forster)   47
152. *B. clavipes* (L.)   36
153. *B. morrisii* Dale   1
154. *B. vallata* (Forster)   10
155. *Pachygaster leachii* Curtis   10
156. *Chloromyia formosa* (Scopoli)   7
157. *Microchrysa cyaneiventris* (Zetterstedt) . .   10
158. *M. flavicornis* Meigen   7
159. *M. polita* (L.) . . . .   550+
160. *Sargus iridatus* (Scopoli)   4
161. *S. splendens* Meigen   28

Xylomyiidae

162. *Solva marginata* (Meigen)   7

Robber flies (Asilidae) are medium-sized to large, narrow-bodied and rather bristly predators of other insects. They have horny and more or less rigid, piercing mouthparts with which they suck the fluids from their prey. Only one individual was captured in the Malaise trap during the period 1980–86, and that was *Dioctria rufipes*, identified by J. Mousley. Many asilids lie in wait for suitable prey to approach, but *D. rufipes* is one of the species that take their prey on the wing, with the front legs extended like a scoop. *Dioctria* species feed particularly on Hymenoptera, especially ichneumonids. Asilid larvae feed on and in decomposing vegetable material such as rotting wood or cow dung, but are found in drier situations than many fly larvae (Colyer & Hammond 1968).

163. *Dioctria rufipes* (Degeer)

Conopidae are medium-sized to fairly large flies, many of which (Conopinae) bear a striking resemblance to solitary wasps and other Hymenoptera. They have large heads with conspicuous eyes, the antennae are long, and the brightly yellow- and brown- or black-banded abdomen is constricted at the base, especially in males. Species of Myopinae look more like typical flies, although the head and eyes are large. Conopid larvae are solitary internal parasites of aculeate Hymenoptera and occasionally of Orthoptera. Eggs are laid directly into the host's abdomen in flight, and develop into larvae that feed on the host's body fluids, while tapping the host's tracheal system for respiration. Pupation takes place within the host's abdomen, which ruptures when the adult conopid emerges. The diurnal adults are often found at flowers, where they feed on nectar (Colyer & Hammond 1968, Smith 1969). Three species of Conopinae have been caught in the Malaise trap and identified by K. G. V. Smith. *Conops quadrifasciata* has been caught on more than one occasion, and has been collected at flowers of goldenrod *Solidago canadensis*. *C. flavipes* is parasitic on *Bombus lapidarius*, *Osmia* spp. and *Vespula rufa*, *C. quadrifasciata* on *Bombus lapidarius*, and *P. rufipes* on several species of *Bombus* and on *V. rufa* (Smith 1969).

164. *Conops flavipes* L.
165. *C. quadrifasciata* Degeer
166. *Physocephala rufipes* (Fab.)

Seventy larvae of nest flies *Neottiophilum praeustum* (Neottiophilidae) were found in July 1976 in a blackbird's nest from which four young had successfully fledged. The fly larvae suck the blood of nestling birds, and are known from the open nests of a variety of passerines.

167. *Neottiophilum praeustum* (Meigen)

Leaf-miners (Agromyzidae) are very small flies whose tiny larvae bore in plant stems or mine the tissue between the upper and lower epidermises of leaves, making characteristic patterns. Moreover, most species are restricted to one or a few closely related species of plants, and so can be identified from their mines. Female *Phytomyza ilicis* lay eggs in June at the base of the midrib beneath holly leaves. The young larva eats its way forwards along the midrib until moving out into the leaf tissue in autumn, where it forms a characteristic blister. The larvae pupate within the mine in March and April, and new adults emerge in May and June. Holly trees in my garden and most in the neighbourhood are conspicuously affected. Holly trees tend to shed mined leaves in the period January–May, whereas normal leaf-fall is in June and July (Owen, D. F. 1978b). Characteristic meandering leaf-mines every year on leaves of honeysuckle *Lonicera periclymenum* are assumed to be those of *Paraphytomyza hendeliana*.

168. *Phytomyza ilicis* Curtis
169. ?*Paraphytomyza hendeliana* (Hering)

Two species of parasite flies (Tachinidae) have been bred from parasitized Lepidoptera larvae found in the garden. The larvae of all tachinids are internal parasites of arthropods, mainly insects, either pupating in the host's skin, or boring out to pupate in the ground. The host lives at least until the tachinid pupates, but eventually dies, and so the tachinids are best considered as

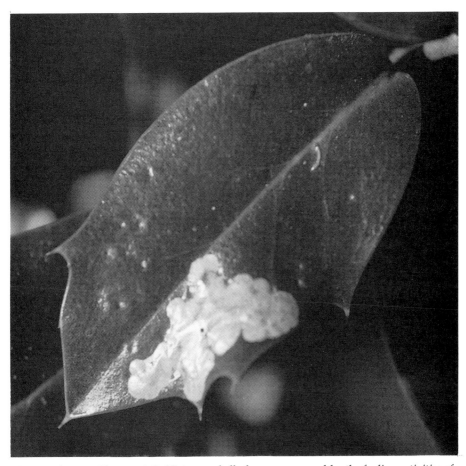

Photograph 11.3. Characteristic blisters on holly leaves are caused by the feeding activities of larvae of the holly leaf-miner Phytomyza ilicis.

parasitoids, similar in host relationships to Ichneumonidae. *Voria ruralis* has been bred on several occasions from caterpillars of the burnished brass *Diachrysia chrysitis*, feeding on *Origanum majorana*, and *Epicampocera succincta* from a caterpillar of the small cabbage white butterfly *Pieris rapae* feeding on *Hesperis matronalis*. The adult flies are black, grey or brown, and rather bristly. They were identified by R. Belshaw.

170. *Voria ruralis* (Fallén)
171. *Epicampocera succincta* (Meigen)

In a campaign well publicized on radio, dipterists at the British Museum (Natural History), anxious to record as fully as possible the status and distribution of blow-flies (Calliphoridae) in Britain, organized a number of National Fish Skin Weeks in 1977, 1978 and 1979. Anyone interested was asked to put out rotting carcasses or fish skins, collect any flies attracted to the bait, and send them to the British Museum for identification. A trap in the garden (Fig. 3.7, p. 58) was duly baited with mackerel heads, skins and offal in July 1978 and blow-flies collected and sent to the British Museum together with those sorted from 1977 and 1978 Malaise trap catches. Flesh flies (Sarcophagidae) were inadvertently included in the samples, but these as well as the Calliphoridae were identified by J.P. Dear, who also provided additional information about the various species. The samples were so large, that the offer of further material in 1979 was politely declined, as it was felt that this locality had been sufficiently covered. The National Fish Skin Weeks were not simply academic exercises, for the distribution of calliphorids is important for veterinary and public health work, and often for forensic medicine. Most blow-flies frequent decaying animal matter and excrement, and females lay eggs on meat and carcasses or in open wounds and sores on live animals; the larvae of other species are external parasites of nestling birds, or parasites of other invertebrates, such as earthworms. The larvae of *Sarcophaga* spp. feed on decaying animal material, and those of other species of Sarcophagidae are found in bee and wasp nests, feeding either on stored food or on the host larvae.

The flesh flies (Sarcophagidae) recorded in the garden, are all common, except *Amobia signatus*, whose larvae are found in nests of solitary and social bees and wasps. The larvae of *Brachycoma devia* occur in nests of *Bombus* spp. and *Vespula* spp., those of *Sarcophaga haemorrhoidalis* and *S. subvicina* in decaying organic matter, and those of *S. incisilobata* in excrement. Of the blow-flies (Calliphoridae), most are carrion-feeders, although *Lucilia ampullacea*, a scarce species generally found in woods, is an occasional internal parasite of toads, and *Phormia terraenovae* feeds and breeds in decaying organic matter. Most are generally common, although the garden record of *Calliphora subalpina*, common in Scotland, is the most southerly record in England, *C. vomitaria* is not found in numbers in cities, *Lucilia caesar* is a common woodland species, and *L. richardsi* is a species of open grassland. In the following list, MT signifies capture in the Malaise trap, and F, capture in trap baited with fish heads, skins and offal.

Sarcophagidae

172. *Amobia signatus* (Meigen) MT
173. *Brachycoma devia* (Fallén) MT
174. *Sarcophaga carnaria* (L.) MT
175. *S. haemorrhoidalis* (Fallén) MT, F
176. *S. incisilobata* Pandellé MT, F
177. *S. subvicina* Rohdendorf MT, F

Calliphoridae

178. *Calliphora subalpina* (Ringdahl) MT
179. *C. vicina* Robineau-Dsesvoidy MT, F
180. *C. vomitoria* (L.) MT, F
181. *Lucilia ampullacea* Villeneuve MT

182. *L. caesar* (L.) MT, F
183. *L. illustris* (Meigen) MT, F
184. *L. richardsi* Collin MT
185. *L. sericata* (Meigen) MT, F
186. *Phormia terraenovae* Robineau-Desvoidy F

The family Muscidae includes many of the flies, such as *Musca domestica*, which commonly enter houses, green-bottle type of flies, flies that cluster around the heads and bodies of cattle and horses, and also the biting flies, *Stomoxys calcitrans* and *Haematobia irritans*. *Mesembrina meridiana*, a large black fly with conspicuously yellow wing-bases, attracted attention in the garden, because it is not only handsome, but also common. Adults feed on nectar from umbellifers and a variety of other accessible flowers, and bask on tree trunks or the ground. The female lays a single large egg, which hatches almost immediately, in cattle or horse dung (Colyer & Hammond 1968).

187. *Mesembrina meridiana* (L.)

SAWFLIES

Sawflies (suborder Symphyta) are distinguished from other Hymenoptera by the lack of a 'waist' at the base of the relatively broad abdomen. Almost all also have two roughened protuberances, called cenchri, on the dorsal metathorax, which engage with rough areas of the forewings when the wings are closed. In most, the ovipositor is toothed like a saw, the pattern of the teeth being characteristic for a particular species, and giving rise to the common name of the group. The saw-like ovipositor is used to cut slits in stems and leaves, inside which the eggs are laid. Some species, however, such as the wood wasps or horntails, have stout, sharp ovipositors, used to bore into wood, where the eggs are laid. The majority of adult sawflies are small to medium-sized insects with biting mouthparts and a relatively long tongue-like labium, used for lapping nectar. They feed at flowers, usually on pollen and nectar, although some species are at least partly carnivorous.

Most sawfly larvae feed on leaves and resemble caterpillars in appearance, although they have more pairs of abdominal pro-legs (at least six). Wood-eating sawfly larvae, those that tunnel in leaves and stems, and the few gall-making species are more like beetle larvae, with no abdominal legs. The larvae of *Orussus abietinus* are unusual in being parasites of timber beetles. The largest British family, with about 390 species, is the Tenthredinidae, most of whose leaf-eating larvae are ecologically very like lepidopteran caterpillars, and some species, such as the gooseberry sawfly, *Nematus ribesii* are commercially important defoliators. The kidney-bean-like galls of *Pontania* sp. are abundant on willow leaves in the garden in some years; they are induced by the egg-laying puncture, and the larva feeds inside its gall. Most species of sawflies complete their development in a single season and overwinter as a pre-pupa, although a few species overwinter as eggs, and wood wasps may remain as larvae for three or four years (Betts 1986, Chinery 1976).

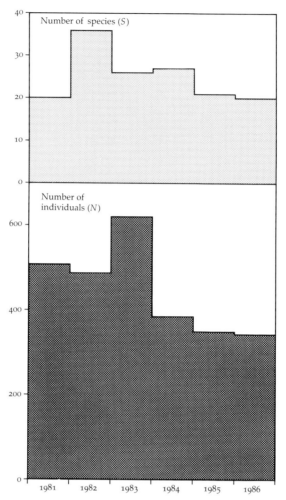

Fig. 11.2 Annual fluctuations in numbers of individuals and species of sawflies caught in Malaise trap, 1981–86.

Sawflies were sorted from Malaise trap catches in the period 1981–86, and were identified by L. R. Cole. A total of 2694 individuals of 51 species were caught, mostly Tenthredinidae, together with a few individuals of two species of Argidae. Over the six-year period, 13 species were represented by only a single individual. Annual catches declined fairly steadily from 508 in 1981 to 345 in 1986, except for a marked rise to 620 in 1983. Numbers of species caught fluctuated with a peak of 36 in 1982 and the smallest number (20) in 1981 and 1986 (Fig. 11.2 and Table 11.1). The average number of individuals per species (N/S) was highest in 1981 and lowest in 1982, when the highest annual total of species was caught; it was also high in 1983, when the largest annual total of individuals was caught. *Protoemphytus carpini* was the commonest species overall and in four of the six years, the proportion it formed of the total catch

Table 11.1 *Size and composition of annual catches of sawflies in Malaise trap, 1981–86*

	Number of individuals (N)	Number of species (S)	N/S	Commonest as % N	Commonest species
1981	508	20	25.4	75.4	*Protoemphytus carpini*
1982	488	36	13.6	25.6	*Cladius pectinicornis*
1983	620	26	23.8	29.8	*C. pectinicornis*
1984	384	27	14.2	58.9	*P. carpini*
1985	349	21	16.6	81.7	*P. carpini*
1986	345	20	17.3	67.6	*P. carpini*
1981–86	2694	51	52.8	49.0	*P. carpini*

varying from 58.9 to 81.7 per cent. *Cladius pectinicornis* was the commonest species in 1982 and 1983, when numbers of species and of individuals, respectively, were at their highest (Table 11.1). The larvae of *P. carpini* feed on a variety of species of *Geranium*, and every year in the garden they reduce the leaves of *G. cinereum* to a network of veins. The larvae of *C. pectinicornis* feed on both wild and cultivated *Rosa* spp. and on strawberries *Fragaria* spp., but have never been evident in the garden. Sawflies may be compared with moths and butterflies because of the ecological similarity of their larvae. Sawflies were like butterflies in being trapped in greater numbers in 1983 than in 1984, whereas moths were trapped in greater numbers in 1984. The summers of both 1983 and 1984 were hot, dry and sunny; neither spring was particularly warm, that of 1983 being wet and dull, and that of 1984, dry with average amounts of sunshine.

Apart from *Protoemphytus carpini*, nine species of sawflies have been found breeding in the garden, six of which were not caught in the Malaise trap. Larvae of three species that were caught in the Malaise trap have been found in the garden: *Aneugmenus padi* on the fern *Dryopteris filix-mas* in 1983, *Nematus olfaciens* on black currant *Ribes nigrum* in 1982, and *Nematus ribesii* on gooseberry *Ribes uva-crispa* in most years. Additional to the Malaise trap list are *Trichiosoma tibiale* on hawthorn *Crataegus monogyna* in 1977 and 1986, *Tenthredo colon* on great willowherb *Epilobium hirsutum* in 1982, *Amauronematus* sp. on crack willow *Salix fragilis* in 1982, *Pontania* sp. on *S. fragilis* in most years, *Croesus septentrionalis* on silver birch *Betula pendula* in most years, and *Nematus spiraeae* on goat's beard *Aruncus dioicus* in 1983.

Diversity of annual catches of sawflies has been calculated as β: the probability that two consecutive random captures will be different. The value of β was high in 1982 and 1983, when numbers of species and of individuals caught, respectively, were high, and particularly low in 1985 (Table 11.2). The series of diversity values, 1981–86, is very markedly heterogeneous ($\chi^2 = 553.16$, df = 5, p<0.001), so the values for any two years can legitimately be compared. The high values of β-diversity for 1982 and 1983 are not significantly different, but the value for 1983 (0.851) is very highly significantly greater (p<0.001) than that for 1984 (0.639). The values for 1981 and 1986,

Table 11.2 *β-diversity of annual catches of sawflies in Malaise trap, 1981–86*

	β-diversity	Standard error	No. of individuals (*N*)	No. of species (*S*)
1981	0.424	±0.027496	508	20
1982	0.856	±0.009780	488	36
1983	0.851	±0.008539	620	26
1984	0.639	±0.027646	384	27
1985	0.332	±0.033269	349	21
1986	0.536	±0.032680	345	20

Note: β-diversity is derived from Simpson's index, λ; $\beta = 1 - \lambda$.

Photograph 11.4. The larvae of the birch sawfly Croesus septentrionalis *rest by day along the edges of leaves and, when disturbed, raise their hind ends from the leaf.*

poor insect years in general, although not in 1981 for sawflies, are highly significantly different (p<0.01), diversity being greater in 1986. This suggests that the garden sawfly community differed considerably in composition from year to year in the period 1981–86.

In the following list of garden species of sawflies, nomenclature and sequence follow Fitton *et al.* 1978. Numbers show total Malaise trap captures,

1981–86, and there are a few additional species collected, or reared from larvae found in the garden or recognized by their galls (*Pontania* sp.).

Argidae

188.	*Sterictiphora geminata* (Gmelin in L.) . . .	2
189.	*Arge cyanocrocea* (Forster)	4

Cimbicidae

190. *Trichiosoma tibiale* Stephens

Tenthredinidae

191.	*Strongylogaster macula* (Klug).	2
192.	*Aneugmenus padi* (L.).	19
193.	*Athalia ?cardata* Lepeletier.	34 females
194.	*A. ?circularis* (Klug).	3 females
195.	*A. ?cornubiae* Benson	52 females
196.	*A. ?liberta* (Klug).	44 females
197.	*A. ?lugens* (Klug)	2 females
	Athalia sp.	134 males
198.	*Empria tridens* (Konow).	2
199.	*Ametastegia glabrata* (Fallén)	7
200.	*Protoemphytus carpini* (Hartig)	1321
201.	*P. tener* (Fallén).	1
202.	*Allantus cinctus* (L.).	61
203.	*A. cingulatus* (Scopoli).	1
204.	*Endelomyia aethiops* (Fab.)	10
205.	*Phytomatocera aterrima* (Klug)	
206.	*Ardis brunniventris* (Hartig)	1
207.	*A. sulcata* (Cameron).	1
208.	*Halidamia affinis* (Fallén)	28
209.	*Metallus gei* Brischke.	5
210.	*Aglaostigma aucupariae* (Klug)	4
211.	*A. fulvipes* (Scopoli).	3
212.	*Tenthredopsis coquebertii* (Klug)	18
213.	*T. litterata* (Geoffroy in Fourcroy).	1
214.	*T. nassata* (L.)	1
215.	*Tenthredo colon* Klug	
216.	*Pachyprotasis rapae* (L.)	2
217.	*Cladius difformis* (Panzer)	96
218.	*C. pectinicornis* (Geoffroy in Fourcroy) .	402
219.	*Priophorus morio* (Lepeletier)	9
220.	*P. pallipes* (Lepeletier)	142
221.	*P. pilicornis* (Curtis).	30
222.	*Trichiocampus viminalis* (Fallén)	1
223.	*Hoplocampa testudinea* (Klug)	1
224.	*Stauronematus compressicornis* (Fab.) . . .	14
225.	*Pristiphora ?alnivora* (Hartig)	1
226.	*P. crassicornis* (Hartig).	24

227. *P. pallipes* (Lepeletier). 91
228. *P. punctifrons* (Thomson) 15
229. *P. saxsenii* (Hartig) 2
230. *Pristiphora* sp. 1
231. *Amauronematus* sp.
232. *Pontania* sp.
233. *Croesus septentrionalis* (L.)
234. *Nematus lucidus* (Panzer). 19
235. *N. hypoxanthus* Foerster 2
236. *N. leucotrochus* Hartig 1
237. *N. myosotidis* (Fab.). 14
238. *N. olfaciens* Benson 2
239. *N. ribesii* (Scopoli). 50
240. *N. spiraeae* Zaddach
241. *N. viridis* Stephens 1
242. *Nematus* sp. 8
243. *Nematus* sp. 2
244. *Nematus* sp. 1
245. *Pachynematus clitellatus* (Lepeletier). . . . 2

ANTS

There is just one large family (Formicidae) of ants, placed in a separate superfamily, Formicoidea, of the Hymenoptera. All are social insects living in colonies, which may be very large. Colonies consist of an egg-laying queen and many female workers, which are usually infertile, although some may lay eggs. Winged males and reproductive females are produced at certain seasons, and leave colonies in great swarms under particular atmospheric conditions. Flights leave nests of the same species in a particular area in synchrony, increasing the chances of out-breeding; mating occurs in the air, and when they land, males soon die and females break off their wings and initiate new nests. Once the first batch of workers has been produced, females devote themselves to egg-laying, and workers take over the tasks of tending the brood, feeding the larvae, and extending and defending the nest.

There are just two species of ants in my garden, although the black ant *Lasius niger* is very abundant with many nests. On sultry afternoons in July, swarms of reproductives rise into the air like plumes of smoke from all the gardens in the neighbourhood.

246. *Lasius niger* (L.)
247. *L. flavus* (Fab.)

In total, 247 species of 'other insects' were recorded in the garden, bugs (Hemiptera) and sawflies (Symphyta) being the most diverse. It is important, however, to emphasize that identification of different groups was very patchy and uneven, perhaps best illustrated by the fact that the house fly *Musca domestica* is not on the garden list, although I am sure that it occurred.

12

Other invertebrates

A s has been indicated, the dominant group in my garden, and indeed in all gardens, are the insects. They are the most conspicuous group of invertebrates, although many other invertebrate groups occur, and some of those in which individual size is small or microscopic, such as mites, are undoubtedly very abundant. Several types of larger invertebrates, particularly spiders, have been collected and identified, providing species lists. Furthermore, myriapods (Chilopoda and Diplopoda) and harvestmen (Arachnida – Opiliones) that have fallen into pitfall traps have been identified, providing comparable annual samples for the seven years 1980–86. Most non-insect invertebrates, however, such as spiders and molluscs, are not amenable to trapping, and so there is no data for these groups comparable with that obtained by Malaise trapping of insects. As with the insects, regular identification and tabulation of samples of other invertebrates, depends upon a specialist being willing and able to deal with them. A visit to the garden by A.J. Rundle, as part of a survey of Leicester's terrestrial invertebrates that he undertook for the Leicestershire Museums, Arts and Records Service, resulted in many additions to the garden species list. Over the years, an extensive species list of invertebrates, other than insects, has been compiled. Much of the systematics and nomenclature used in this chapter is from an unpublished systematic list of selected invertebrate groups prepared by A.J. Rundle in 1986, a copy of which is held by Leicestershire Museums.

FLATWORMS

The flattened, simple worms of the phylum Platyhelminthes, which lack coelom and circulatory system and have only one opening to the gut, are known as flatworms. Only the class Turbellaria are free-living, the other classes being parasitic, and the only suborder represented in the garden is the Tricladida, relatively large planarians, with a three-branched intestine. They are carnivorous, small creatures being caught by the protrusible pharynx, but will also eat freshly dead animals. Prey too large to be ingested whole is partially digested externally by enzymes poured onto it from the pharynx, and small particles are then sucked in. Flatworms are covered with cilia which, together with an investing film of mucus, enable them to glide over surfaces, including, in the case of aquatic forms, the underside of the surface film of water. *Polycelis nigra* is super-abundant in the pond, and *Dendrocoelum lacteum* is common. Several individuals of the terrestrial flatworm *Microplana terrestris* were found under damp stones. Flatworms were identified by U. Bowen and A. J. Rundle.

Planariidae

1. *Polycelis nigra* (Müller)

Dendrocoelidae

2. *Dendrocoelum lacteum* (Müller)

Rhynchodemidae

3. *Microplana terrestris* (Müller)

SNAILS AND SLUGS

Snails and slugs are molluscs of the class Gastropoda; most, including all those found in the garden, belong to the subclass Pulmonata, in which the mantle cavity (beneath the shell if there is one) has become a vascular lung for air-breathing. Snails have a hard, spirally coiled shell into which they can withdraw at times of adversity; in most slugs, the shell is reduced to a thin, internal plate or a few calcareous granules, but others have a small shell perched on the hind end of the body. The outer skin is soft and wet, making slugs and snails prone to desiccation, and consequently their activity is normally confined to night-time or to wet days. The skin is coated with mucus, which also lubricates the sole of the muscular creeping foot, so that they leave a characteristic slime trail. Slugs, in particular, can burrow into soil, moving down more than a metre in dry weather.

Pulmonate slugs and snails are hermaphrodite but cross-fertilization, after a more or less elaborate courtship and mating ritual, is the general rule. Clutches of eggs, often pearl-like or translucent, but sometimes in a hard, calcareous shell, are laid in the soil or under logs or stones. The small newly hatched young are vulnerable, and mortality is high, but the larger species may be long-lived, up to ten years or longer for some individuals.

The aspect of snails and, in particular, slugs that most frequently draws them to the attention of a gardener is their feeding habits. Most feed on rotting vegetation, on Fungi, algae and lichens, or on underground storage organs, such as radishes and potatoes, and healthy green plants are not attacked. Many species also eat carrion, and some are carnivorous. A few species that normally eat decaying matter are, however, pests, preferentially eating cultivated plants, which tend to be softer and more nutritious than

Photograph 12.1. Lush garden vegetation shelters snails, spiders and other invertebrates.

wild ones. Slugs are a particular problem in my garden, probably because of the large amounts of decaying vegetation on the soil surface. The feeding equipment of slugs and snails is unusual in that they have a single horny jaw in the roof of the mouth, and a strip of continuously growing horny tissue, the radula, bearing numerous rows of tough, sharp teeth, in the floor of the mouth. Jaw and radula are scraped and rasped over food, tearing off particles which become entangled in mucus and move back into the oesophagus.

In 1972, 724 *Cepaea nemoralis* were introduced to the garden from a locality in Sweden where there are no song thrushes, the usual predator in England. The grove snail *C. nemoralis* occurs in many colour forms, the background colour of the shell varying from yellow through brown to pink. There is also great variation in the number and density of dark bands on the shell. The relative frequency of different colour forms varies in different habitats and the object of the exercise was to see what would happen in a garden under selection pressure from thrushes. In the first year, garden song thrushes ate large numbers of the introductions, although the rate of predation gradually fell to nothing as snail numbers declined (Owen, D.F. 1980c). In 1978 there were still a few *C. nemoralis* in the garden, but none has been seen since. As a planned introduction, *C. nemoralis* is not included in the species list.

The great pond snail *Lymnaea stagnalis* and the ramshorn *Planorbis planorbis* have been common in the pond since its re-filling in 1984, when they were almost certainly re-introduced with aquatic vegetation. Seven species of slugs occur in the garden, and all are common: the four *Arion* species are common and widespread in the British Isles; *Tandonia budapestensis* and the two *Deroceras* species are common in association with man and his cultivation, and *T. budapestensis* and *D. reticulatum* are generally regarded as pests. Five species of snails are common in the garden: tiny *Cochlicopa lubrica* with its tall shell, tiny *Vallonia costata* with a flat, spiral shell, small *Oxychilus draparnaudi* and *Trichia striolata* with flat spiral shells, and the large garden snail *Helix aspersa* with its intricately patterned shell. Tiny *Vitrea contracta* and small *Aegopinella nitidula* and *Oxychilus alliarius*, all with flat spiral shells, have occasionally been collected. All eight species of snails are common and widespread in many parts of the British Isles, and *Oxychilus draparnaudi*, *Trichia striolata* and *Helix aspersa* are particularly associated with gardens. Interestingly, some of the snail species, such as *Cochlicopa lubrica* and *Aegopinella nitidula*, usually occur in moderately damp habitats, whereas others, such as *Vallonia costata* and *Vitrea contracta*, favour drier, more open places (Kerney & Cameron 1979). Garden snails and slugs, 17 species in all, have been variously identified by A.J. Rundle and D. Whiteley.

Lymnaeidae

  4. *Lymnaea stagnalis* (L.)

Planorbidae

  5. *Planorbis planorbis* (L.)

Cochlicopidae

  6. *Cochlicopa lubrica* (Müller)

Vallonidae

  7. *Vallonia costata* (Müller)

Arionidae

  8. *Arion ater* (L.) agg.
  9. *A. hortensis* Ferussac
10. *A. distinctus* Mabille
11. *A. fasciatus* (Nilsson)

Zonitidae

12. *Vitrea contracta* (Westerlund)
13. *Aegopinella nitidula* (Draparnaud)
14. *Oxychilus draparnaudi* (Beck)
15. *O. alliarius* (Miller)

Milacidae

16. *Tandonia budapestensis* (Hazay)

Agriolimacidae

17. *Deroceras reticulatum* (Müller)
18. *D. caruanae* (Pollonera)

Helicidae

19. *Trichia striolata* (Pfeiffer)
20. *Helix aspersa* (Müller)

EARTHWORMS

Worms of the family Lumbricidae (Annelida – Oligochaeta) are the familiar earthworms, so abundant in garden soil. The annulated appearance arises because the body consists of many, serially repeated segments, in each of which bristle-like external chaetae, muscles, excretory organs and nerve branches are the same. Earthworms require damp surroundings, and quickly desiccate if exposed to dry air. The unprotected skin must remain moist for respiration, water is necessary for excretion, and movement is achieved by a hydrostatic skeleton based on the pressure of fluids in the cavities of the segments, which are separated by septa. Earthworms are hermaphrodite, the reproductive organs being restricted to a few segments, but cross-fertilization is ensured by an elaborate mechanism whereby there is mutual exchange of sperm between two worms bound together by mucus. In most species, a prominent saddle or clitellum produces an envelope of mucus which receives eggs and stored sperm (from another individual) as the worm wriggles out of it, and which forms a cocoon for the fertilized eggs.

It has long been recognized that earthworms enhance the organic content of soil and improve its texture and quality. They do this because of the way they move and eat. Although they may burrow through loose soil by alternately contracting and extending, then, as the body is anchored in position by the chaetae, expanding and shortening, in compact soil many species are also eating excavators. Soil is sucked in at the mouth, the particles cleaned by digestion of all organic matter, and then defaecated. Some species defaecate in their burrows, but others back up to the soil surface, where they deposit characteristic worm casts. Charles Darwin calculated that this amounts annually to a 5 mm layer of fine soil deposited on the surface. Some species, such as the familiar, large *Lumbricus terrestris*, which may burrow to depths of several metres to escape unfavourable surface conditions, drag leaves and other vegetable material into the entrances of their burrows, and eat it when it is decomposed. Earthworms promote the overall soil-forming process by influencing soil pH by the alkalinity of their faeces, by breaking down the organic content of soil and enhancing microbial activity, by promoting humus formation, by improving soil texture as organo-mineral complexes or 'crumbs' are formed in the gut, by mixing soil and improving aeration and drainage, and by enriching the soil through conversion of minerals to forms that are available to plants. The presence of earthworms in garden soil is undoubtedly beneficial to the gardener. Five species have been found in the garden, the brandling *Eisenia fetida* particularly in the compost heap; most were identified by A. J. Rundle.

Lumbricidae

21. *Allolobophora chlorotica* (Savigny)
22. *Aporrectodea longa* (Ude)
23. *Dendrilus rubidus* (Savigny)
24. *Eisenia fetida* (Savigny)
25. *Lumbricus terrestris* L.

LEECHES

One species of leech was found in the pond and identified by A. J. Rundle. Leeches (Hirudinea), like earthworms, are annelids, but have a sucker at each end. Most have developed as blood-suckers, using suckers to attach, usually temporarily, to their prey, and may be regarded either as specialized predators or as ectoparasites. *Glossiphonia* spp. feed on aquatic snails, whose body fluids and soft parts they suck out, using the eversible proboscis.

Glossiphoniidae

26. *Glossiphonia heteroclita* (L.)

WOODLICE

The crustacean order Isopoda, members of which are dorso-ventrally flattened, with seven pairs of similar walking legs on the thorax, includes the terrestrial woodlice. On the head, they have a pair of long antennae, with a second, rudimentary, pair between them, and usually, at the sides, a pair of eyes, which in most species are compound. The mouthparts, which resemble biting and chewing insect mouthparts, include a pair of stout mandibles, and two pairs of maxillae and a pair of maxillipeds which manipulate the food. Behind the thorax or pereion, with its seven pairs of walking legs, is the abdomen or pleon, which, on all but the last segment, bears biramous appendages modified as gills. On the last segment are a pair of flattened uropods. The first two pairs of pleopods in males are modified for sperm transfer, fertilization being internal, as is typical for land animals. Females retain eggs and newly hatched young in a brood pouch that develops beneath the thorax from overlapping outgrowths of cuticle.

Woodlice, being Crustacea with gills for respiration, are particularly susceptible to desiccation. Only in the species most highly adapted to terrestrial life is there a waterproofing layer in the cuticle, but this is not as efficient as the waxy covering of insects. There are consequently continual risks of water loss, which is avoided by behavioural adaptations. Woodlice move along

Photograph 12.2. Buddleia davidii, marguerites Chrysanthemum frutescens and other plants contribute to the richness and variety of a garden.

humidity, light and temperature gradients into damp, dark, cool places, and they also respond positively to contact, including contact with other wood-lice. They tend consequently to aggregate in cool, dark, confined places, which on balance will also be humid, and are usually only active at night. The permeable body covering facilitates water absorption by osmosis as well as water loss, so they are also subject to waterlogging, and move away from a very wet place. They, to some extent, make good water loss from their food, by drinking, by absorbing water through the body surface, particularly the pleopods, and by anal drinking, achieved by capillary movement of water along the closely adpressed uropods to the anus. Excess water is lost by transpiration through the cuticle, with the faeces, and through the uropods by a reverse of the capillary action used in anal drinking (Sutton 1972).

Woodlice vary in the extent to which they have become independent of water, showing increasing specialization to life on land in the series Trichoniscidae, Oniscidae, Porcellionidae and Armadillidiidae. In those most highly adapted to terrestrial existence, there is a mass of minute air-channels or pseudotracheae in the outer ramus (exopodite) of the pleopods, connecting to the outside through a small pore. The Armadillidiidae includes the common pill-bug *Armadillidium vulgare*, which is often seen in the sun, and is restricted to calcareous soils where it can take in enough calcium to build its massive exoskeleton.

Woodlice generally feed on dead plant material, but in some circumstances

315

living plants may be eaten, particularly by *A. vulgare*. They may also feed extensively on animal remains and dung, and die if they are prevented from eating their own faeces. The reason for coprophagy is the need to maintain high levels of copper, which is a constituent of the respiratory pigment haemocyanin. Copper compounds are present in most foods, but copper is not available for absorption until released by bacterial activity in the faeces.

Woodlice abound in my garden, almost certainly because large amounts of rotting plant material are allowed to accumulate on the soil surface. Lifting any stone reveals a seething mass of little bodies struggling to escape the light. *Oniscus asellus, Porcellio scaber, Trichoniscus pusillus* and *T. pygmaeus* are common, *Philoscia muscorum* and *Armadillidium vulgare* are occasionally found, and A.J. Rundle found one *Androniscus dentiger* beneath a loose paving slab by the pond, and several individuals of the white, blind *Platyarthrus hoffmann-seggi* in a nest of the ant *Lasius niger*. Garden woodlice were variously identified by A.J. Rundle and S.L. Sutton. Taxonomy and sequence follows Harding & Sutton (1985).

Trichoniscidae

27. *Androniscus dentiger* Verhoeff
28. *Trichoniscus pusillus* Brandt
29. *T. pygmaeus* Sars

Oniscidae

30. *Oniscus asellus* L.

Philosciidae

31. *Philoscia muscorum* (Scopoli)

Platyarthridae

32. *Platyarthrus hoffmannseggi* Brandt

Armadillidiidae

33. *Armadillidium vulgare* (Latreille)

Porcellionidae

34. *Porcellio scaber* Latreille

CENTIPEDES

Centipedes form the class Chilopoda, one of four classes of terrestrial arthropods with a distinct head bearing jaws, and possessing more than eight pairs of legs, which are generally known as myriapods. Centipedes are distinguished from the other three groups by being flattened in form, and being predatory, with a pair of poison claws on the first body segment; in contrast to millipedes, there is but one pair of legs on each body segment. Three orders of centipedes, differing in form and structure, occur in Britain and are represented in the garden.

Geophilomorph centipedes are long, thin, blind, burrowing animals, with up to 101 pairs of short legs. The large number of intersegmental joints is augmented by a subsidiary joint in each segment giving great flexibility, such that movement of three segments can produce a body flexion of 180 degrees. The numerous joints also enable both shortening of the body, as the segmental sclerites override each other at the joints, and also a thickening as the loosely articulated sides of the body expand; the converse produces lengthening and narrowing. Consequently, geophilomorphs can burrow as do earthworms, by alternately extending and contracting the body, the numerous legs acting as anchors in the soil. Lithobiomorph centipedes, on the other hand, are shorter and more compact, with only 15 pairs of powerful legs; these are the familiar, active, brown centipedes. They have no subsidiary segmental joints, and, moreover, lateral flexion is limited because long and short tergites alternate along the body, so that inter-tergal joints do not lie immediately above inter-sternal joints. Lithobiomorphs have one or more ocelli on each side of the head. Scolopendromorph centipedes are in many ways intermediate to the other two orders, being blind, with subsidiary joints in the sternites, although not in the tergites, which show some tendency to alternate in length along the body. They are small, active animals, with 21 pairs of legs, and although their legs are not so long, they can run almost as fast as Lithobiomorpha (Eason 1964).

All types of centipedes have long antennae, which are assumed to have a tactile function. The mouthparts consist of a pair of mandibles and two pairs of maxillae, which move food towards the mandibles. Immediately behind the head, but curving forwards in front of the mouth, are the poison claws with which prey is grasped and paralysed or killed; these are the terminal part of the much modified appendages of the segment immediately behind the head, and move and grasp in a horizontal plane. A poison gland opens just behind the tip of each claw, and in some large, tropical centipedes, the claws can inflict a venomous and painful bite to humans. Centipedes breathe by means of a ramifying tracheal system, they have a dorsal tubular heart, and the reproductive openings of both males and females are at the hind end of the body.

Reproductive processes are poorly known, but it is believed that there is copulation and internal fertilization in some species, whereas in others, males deposit spermatophores on specially constructed webs from which females pick them up with the appendages of the genital segments. Female Geophilomorpha lay 30–40 eggs in an excavated cavity in the soil and remain coiled around eggs and young for some time. If the female is removed, the eggs become infected by fungus, which she normally seems to keep at bay with fungicidal secretions from her mouth. Newly hatched geophilomorph young have the adult number of legs, although these are not fully developed, but they lack such structures as tracheae. Lithobiomorph females, however, lay a number of eggs separately, coating each with mucus and soil particles. The newly hatched young are active and independent, able to feed and breathe, but have fewer than the adult number of segments and legs, which are progressively added at each moult.

Centipedes require moist surroundings, having no waxy, waterproof covering. They also avoid light, and tend to creep into crevices and crannies where both dorsal and ventral surfaces are in contact with something firm. Consequently, they are found under stones, beneath the bark of decayed logs, and in similar situations, and are most active at night, when they do their hunting.

Thirty-two species of centipede are native to the British Isles. Five species fell into pitfall traps in the garden, and have been identified as annual samples by D. T. Richardson. The annual catches in the seven years 1980–86 were 73, 90, 94, 111, 130, 82 and 86, consisting largely of *Lithobius* spp., particularly *L. forficatus*. 1984 was thus the year of maximum capture (130); this was a year when summer was hot, dry and bright, and many insect groups also did well in the garden.

In the following species list, numbers indicate total pitfall captures, 1980–86. *Schendyla nemorensis* is not one of the commonest British species, but is found in a range of microhabitats, including under bark and leaf litter, usually in woodland, but also in gardens. Of the other species, several *Haplophilus subterraneus* were found by A. J. Rundle under stones, and he also found one *Cryptops hortensis* under a paving slab. *H. subterraneus* is the only British centipede which sometimes turns vegetarian and damages root crops.

ORDER: GEOPHILOMORPHA

Himantariidae

35. *Haplophilus subterraneus* (Shaw)

Schendylidae

36. *Schendyla nemorensis* (Koch)   8

Geophilidae

37. *Geophilus insculptus* Attems.   2
38. *Necrophloeophagus longicornis* (Leach) . . .   2

ORDER: SCOLOPENDROMORPHA

Cryptopsidae

39. *Cryptops hortensis* Leach

MILLIPEDES

Millipedes, like centipedes, are what are generally called myriapods, but they form a separate class, Diplopoda, characterized by two pairs of legs on each of the body segments, each of which is in reality a double segment. They are vegetarians, eating a wide range of plant material, especially when it is soft and decaying, although they sometimes also eat dead animals, such as worms and insects. The mouthparts include mandibles and behind them, a broad plate, the gnathochilarium, formed from a pair of appendages equivalent to maxillae. The spotted snake millipede *Blaniulus guttulatus* is frequently a pest of arable crops, but it is doubtful whether their weak jaws can effect entry to (say) potatoes, in which they often occur in numbers, and their attack almost certainly follows damage by fungus or another animal.

Most millipedes, but not *B. guttulatus*, have ocelli on the head; they have only short antennae, but these constantly tap the substrate ahead as they move. Many are cylindrical in shape, the tergites much larger than the sternites, so that the many little legs seem to originate almost centrally beneath the body. The backstroke of the legs is of longer duration than the forward stroke, so that many legs are involved in a metachronal wave of movement, and they have a marked ability to push forwards using the motive force of the legs. They are essentially animals of the soil, and the rounded head and the massive tergite (collum) behind it expedite pushing into soil or litter. Other species have lateral or dorso-lateral keels on the tergites, or are fringed with bristles so that they look almost chiton-like, and some of these are very small.

Many species have a pair of repugnatorial glands on each segment, and in *Blaniulus guttulatus* the contents of these appear as a row of bright red spots down each side. The glands produce a noxious mixture of substances, chiefly hydrocyanic acid, iodine and quinine, and chlorine often gives a characteristic odour (Cloudsley-Thompson 1968). The larger tropical species seem able to walk about with impunity, being presumably distasteful, although many predators do eat millipedes. Many species are also protected by a tough cuticle, impregnated with calcium as well as phenolic tannins. The pill-millipedes, family Glomeridae, have massive calcareous exoskeletons and roll into protective balls so that they resemble pill-bugs (woodlice). Most millipedes, however, are prone to desiccation; they are usually found in damp places, avoid light, and are more active at night.

The genital apertures are on the third segment, near the front end of the animals. Males of most species have special appendages, called gonopods, on the seventh segment, which they charge with sperm and use as intromittent organs; in other species, the hindermost legs are used to introduce sperm into females. Females of some species coat their eggs with soil or excrement and leave them in soil crevices, but others make nests of soil particles moistened with saliva, and line them with excrement. In some species, the female stays tightly coiled around the nest for several days, and, as in centipedes, her actions seem to protect the eggs from fungal infections. In some julid millipedes in cold climates, males revert in summer to forms without gonopods and in which the testes are inactive; this prolongs their lives to two or more years. Millipedes moult their skin several times as they grow, adding segments and legs each time, and many species construct special moulting chambers of soil or, in some cases, silk for this purpose. The cast skin is eaten, thus restoring lost calcium supplies, and further development seems to depend on this.

Millipedes are not especially common in my garden, but two species were taken in pitfalls during the period 1 April – 31 October 1980–86, the numbers trapped being given in the species list. One female *Archiboreoiulus pallidus* was caught in a pitfall during the winter 1985/86; this is one of the less common British species, but had already been recorded for Leicestershire. A. J. Rundle found *Macrosternodesmus palicola* commonly under stones, and found a single *Blaniulus guttulatus* beneath a paving slab by the compost heap. Millipedes trapped in pitfalls were identified by D. T. Richardson.

Polydesmidae

42. *Polydesmus gallicus* Latzel   15

Macrosternodesmidae

43. *Macrosternodesmus palicola* Brolemann

Blaniulidae

44. *Blaniulus guttulatus* (Fab.)
45. *Archiboreoiulus pallidus* (Brade-Birks)

Julidae

46. *Ophyiulus pilosus* (Newport)   9

SYMPHYLA

Symphyla are minute, pale, many-legged animals, with long antennae. They were once included in the group 'Myriapoda', but are now recognized as a separate class, and are perhaps closer to ancestral insects than other myriapods. Superficially they look like tiny centipedes, the largest British species being only about 8 mm long, but they lack poison claws. They have 11 or 12 pairs of legs, but many more tergites, up to 24. Symphylans live in soil, in moss, under bark and in organic debris in a variety of habitats, but are a poorly known group. They are herbivorous, feeding mainly on soft or decaying plant material, but often attacking delicate plant root systems, and are sometimes so abundant as to be serious economic pests, densities of 22 million per acre (roughly 54 million per hectare) out-of-doors and 90 million per acre (roughly 222 million per hectare) in glasshouses being reported (Cloudsley-Thompson 1968). A. J. Rundle collected two symphylans from the garden, but these have not as yet been identified.

FALSE-SCORPIONS

Tiny, retiring false-scorpions, with their relatively enormous pincers, or pedipalps, form the Pseudoscorpiones, one of the orders of the class Arachnida, all of whose members have four pairs of walking legs; other orders of Arachnids include scorpions, harvestmen and spiders. False-scorpions look somewhat like scorpions on a vastly reduced scale, the largest British species being less than 4 mm long, but they lack the tapering tail and 'sting', the body instead ending bluntly. They are predators of other tiny animals for which they lie in wait, pedipalps extended, grabbing with them whenever the long tactile setae covering the appendages are stimulated. The prey is injected with toxins from glands that open on the pedipalp claws, and is transferred by them to the jaws, or chelicerae. After these make a wound, digestive enzymes are introduced, and fluid food sucked out.

False-scorpions produce silk from glands that open on the chelicerae, and this is used to spin chambers for moulting, for overwintering and for brooding young. They have elaborate courtships involving male and female grasping each other's pedipalps and walking to and fro while, in some species, the male displays with the 'ram's-horn organ', an eversible sac between the second and third sternites of the abdomen, close to the genital aperture; the male produces a spermatophore over which he guides the female while she takes it up. Females lay eggs into a membranous sac extruded from the oviduct; the eggs hatch in this, and the young feed on 'milk' produced from their mother's degenerating ovaries. When, after four or five weeks, they leave the sac, the young remain for some time clinging to their mother's abdomen (Cloudsley-Thompson 1968). Dispersal seems to be by phoresy, as false-scorpions are frequently found clinging onto flies and other insects, harvestmen, or other animals. They live in such places as leaf litter, in moss, and beneath bark, and A. J. Rundle found two individuals of *Lamprochernes nodosus* in the drier top layers of the compost heap.

Chernetidae

47. *Lamprochernes nodosus* (Schrank)

HARVESTMEN

The familiar harvest-spiders or harvestmen (order Opiliones) are arachnids characterized, among other features, by the uniting of cephalothorax and abdomen, so that the body appears as a single, often rounded, entity, by conspicuously long, often gangling, legs, especially the second pair, and by the absence (in contrast to spiders) of silk glands. Most have a pair of simple eyes, usually placed one either side of a prominent tubercle, the ocularium, near the middle of the cephalothorax, and near the attachment of the first pair of legs are a pair of odoriferous glands, whose secretion makes harvestmen distasteful to most invertebrate predators. The chelicerae are pincer-like, and there is a pair of short, leg-like pedipalps, which are chiefly sensory and help in grasping food.

Harvest-spiders, although primarily carnivorous, are omnivores, often scavenging from bird droppings and dead animal material; some species feed on snails. The sensitive second legs are particularly important in locating and recognizing prey, which is held down by the legs and palps, and torn to pieces by the pincers on the chelicerae. Harvestmen regularly clean legs and pedipalps by pulling them through the chelicerae, cleanliness of all append-ages being essential to their hunting and feeding behaviour. In adversity, a harvestman will shed a leg to escape, four-legged individuals being able to survive as long as they retain one of the long, sensitive second legs; without them, they presumably cannot feed. Harvestmen need to drink at least every two days, although they can fast for two or more weeks. As there is no lipid layer in the cuticle, they require humid surroundings, and they are normally nocturnal in their activity.

Mating is a casual, brief affair, often repeated with other individuals, and involves copulation, the male having a relatively long penis which, like the opening of the female's oviduct, is on the first sternite of the abdomen, beneath the genital plate. Females use their long ovipositors to lay eggs in crevices in soil, under stones, beneath bark, and in other moist places, the number varying from a few to several hundred in *Phalangium opilio*. Newly hatched young, each about 1 mm in length, resemble the adults in most features, but moult several times as they grow (Cloudsley-Thompson 1968).

All British harvest-spiders belong to the suborder Palpatores, but not all have the familiar long legs and round bodies; Trogulidae, for instance, which feed on snails, are flattened slow-moving species with shorter legs and a longish body often covered with earth particles. Some species are active all year round, and take more than one year to complete their life cycle, a few are annuals overwintering as immatures, and many, such as *Mitopus morio*, are annuals that overwinter as eggs (Williams 1963). Todd (1948) found the greatest seasonal density of harvestmen, about 50 per square metre of ground litter, in deciduous woodland in June, but these were mostly immature individuals, and the biomass was at its greatest in July. She found that many woodland species hatch on the ground, but then move vertically up to the field layer and then to the tree trunks, even in some species up to the

Photograph 12.3. Overgrown shrubbery increases the number of living spaces for harvest-spiders and spiders as well as providing nesting sites for birds.

branches, as they get older. Different species typically live at different levels in accord with their humidity preferences as tested in the laboratory: *Nemastoma lugubre* is found on the ground, where relative humidity (RH) is 85–100 per cent, *Oligolophus tridens* in the field layer (RH 70–80 %), *Leiobunum rotundum* on tree trunks (RH 60–75 %), and *Paroligolophus agrestis* on tree branches (RH 50–60 %). Todd (1948) also found that *Odiellus spinosus*, which often occurs in gardens, has a much higher range of temperature preference than woodland species.

Ten species of harvestmen were trapped in pitfalls between 1 April and 31 October in the seven years 1980–86, and identified by J. H. P. Sankey. Many harvestmen are not available for capture on the ground, because they are higher up in the vegetation, and it is significant that four species were also taken in the Malaise trap, which would have entailed a climb of nearly 2 metres; all are species that are usually found well above the ground. Three of them, *Paroligolophus agrestis*, *Phalangium opilio* and *Leiobunum blackwalli*, were also beaten from vegetation, and several females of *L. rotundum*, only once taken in a pitfall, were found by A. J. Rundle in the compost heap. *Odiellus spinosus*, *Opilio saxatilis*, and *Leiobunum blackwalli* were also trapped in pitfalls during winter 1985/86. The most interesting aspect of the pitfall captures was the large catch of *Odiellus spinosus*, the largest British harvestman and one of the less common species, which has a southeastern distribution in the British Isles and inhabits relatively dry and warm places, especially on walls, fences and under window ledges in urban situations (Sankey 1988).

The annual total pitfall captures of harvestmen in the seven years 1980–86 were 44, 50, 18, 50, 38, 70 and 75; thus the poorest year (18) was 1982 and the best (75) was 1986, the reverse of the annual catches of most groups of insects. There is no reason why catches of harvestmen on the ground should follow those of flying insects, and, indeed, the trapping results demonstrate that harvestmen respond differently from flying insects to weather factors; 1986, when most were captured, was a damp summer with only average temperatures and amount of sunshine. In the following species list, nomenclature and sequence follow Sankey (1988), numbers refer to total pitfall captures 1 April to 31 October 1980–86, and MT indicates that the species was also caught in the Malaise trap.

Phalangiidae

48. *Oligolophus tridens* (Koch)   3
49. *Paroligolophus agrestis* (Meade) . . . .   7 MT
50. *Odiellus spinosus* (Bosc) 227
51. *Mitopus morio* (Fab.)   1
52. *Phalangium opilio* L. . . . .   11 MT
53. *Opilio parietinus* (Degeer)   3
54. *O. saxatilis* (Koch).   62

Leiobunidae

55. *Leiobunum rotundum* (Latreille) . . . .   1 MT

SPIDERS

Spiders (order Araneae) are distinguished from other arachnids by the possession of silk glands that open through spinnerets at the end of the abdomen. They are an abundant and diverse group found in an astonishing range of terrestrial habitats. More than 600 species are found in the British Isles, and can be used to illustrate the diversity of the group. The cephalothorax is separated by a narrow pedicel from the abdomen, which may be tubular, as in *Harpactea* sp., or spherical, as in Theridiidae. *Segestria florentina* females are up to 22 mm long, but some Linyphiidae are less than 2 mm. The eyes, always simple ocelli, number six or, more usually, eight, and are sometimes on tubercles, as in crab-spiders, Thomisidae. They have four pairs of legs usually of similar length, although in Thomisidae the first and second pair are very long and robust for prey capture. The tarsi usually end in two toothed claws, occasionally in three, and the orb-web builders, Araneidae, have claw-like hairs at the end of the tarsi for gripping the silk of their webs. Many hunting spiders also have dense tufts of hairs, scopulae, under their claws, that help maintain a grip on vertical or overhanging surfaces. Most spiders are nocturnal in their activity, but some conspicuous species, such as Salticidae, are diurnal.

Spiders have one or, in Atypidae (Mygalomorpha), two pairs of spiracles ventrally on the forepart of the abdomen leading to book lungs, and one posterior spiracle, occasionally four, leading to tubular tracheae which branch throughout the abdomen and, in some species, also through the cephalothorax. The body cavity is filled with haemolymph containing the copper-based respiratory pigment haemocyanin, and there is an open-ended tubular dorsal heart. The book lungs consist of many horizontal air pockets in contact with haemolymph-filled lamellae, and most of the blood goes through them on its way back to the heart. The tracheae are all open-ended, rather than contacting a cell, so haemolymph is always responsible for the final delivery of oxygen. Mygalomorphs have only book lungs, but the fastest-moving spiders have only tracheae, presumably the most efficient means of oxygenating the tissues (Foelix 1982).

Abdominal silk glands open on many little 'spigots' on three pairs of mobile spinnerets situated just below the anus. Silk, a proteinaceous substance, is extruded as a liquid but solidifies as it is stretched and exposed to the air. Hunting spiders have four types of silk glands producing different types of silk for attachment discs, for drag lines, for wrapping prey and making a sperm web or outer egg sac, and for making the inner layer of egg sacs. Orb-weavers have two additional types, producing silk for making the spiral threads of their orb-webs, and for producing the viscous liquid that coats the sticky spiral that ensnares prey (Jones 1983). An oily coating to their legs prevents them becoming caught themselves. Many spiders are wanderers and do not make webs, but the more sedentary species construct webs of various designs for catching their prey. For instance, those of Agelenidae are slightly concave mats, suspended by vertical threads, with a funnel-shaped retreat at one end, Linyphiidae build convex sheets also with vertical trip-lines, Theridiidae make irregular meshes, Araneidae construct intricate orb-webs, and Amaurobiidae weave a tubular retreat from which radiate signal lines and catching threads. The catching threads of *Amaurobius* spp. and other cribellates are flocculent, bluish ribbons which are made by combing with the calamistrum (comb-like series of curved spines along the metatarsi of the fourth legs) the fine silk that emerges from a sieve-like plate (the cribellum) crowded with spigots in front of the spinnerets, so that it forms a dense ribbon around the web threads.

The body and limbs of all spiders are covered with hairs and bristles, and on the legs are fine trichobothria that are sensitive to air currents and vibra-tions. These play a part in alerting spiders to prey, usually insects, although hunters, such as Lycosidae, rely heavily on sight. Almost all spiders have poison glands opening just behind the tip of the fangs that form the second segment of the chelicerae. The venom is a mixture composed mostly of neuro-toxic polypeptides, but less than 0.1 per cent of the world's spiders are dangerously poisonous to man. Apart from *Scytodes thoracica*, which shoots ensnaring silk to disable its prey from a distance, hunting spiders bite and inject poison at the first strike, but most other spiders ensnare and wrap their prey in silk before biting it. The chelicerae with their fangs strike vertically in Atypidae and other Mygalomorpha, but work horizontally in other spiders. In most species there are cheliceral teeth that, together with the massive coxae of the pedipalps, macerate the prey beyond all recognition while digestive enzymes are poured onto it. The resulting nutritious liquid is then sucked in, for no spider can ingest solid food. Thomisidae and Theridiidae, however, have no cheliceral teeth and inflict only a small wound on their intact prey from which they suck the contents liquefied by extra-oral digestion.

The pair of six-segmented pedipalps, between the chelicerae and the first pair of walking legs, have a special function in males. The last, bulbous segment is variously modified to receive sperm from a drop that the male extrudes onto a specially-made little silken web. Males then use their charged pedipalps as intromittent organs when they copulate with females. Males are vagabonds after their final moult, wandering around in search of females, which are usually larger; in most species, courtship involves the male dis-suading the female from treating him as food. In some spiders, such as Thomisidae, direct contact is required for courtship to ensue; in others, female pheromones stimulate the males and some Araneidae, for instance, mate immediately after the females final moult. In Salticidae, however, visual signals are the primary releaser of courtship. In nocturnal hunters there is little courtship, but in Salticidae and Lycosidae there is an elaborate visual display often involving much semaphoring with the legs, and in many web-makers the approach is tactile involving coded vibrations of the web. Many males have stridulatory organs; in Theridiidae, for instance, males vibrate their abdomens so that spurs on the front of the abdomen rub against ridges on the back of the thorax. A few females attack and eat males after mating but many, such as *Tegenaria* spp., co-exist in the same web, and the male is only eaten later, when he dies. Male *Pisaura mirabilis* present the female with a silk-packaged prey item, and mate while she is occupied with it, and in other species, males copulate while the female is feeding. In *Linyphia triangularis*, however, the male meets no resistance, and he destroys most of the female's web during courtship so that she can be interrupted neither by feeding nor by the attentions of competitors (Foelix 1982).

Eggs are laid in a silken sac containing in some cases, such as Oonopidae, only two, in others, such as *Dolomedes*, more than 2000 eggs. The sac may be suspended in the web, lodged in the retreat, when it may be stalked, as in Clubionidae, or is carried around by the female, either attached to her fangs, as in Pisauridae, or attached to her spinnerets, as in Lycosidae. Early larval stages, which lack silk and poison glands, subsist on yolk, but these moult into cannibalistic spiderlings, resembling adults, which soon leave the egg sac. Female Theridiidae feed their young from their mouths or on specially caught and macerated prey, and Lycosidae carry their young around on their backs, but most spiderlings soon disperse either to make tiny webs for prey capture, or to take to the air. 'Ballooning' is the usual dispersal method of small species and of young. The little spider climbs to an elevated position, stands on tip-toe with the abdomen raised, pays out silk, and, when a breeze catches the threads, releases its foothold to sail away.

As spiders grow they moult their skins, sometimes, as in Salticidae, in a specially prepared moulting chamber. Lost limbs can usually be regenerated at a subsequent moult. Different species mature and mate at different seasons, so that mature and active spiders can be found in all months. *Araneus diadematus*, for instance, mate in August, and in September or October lay eggs which overwinter and hatch the following spring; it is usually the following summer, more than a year later before they mature (Bristowe 1971). Large spiders often take two, three or even four years to mature, and females of *Steatoda bipunctata* may live several years indoors, although the life cycle of most species occupies only a year.

A total of 64 species of garden spiders were identified by R. F. Owen using Locket & Millidge (1951); I. M. Evans confirmed the identification of a few of the more difficult species. Taxonomy, nomenclature and sequence in the following annotated list follows Locket, Millidge & Merrett (1974). All 63 species had previously been recorded in Leicestershire. MT refers to capture in the Malaise trap, and P to capture in pitfalls.

Amaurobiidae

58. *Amaurobius fenestralis* (Stroem). On house wall and fences.
59. *A. similis* (Blackwall). Common on house wall and fences.
60. *A. ferox* (Walckenaer). Frequent in dark crevices.

Dictynidae

61. *Dictyna arundinaceae* (L.). Common on tops of plants.
62. *Lathys humilis* (Blackwall). One male in litter beneath *Chamaecyparis*.

Oonopidae

63. *Oonops pulcher* Templeton. One under bark of apple tree.
64. *O. domesticus* de Dalmas. In house.

Dysderidae

65. *Harpactea hombergi* (Scopoli). Common hunting spider.

Gnathosidae

66. *Drassodes lapidosus* (Walckenaer). Common in garage.
67. *Herpyllus blackwalli* (Thorell). On house walls, not common.

Clubionidae

68. *Clubiona corticalis* (Walckenaer). MT
69. *C. pallidula* (Clerck). On vegetation. MT
70. *C. terrestris* Westring. On vegetation.
71. *C. lutescens* Westring. MT
72. *C. compta* Koch. MT
73. *C. brevipes* Blackwall. MT

Thomisidae

74. *Misumena vatia* (Clerck). One on flower *Chrysanthemum frutescens* 1980.
75. *Xysticus cristatus* (Clerck). Common on vegetation and flowers. MT
76. *Philodromus aureolus* (Clerck). Common on vegetation. MT
77. *Tibellus oblongus* (Walckenaer). On vegetation.

Salticidae

78. *Salticus scenicus* (Clerck). Fairly common.

Lycosidae

79. *Pardosa pullata* (Clerck). P
80. *P. prativaga* (Koch). Common, including females with egg sacs.
81. *P. amentata* (Clerck). Common. P
82. *Alopecosa pulverulenta* (Clerck). Common.

Pisauridae

83. *Pisaura mirabilis* (Clerck). MT

Agelenidae

84. *Textrix denticulata* (Olivier). Common under stones.
85. *Tegenaria saeva* Blackwall. P
86. *T. domestica* (Clerck). Common; occasionally in P.

Photograph 12.4. Privet bushes Ligustrum ovalifolium *cut back in the winter of 1985/86 were growing vigorously in August 1986 and soon restored dense cover.*

Theridiidae

87. *Steatoda bipunctata* (L.). Common in house and garage.
88. *Theridion varians* Hahn. Common on vegetation, buildings and fences.
89. *T. melanurum* Hahn. Common on vegetation.
90. *T. cinctum* (Walckenaer). Commonly beaten from *Chamaecyparis*.
91. *T. bimaculatum* (L.). Common on vegetation.
92. *Enoplognatha ovata* (Clerck). Common on vegetation.

Nesticidae

93. *Nesticus cellulanus* (Clerck). Common on buildings.

Tetragnathidae

94. *Tetragnatha extensa* (L.). One female beaten from *Prunus serratula*.
95. *T. montata* Simon. Immatures commonly beaten from trees.
96. *Meta segmentata* (Clerck). Common; often in MT.

Araneidae

97. *Araneus diadematus* Clerck. Common.
98. *A. umbraticus* Clerck. Common on fences.

99. *A. cucurbitinus* Clerck. Fairly common on vegetation.
100. *Zygiella x-notata* (Clerck). Fairly common on buildings.

Linyphiidae

101. *Ceratinella brevis* (Wider). On vegetation.
102. *Gongylidium rufipes* (Sundevall). MT
103. *Hypomma cornutum* (Blackwall). MT
104. *Oedothorax agrestis* (Blackwall)
105. *O. retusus* (Westring). MT
106. *Tiso vagans* (Blackwall). MT
107. *Micrargus herbigradus* (Blackwall). P
108. *Erigone dentipalpis* (Wider). Very common aerial species. MT
109. *E. atra* (Blackwall). MT
110. *E. promiscua* (O.P.-Cambridge). On vegetation.
111. *Porrhomma convexum* (Westring). On vegetation. P, MT
112. *Meioneta rurestris* (Koch). MT
113. *Centromerus dilutus* (O.P.-Cambridge). P
114. *Diplostyla concolor* (Wider). Very common in P.
115. *Stemonyphantes lineatus* (L.). Commonly in P. MT. Identification confirmed by British Museum (Natural History).
116. *Lepthyphantes nebulosus* (Sundevall). Very common in garage.
117. *L. leprosus* (Ohlert). Common in garage; in litter below *Chamaecyparis*. P
118. *L. minutus* (Blackwall). MT
119. *L. tenuis* (Blackwall). Commonly in P.
120. *Linyphia montana* (Clerck). Common in 1982, males only. MT
121. *L. clathrata* Sundevall. P

The 121 species of invertebrates, other than insects, identified from the garden by no means form a definitive list. Special techniques are needed to find some groups, such as false-scorpions, diligent searching of vegetation at all levels is necessary to find harvestmen and spiders, and groups such as Crustacea in the pond and soil-dwelling mites have not been investigated at all. Nevertheless, the varied list gives a flavour of the diversity of the fauna of a suburban garden.

13

Vertebrates

Vertebrates, unlike insects and some other invertebrate groups, have not been trapped in the Leicester garden, but continual observation has led to accumulation of many records, particularly of birds. They are conspicuous residents and visitors since there is dense vegetation for shelter and little threat from cats, which are discouraged. Indeed, over the years, an astonishing variety of birds has visited the garden or flown over it. The relatively small size of the garden and its location on a busy corner make it unsuitable for most mammals, but the pond, small though it is, attracts amphibians. In all, three species of amphibians, 49 of birds and seven of mammals have been recorded.

Amphibians

Five species of amphibians occur in the county of Leicestershire: common frog, toad, smooth newt, palmate newt and crested newt. The common frog and the smooth newt are generally distributed and common in many places, and toads are widespread although not as universally distributed as frogs. The palmate newt occurs in a few localities on Charnwood Forest and in east Leicestershire, and the crested newt is widely although patchily distributed. In the ten years 1965–75, installation of garden ponds was common and widespread, causing a rapid expansion of urban standing water, although most ponds were of less than two cubic metres in capacity. Frogs and smooth newts successfully invaded garden ponds, although toads were slower to establish themselves and crested newts have had little success in spreading to urban development (Mathias 1975).

Garden ponds are evidently important for the maintenance of breeding populations of amphibians, which have suffered countrywide from the drainage, pollution or infilling of field ponds and in the 1940s and 1950s from large-scale collection of frogs for dissection in school biology classes, when this was regarded as an essential part of everyone's education. Certainly, frogs and smooth newts are now common in the city of Leicester, although toads are less common as they normally maintain larger breeding populations than the typical small garden pond can support. There are also many ponds

Photograph 13.1. Garden ponds are important for the maintenance of breeding populations of amphibians, particularly common frogs Rana temporaria *and smooth newts* Triturus vulgaris. *Lush garden vegetation, such as the valerian* Centranthus ruber *beside the pond in the Leicester garden, ensures that their invertebrate food away from the water is plentiful.*

in the city on public land, such as parks and golf courses, and although old field ponds have been lost by urban development, many new ponds have been created. Some of these ponds are excellent for wildlife, including amphibians, and even crested newts breed in a few of the larger ponds on golf courses. It seems that all city ponds are eventually colonized by amphibians, particularly frogs, whereas only about half of ponds on agricultural land are occupied. This is almost certainly because of the terrestrial requirements of amphibians. Garden ponds are particularly important not only because there are so many of them, well distributed, but also because lush, diverse garden vegetation provides a good habitat for amphibians away from the water, as their invertebrate food is plentiful.

I have recorded common frog, smooth newt and toad in my garden. Before the pond was drained in 1976, frogs were repeatedly seen both in the pond in spring and summer and in all other parts of the garden in summer and autumn, although there was no evidence of breeding. When, however, the pond was re-established in early 1984, they quickly moved in and bred successfully. They probably recolonized so quickly because they were already in the garden, two being found in the rubble filling the pond well, when it was cleared away prior to reinstating the pond. In October 1984, a small frog, presumably bred that year, was found in a pitfall trap. Smooth newts, too,

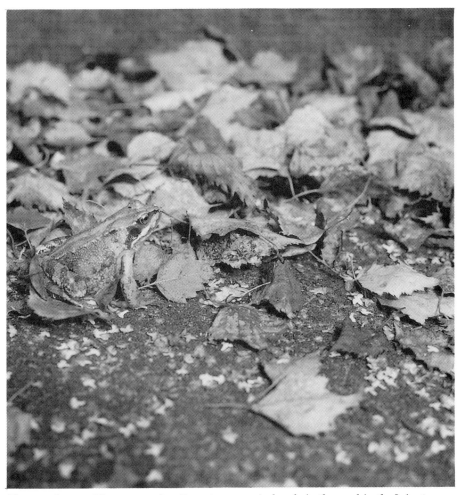

Photograph 13.2. The common frog Rana temporaria *breeds in the pond in the Leicester garden.*

were present in the pond before 1976, although breeding was not proved. Individuals were found hibernating in the rockery beside the pond, even in the years when the pond was dry. When it was re-established, smooth newts quickly recolonized and bred successfully, and have occasionally been found in June beneath a spreading clump of *Iberis sempervirens* close to the pond. The only record of a toad was of a semi-torpid individual found in a heap of dead leaves in August 1986. Small though the pond is, it is evidently a suitable breeding site for frogs and smooth newts, and the decision to reinstate it is amply justified in terms of wildlife.

1. Toad *Bufo bufo* (L.)
2. Common frog *Rana temporaria* L.
3. Smooth newt *Triturus vulgaris* (L.)

BIRDS

Leicestershire, being very much an inland county, does not have a bird list as large as those of some coastal counties on migration routes, but it does have a variety of habitats, including the internationally important Rutland Water; in 1986, 194 species were recorded in the county (including Rutland) by the Leicestershire and Rutland Ornithological Society (LROS 1987). In the period 1983–87, 108 species were reported within the city boundary, including a fair number of ducks and other water birds (Lomax 1987). In the 1950s, my father, F. A. Bak, trapped and ringed 55 species in his 1000-square-metre garden less than 2 km from Leicester city centre in an area of spacious gardens with many mature trees. The diverse traps he used, including a large one enclosing soft fruit bushes and baited with dripping water, caught unexpected visitors to a suburban garden, such as sedge warbler *Acrocephalus schoenobaenus*, grass-hopper warbler *Locustella naevia*, nightingale *Luscinia megarhynchos*, brambling *Fringilla montifringilla*, and the first marsh warbler *A. palustris* to be recorded in the county. (The only other record is of one ringed north of Leicester in 1981.) Many additional species passed over the garden, or, like the local house martins and swifts, were not amenable to trapping.

The trapping list from my father's garden well illustrates the feature of birds that leads to them doing so well in gardens even in a city: birds are supreme opportunists. Since they investigate every feeding possibility and are relentlessly curious, they are readily caught by traps that provide food, whether it be soft fruit, grain or meal-worms, and the sound of dripping water is a magnet to warblers, in particular, during hot summer weather. My father even brought black-headed gulls down onto his garden lawn by pro-vision of sprats.

Blue tits are adroit at levering open the blisters on holly leaves caused by holly leaf-miners to extract the larvae, and have turned this skill to acquiring another food, for they tear open cardboard milk-bottle tops and hammer holes in foil ones to drink the cream. The habit was first noticed in the 1920s, but spread rapidly and now occurs in most parts of the country. Although there was undoubtedly some copying, the milk-stealing habit appears to have been acquired independently many times, and species other than blue tits have been involved. Greenfinches, too, quickly acquire new eating habits, learning in the early nineteenth century to eat *Daphne mezereum* seeds in mid-summer while they are still soft, a habit that then spread at a rate of about 3 km a year (Murton 1971).

I have seen countless examples of opportunism by birds in my garden, such as a blue tit investigating a new nest-box within hours of it being fixed. Robins are also opportunist, landing close by me to snatch worms whenever I start to dig, pecking at large toadstools, presumably for insects, searching the timbers of a dismantled old oak gate for spiders and insects, and drinking melt-water dripping from icicles on the roof when a thaw set in after a prolonged cold spell. When I pulled up a root of *Chrysanthemum frutescens* enclosed in an ant nest exposing to view not only the ant brood but also unpigmented aphids on

Photograph 13.3. Well-stocked bird-tables are particularly important to birds in hard winter weather when fewer insects, berries and other foods are available.

the root, a song thrush quickly moved in to eat both ants and aphids. Dunnocks exploit the untidy feeding activities of other birds, regularly gleaning morsels beneath the bird-table when coal tits peck somewhat ineffectually at a bag of nuts, and, one winter's day, picking up seeds dropped on the snow by a bullfinch clumsily feeding at the seed-heads of antirrhinums. Greenfinches pick seeds from newly spread compost, in hard weather wood pigeons strip broccoli plants to the mid-ribs of the leaves, black-headed gulls, although normally not much in evidence in summer, circle high in the air feeding on flying ants whenever nests of the black ant *Lasius niger* produce swarms of reproductives, and house sparrows pick up birch seeds dropped from the tree by the feeding activities of the more agile redpolls. House sparrows will optimistically have a go at anything that might be edible, and one chased and snapped without success at a large *Aeshna* dragonfly crossing the garden.

Perhaps, however, the most striking instance of opportunism in birds is the speed and success of the colonization of English suburbia by collared doves, suggesting that there were hitherto unexploited opportunities for a small, seed-gleaning dove. They reached Britain in 1955, and in the space of 15 years numbers increased to between 15 000 and 25 000 breeding pairs. They were first recorded in Leicestershire in 1961, by 1964 were widespread, first bred in the city in 1975 (Hickling 1978), and by 1985 were designated as common resident breeders in the county (LROS 1986).

The basis of my garden bird list is species that have bred in the garden, some of them every year 1972–86, others occasionally: wood pigeon, collared dove, dunnock, robin, blackbird, song thrush, blue tit, starling, greenfinch and linnet. Wood pigeon nests were found in three years, twice in Lawson's cypresses, once in a holly tree, and pairs were seen courting in other years; collared dove nests were found in Lawson's cypresses in three years, and pairs were seen courting every year. Blackbird and song thrush nests were located nearly every year, blue tit (in a nest-box), greenfinch and linnet nests occasionally, and starlings nested one year in a hole in the foulstack on the side of the house. Neither dunnock nor robin nests were ever located, although they certainly bred in the garden. The constant presence and behaviour during the nesting season of goldcrests, redpolls (since 1977) and bullfinches suggested that they too were breeding, although no nests were found. It is tempting also to include on the breeding bird list the wren which built a large unoccupied nest behind the honeysuckle against a fence, and the mallard, presumably one from the nearby park, which laid an egg in the strawberry patch. The status of all, other than redpoll, in the county is either common or abundant, i.e. 1000–10 000 pairs or >10 000 pairs (LROS 1974–1987). The redpoll is now a fairly common resident breeder in the county, i.e. 100–1000 pairs, after years when it was known only as a winter visitor. Nests with young were first recorded in 1971, but by then it was clearly breeding over a large area (Hickling 1978), and it was first recorded breeding in the city in 1975 (LROS 1977).

Birds of many other species visited the garden, hunted in its air-space or perched on the chimney-pots or television aerial. Regular visitors included swift, house martin, mistle thrush, willow warbler, long-tailed tit, coal tit, great tit, carrion crow, house sparrow, chaffinch and goldfinch. Mistle thrushes visited at all times of year; a family party passed through one June, and one winter a pugnacious individual defended holly berries from blackbirds. Long-tailed tits, often in family parties, great tits, carrion crows on the chimney-pots, and house sparrows were frequent all year round. Swifts and house martins nested nearby every year, and in summer were major predators of the aerial plankton rising from the garden. Willow warblers were active in most years, feeding and often singing, particularly in early and late summer months, and are presumed to have bred in adjacent gardens. Chaffinches and goldfinches, too, were mainly summer visitors, the latter taking seeds of Oxford ragwort and perennial cornflower but also probably eating aphids on the still green cornflower seed-heads. Coal tits, however, were mainly winter visitors, although they were seen at all times of year. All these regular visitors to the garden are regarded as common or abundant in the county.

Kestrel, red-legged partridge, tawny owl, lesser spotted woodpecker, fieldfare, redwing, chiffchaff, marsh tit, treecreeper and jackdaw were irregular visitors to the garden. Kestrels were repeatedly seen overhead, and on separate occasions perched on the television aerial and hunted low over the garden. Red-legged partridges were twice disturbed from vegetable patches where there had been previously unexplained damage to lettuces and cabbages. Tawny owls were occasionally heard calling from the chimney-pots, lesser spotted woodpeckers sometimes flew around the garden in September or October, and fieldfares and redwings quite often visited in winter to feed on holly berries or rootle in leaf litter. Single chiffchaffs visited in April, August and September in different years, marsh tits were occasionally seen at different times of year, and in July 1982 a fledgling treecreeper was found on the back-door-step. Jackdaws occasionally perched on the chimney-pot, and in June 1984 a young one spent some time walking around the garden and inspecting the pond. Red-legged partridge, treecreeper and jackdaw are regarded as common resident breeders in the county (i.e. 1000–10 000 pairs); fieldfares and redwings are common winter visitors (i.e. 1000–10 000 birds). Kestrel, tawny owl, lesser spotted woodpecker, chiffchaff and marsh tit are fairly common resident breeders (i.e. 100–1000 pairs), and both tawny owl and lesser spotted woodpecker regularly nest in urban or suburban areas (LROS 1987). Kestrels quickly adapted to hunting on the broad grassy verges of motorways, and for at least 25 years have bred in the city, in distinctly urban localities (Hickling 1978).

Other birds passed over the garden in their foraging, roosting or dispersal movements; these included heron, lapwing, black-headed gull, common gull, swallow, pied wagtail, jay, magpie and rook. Skylarks and grey wagtails sometimes flew over, evidently on migration, and cuckoo and green woodpecker were each seen once. The county status of these birds seen over the garden varies greatly, and only one, the skylark, is classed as abundant. Swallow, pied wagtail, magpie and rook are regarded as common breeding birds (i.e. 1000–10 000 pairs). Nationally, numbers of rooks seem to have been declining since the mid- to late-1970s, probably because of the loss to Dutch elm disease of mature elms, in which they traditionally nested, and also because large areas of rough pasture, where they once fed, have been put under cereals. During the 1980s, some rooks were recorded in the county using electricity pylons for nesting; two rookeries were established in the city in the early- to mid-1980s, but have since been abandoned. Since at least 1976, there has been a large winter roost of pied wagtails in trees on a busy thoroughfare in the centre of the city of Leicester, and in 1986, this was estimated to contain 1500 birds. The suburban population of magpies appeared to be increasing in the early 1980s.

Lapwings, cuckoos, green woodpeckers and jays are classed as fairly common breeding birds in the county (i.e. 100–1000 pairs), and the lapwing is abundant in winter, when local birds are joined by immigrants. The heron is an uncommon resident breeder (i.e. 10–100 pairs), with three heronries in the county in 1986. Black-headed and common gulls are present in the county all year, and abundant in winter, when there are large flocks on parks and playing fields in the city; black-headed gulls bred in the county three times in the 1940s and 1950s. The locally least common species on the garden list is the grey wagtail, which is regarded as a rare to scarce breeder in the county (i.e. breeding less than annually with less than ten pairs), and an uncommon winter visitor (LROS 1987).

The total garden bird list is thus 49 species, which could marginally be boosted to 54 if racing pigeons and escaped cage-birds (three species of parrot and the cockatiel) are included. Sequence and nomenclature follow *A List of Recent Holarctic Bird Species 1977* by K.H. Voous, as used in *The Birds of Leicestershire and Rutland – 1986* (LROS 1987); authorities for the scientific names are from Witherby *et al.* (1946), except for that of the collared dove, taken from Coombs *et al.* (1981). In the cases of pied wagtail and carrion crow, subspecies is recorded, as other, visibly different, subspecies are known in the county.

4. Heron *Ardea cinerea* L.
5. Mallard *Anas platyrhynchos* L.
6. Kestrel *Falco tinnunculus* L.
7. Red-legged partridge *Alectoris rufa* (L.)
8. Lapwing *Vanellus vanellus* (L.)
9. Black-headed gull *Larus ridibundus* L.
10. Common gull *L. canus* L.
11. Wood pigeon *Columba palumbus* L.
12. Collared dove *Streptopelia decaocto* (Frivaldsky)
13. Cuckoo *Cuculus canorus* L.
14. Tawny owl *Strix aluco* L.
15. Swift *Apus apus* (L.)
16. Green woodpecker *Picus viridis* (L.)
17. Lesser spotted woodpecker *Dendrocopus minor* (L.)
18. Skylark *Alauda arvensis* L.

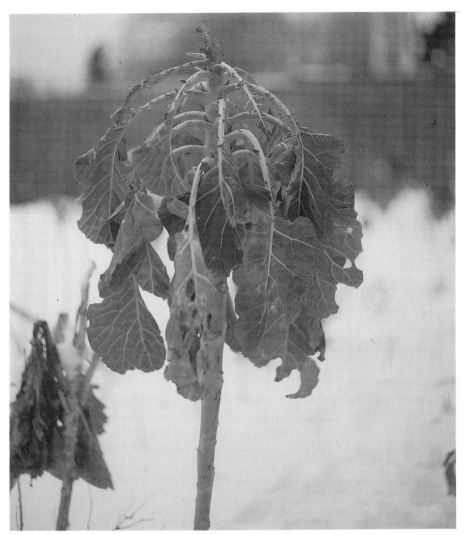

Photograph 13.4. In hard winters, particularly when snow lies on the ground, cruciferous plants such as Brussels sprouts are damaged by feeding wood pigeons Columba palumbus.

19. Swallow *Hirundo rustica* L.
20. House martin *Delichon urbica* (L.)
21. Grey wagtail *Motacilla cinerea* Tunstall
22. Pied wagtail *M. alba yarrellii* Gould
23. Wren *Troglodytes troglodytes* (L.)
24. Dunnock *Prunella modularis* (L.)
25. Robin *Erithacus rubecula* (L.)
26. Blackbird *Turdus merula* L.
27. Fieldfare *T. pilaris* L.
28. Song thrush *T. philomelos* Brehm
29. Redwing *T. iliacus* L.

30. Mistle thrush *T. viscivorus* L.
31. Chiffchaff *Phylloscopus collybita* (Vieillot)
32. Willow warbler *P. trochilus* (L.)
33. Goldcrest *Regulus regulus* (L.)
34. Long-tailed tit *Aegithalos caudatus* (L.)
35. Marsh tit *Parus palustris* L.
36. Coal tit *P. ater* L.
37. Blue tit *P. caeruleus* L.
38. Great tit *P. major* L.
39. Treecreeper *Certhia familiaris* L.
40. Jay *Garrulus glandarius* (L.)
41. Magpie *Pica pica* (L.)
42. Jackdaw *Corvus monedula* L.
43. Rook *C. frugilegus* L.
44. Carrion crow *C. corone corone* L.
45. Starling *Sturnus vulgaris* L.
46. House sparrow *Passer domesticus* (L.)
47. Chaffinch *Fringilla coelobs* L.
48. Greenfinch *Carduelis chloris* (L.)
49. Goldfinch *C. carduelis* (L.)
50. Linnet *C. cannabina* (L.)
51. Redpoll *C. flammea* (L.)
52. Bullfinch *Pyrrhula pyrrhula* (L.)

A larger garden with meadows, scrub and a spinney, especially one near the coast or a large body of inland water, would have a far longer bird list, but the garden list is impressive for a relatively small, suburban garden. Nearly half the species recorded recently in the city of Leicester have been seen in or flying over the garden, and the list compares well with the 103 species recorded in the period 1951–74 in the Outer London suburb of Dollis Hill, an area that included parks and other open areas as well as gardens (Simms 1975). Were the garden larger, more species would undoubtedly breed, for even Regent's Park in Inner London had 25 breeding species in 1959 (Simms 1975). Birds of many sorts do well in suburbia, and in large part this can be attributed to their opportunism.

MAMMALS

For reasons of size and locality, the Leicester garden is not an ideal habitat for mammals, and only seven have been recorded, none of them frequently. Nevertheless, in the city as a whole, 20 species were recorded during the period 1983–87 (Lomax 1987), so other species might be expected occasionally in the garden.

Hedgehogs were frequently seen in the garden in the 1970s, including a female with young in 1972. In the period 1980–86, they were rarely seen, the latest record being of one squashed on the road outside in June 1985. They are believed to be common, although patchily distributed, in rural parts of the county, but numbers have been declining in the city since about 1980. Bats, that were occasionally seen flying above the garden at dusk during the period 1972–86, were assumed to be pipistrelles. This, the commonest bat species in the county and nationally, is reasonably common in the city, having taken over roof- and wall-spaces of many modern houses for roosting. Nine species of bats, three of them rare, have been recorded recently in the county, and at least three – pipestrelle, noctule and Daubenton's – occurred in the city 1983–87 (Lomax 1987).

Grey squirrels were irregular but quite frequent visitors to the garden, presumably having come from nearby parks, where they are abundant. This introduced species was first recorded in the county in the late 1920s but is now widespread, particularly in urban and suburban areas, and is doing well. The field vole, common and widespread in the county on grassland and roadside verges, is not likely to visit a suburban garden, but, in August 1981, one was found damaged and dead, perhaps having been dropped by a kestrel. Its exposed flesh was attacked by large numbers of wasps. A dead wood mouse was found on a path in 1980, and in April 1981, in thick snow, there were two sightings in the yard on the same day of what was presumably one individual. The wood mouse is the most abundant mammal in city and county, being even commoner than the house mouse. A dead house mouse was found in the garage, and three were disturbed eating stored wild-bird seed in an out-house. The house mouse is common and widespread in the county, although probably under-recorded. That small mammals are more often found dead than seen alive, is perhaps an indication that they are more abundant in the garden than I suppose, but are rarely observed because of their secretive habits. The wood mouse and the bank vole *Clethrionymus glareolus* are common in Oxford gardens (Dickman & Doncaster 1987), and, in a survey of patches of semi-natural and disturbed vegetation in Oxford city, Dickman (1987) found that the number of mammal species increased with increasing density of vegetation in the layer 21–50 cm above the ground. Foxes have never been seen in my garden, although on two occasions their rank odour was unmistakable. They became established in many British towns and cities in the inter-war years when there was a boom in building low-density housing with medium-sized gardens (Harris & Rayner 1986b). They are common and widespread in the county of Leicestershire, and during the 1980s were regularly seen in the city, some garden-owners having resident foxes, which they supplied with food.

Dogs rarely enter the garden, but cats are frequent, and constantly have to be chased away, as they pose a threat to the garden's birds. The 70 domestic cats of a Bedfordshire village of 173 houses were judged in 1981–82 to be major predators, 1090 prey items (22 species of birds and 15 of mammals) being taken in a one-year period, the most important items being wood mice (17%), house sparrows (16%) and field voles (14%) (Churcher & Lawton 1987). If other mammals are to be added to the garden list, the most likely candidate is the common shrew. Moles are quite common in the city, but in most gardens, the soil is disturbed too much by digging for them to feel at home. Muntjac, small non-native deer, are occasionally reported in the city, and one has been seen in the Humberstone area, but the position of my garden, with roads on two sides, makes it unlikely that one will visit. A mink, too, was recorded on Scraptoft Lane in the 1980s, but I regard that as a most unlikely garden visitor.

53. Hedgehog *Erinaceus europaeus* L.
54. Pipistrelle *Pipistrellus pipistrellus* (Schreber)
55. Grey squirrel *Sciurus carolinensis* Gmelin
56. Field vole *Microtus agrestis* (L.)
57. Wood mouse *Apodemus sylvaticus* (L.)
58. House mouse *Mus musculus* L.
59. Fox *Vulpes vulpes* (L.)

The mammal list is short, and other suburban gardens, adjacent to parks, golf courses or open countryside, have more species. It is unlikely, however, that more species will be recorded in my garden. This illustrates particularly well that it is a medium-sized, ordinary, unsecluded, suburban garden. Insects, other invertebrates, and birds are diverse and abundant, but the habitat is neither extensive enough nor sufficiently undisturbed for there to be many mammals.

14

The garden habitat

In the 15-year period 1972–86, an astonishing diversity of animal species was recorded in the Leicester garden. It does not accord with the usual idea of a wildlife garden, but instead contains the elements of a conventional garden and is in no sense a wilderness. Neither do I believe it to be unusual in the richness of its animal life. There is every reason to suppose that the majority of gardens have an equally varied fauna. In other words, the diverse and abundant animal life of my garden is that of gardens in general. Why are gardens so rich? What is it about gardens that enables them to support so many different animals? The answer lies in the nature of the process of gardening, which creates a varied, bountiful habitat, yet one which is in a continual state of flux. The essence of gardening is the creation and maintenance of high plant and structural diversity, while maintaining the habitat in an early successional stage, susceptible to colonization and change.

CONTRIVED PLANT DIVERSITY

A gardener crowds together in one place a far greater diversity of plants than is ever found in one place in the wild. Even in tropical rainforest, the habitat generally agreed to have the highest plant diversity, an area equivalent to that of a typical garden would not contain so many different plant species. In 1984, my garden contained 264 species of flowering plants, excluding grasses, 197 of them intentionally cultivated, in a planted area of 741 square metres, a species density of 3563 species per hectare. It is difficult to compare this with species diversity in tropical rainforest, because figures given for the latter usually refer only to trees. A typical 2.5-acre (approximately 1 hectare) plot of West African rainforest includes 20–25 species of trees of 16 inch (40.6 cm) minimum diameter, 45–60 species of 8 inch (20.3 cm) minimum diameter and 50–70 species of 4 inch (10.15 cm) minimum diameter (Hopkins 1965). The number of tree species in Malaysian or Brazilian rainforest plots of equivalent area might well be considerably greater, as there are many more species of flowering plants there than in West Africa. It is unlikely that inclusion of all types of plants would bring the species diversity of West African rainforest up to 3563 species per hectare, and certainly, I was not aware of anything like

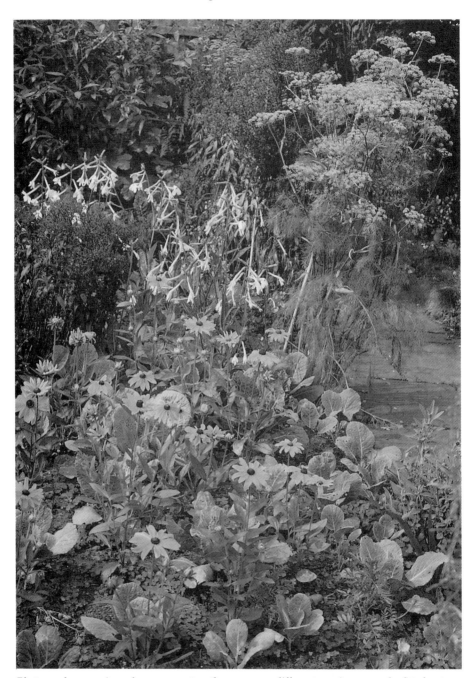

Photograph 14.1. A gardener grows together as many different species as can be fitted onto a plot, producing contrived plant diversity such as this mixture of cabbages Brassica oleracea, *black-eyed Susan* Rudbeckia hirta, *African marigolds* Tagetes erecta, *sage* Salvia officinalis, *flowering tobacco* Nicotiana alata *and fennel* Foeniculum vulgare.

this degree of small-scale diversity in West African forests. It might be argued that most gardeners grow more or less the same thing, so that the floristic composition of one garden is repeated again and again, making the average species diversity per hectare smaller, the larger the area considered. True, the composition of the plant diversity of one garden may be similar to that of other gardens in the same area of suburbia, but not identical, for different gardeners grow different species and manage their land in different ways. Collectively, the gardens of suburbia are an area of incredibly high plant diversity.

I have called this diversity contrived plant diversity because it is achieved by the decision of the gardener to grow as many different plants as he can fit onto his plot. Aliens originating in many different parts of the world, together with native plants, are cultivated side by side. For instance, in the 14.5 square metre mixed lawn bed in my garden, runner beans *Phaseolus coccineus* from South America and sweet peas *Lathyrus odoratus* from Sicily surround a medley of tomatoes *Lycopersicon esculentum* from South America, courgettes *Cucurbita pepo* from North America, spinach *Spinacia oleracea* from southwest Asia and native cabbages *Brassica oleracea* interplanted with flowering tobacco *Nicotiana alata* from southern Brazil, native cornflowers *Centaurea cyanus*, pot marigolds *Calendula officinalis* from southern Europe, asters *Callistephus chinensis* from China and Japan, African and French marigolds *Tagetes erecta* and *patula* from Mexico, dwarf convolvulus *Convolvulus tricolor* from southwest Europe, *Clarkia unguiculata* from California, viper's bugloss *Echium plantagineum* and borage *Borago officinalis* from southern Europe, snapdragons *Antirrhinum majus* from the Mediterranean and poppies *Papaver alpinum* from the Alps. Growing between them are native weeds, such as chickweed *Stellaria media*, and the persistent alien weed, *Oxalis corymbosa* from South America. This extraordinary assemblage of species, diverse taxonomically as well as in the country of origin, is typical of the plant diversity that a gardener contrives in a small space.

Since green plants are the starting points for all terrestrial food chains, and many, perhaps a majority, of insect herbivores are confined to one foodplant species, or at least a group of related species, diversifying the plants generates many different food chains. Plant diversity has been shown to be linked to diversity of herbivorous insects (e.g. Murdoch, Evans & Peterson 1972), which in turn is almost certainly linked to diversity of predatory insects. Latitudinal gradients in animal diversity are generally assumed to stem, at least in part, from latitudinal gradients in plant diversity. Because tree diversity, and by inference plant diversity as a whole, is greatest in the tropics, resources for herbivores are greater, and hence so are resources and opportunities for predators. While the situation is not identical in the artificial confines of a garden in the temperate region, it seems clear that high plant diversity, even if contrived, should generate high animal diversity by increasing consumer food chains. The abundant opportunities for herbivores are illustrated by the 68 species of moths that bred in the garden during the period 1972–86, and the diversity of top consumers by the 529 species of ichneumonids trapped in the three years 1972–74.

PERMANENT SUCCESSION

The activity of gardening entails regular clearing, digging, working and watering the soil, which render it continually receptive to the establishment of native and naturalized plant species as weeds. Ninety-four species recorded in the Leicester garden during the period 1975–86 were plants that had come in of their own accord. It is clear that while land is managed as a garden, contrived plant diversity is augmented by continual spontaneous introductions from outside that fill spaces as soon as they are created. Most gardeners never let their garden alone, but are always clearing and re-planting. The cleared land is cultivated, watered, and often treated with fertilizers of one sort or another, so that the new plantings usually have to compete with a range of spontaneous introductions (weeds).

Were my garden abandoned, in about 30 years it would almost certainly become a dense mixed scrub of birch *Betula pendula*, holly *Ilex aquifolium*, ash *Fraxinus excelsior*, elder *Sambucus nigra* and hawthorn *Crataegus monogyna*, with brambles *Rubus fruticosus*, bittersweet *Solanum dulcamara*, nettles *Urtica dioica*, several sorts of grass, and perhaps thistles *Cirsium* spp., smooth hawk's-beard *Crepis capillaris* and yarrow *Achillea millefolium* in the more open places. Eventually, the natural processes of plant succession would convert it to mixed deciduous woodland, the characteristic climax community of central England. As it is, maintaining the land as a garden continually interferes with the natural progress of plant succession, preventing the establishment of a climax community, and keeping the plot in an early successional stage, receptive to invasion from outside. Without my efforts as a gardener, the apparent stability of ordered but contrived plant diversity would soon be lost, and the floristic richness would be reduced to perhaps only a twentieth the number of species that are maintained or arrive at present.

An aspect of gardening that is as significant for the garden community as making introductions, is the removal of unwanted plants. Not only are plants that have come in of their own accord treated as weeds and destroyed, but in addition, annuals and biennials are uprooted as soon as they have served their purpose as flowers or vegetables, and perennials and woody plants are often removed as the gardener's tastes or preferences change. Continual introductions to a garden, whether by deliberate planting or invasion, are constantly countered by changes of policy and weeding. At the same time, both cultivated plants and weeds die out naturally for reasons beyond the gardener's control, either because conditions were unsuitable from the outset, or because they have changed.

In my garden, alien plants grown as vegetables, such as runner beans, or flowers, such as African marigolds, cannot maintain themselves and are replanted every year. Some cultivated plants, however, such as native forget-

Photograph 14.2. In May, the lawn bed is bare, ready for planting annual flowers and vegetables. The trellises are for sweet peas and runner beans. Regular clearing of vegetation and working the soil leave a garden continually receptive to invasion by plants that come in of their own accord, and maintain a garden in a state of permanent succession.

me-not *Myosotis alpestris* and alien goldenrod *Solidago canadensis*, seed themselves so freely and effectively that they have to be rigorously controlled. Trees, such as almond *Prunus dulcis* and walnut *Juglans regia*, were grown from seed and introduced, but did not survive. The herbaceous border perennials sneezeweed *Helenium autumnale* and *Heliopsis helianthoides*, present when the garden was taken over in 1971, eventually died; both were subsequently grown from seed and replanted, but both eventually died again. Of annual flowers, cornflowers were grown in only three years, *Petunia hybrida* in five; among vegetables, broad beans *Vicia faba* were grown in only one year, asparagus peas *Lotus tetragonolobus* in six.

Unplanned garden introductions, or weeds, similarly arrive and quickly disappear or persist. Hairy tare *Vicia hirsuta* appeared in only one year, scarlet pimpernel *Anagallis arvensis* in three, but wood avens *Geum urbanum*, which appeared in 1976 and was encouraged as an interesting addition to the flora, has since become one of the most persistent and aggressive weeds. Colt's-foot *Tussilago farfara*, initially well-established in a herbaceous border, died out in 1977, perhaps as accumulation of dead plant material and dense soil cover created conditions in which this early successional species was a poor competitor. Stinking chamomile *Anthemis cotula* and scentless mayweed *Matricaria perforata* appeared in 1975 in the newly created lawn bed, where the soil was sandy and lacking in humus, but as it changed in texture and fertility with repeated cultivation, these species disappeared, *M.perforata* after four years

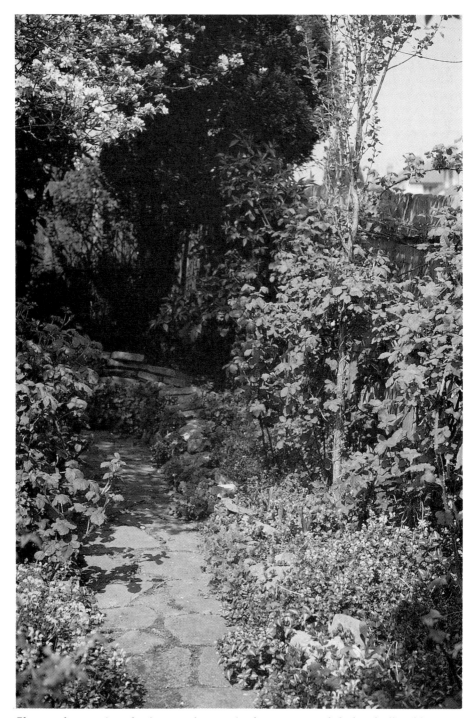

Photograph 14.3. A garden is a complex mosaic of open space and shade, of tall and low vegetation, forming an intricately structured and patchy environment.

and *A.cotula* after only two, although in 1986 one plant of *A.cotula* appeared elsewhere in the garden growing between paving slabs.

The plant community of a garden is never allowed to stabilize; it is continually interfered with and remains receptive to new arrivals from outside. Dominants are suppressed and prevented from forming an unbroken canopy that would shade the ground, which is consequently repeatedly invaded by early successional species. The outcome is that a garden is maintained in a state of permanent succession, and every step towards establishment of a climax community is countered. Indeed, a gardener's prowess may be assessed in terms of his ability to check the normal processes of plant succession. This continual state of flux in a garden, and the constant arrival of their own accord of new species creates yet more feeding opportunities for herbivores and so generates more food chains. Abundance of food resources combined with instability of the habitat, makes a garden highly susceptible to temporary colonization. The arctiid moths, the muslin *Diaphora mendica* and the ruby tiger *Phragmatobia fuliginosa*, although uncommon, were regularly captured at light in the early 1970s, but no *P.fuliginosa* were seen after 1975, and no *D.mendica* after 1976. The hoverfly *Cheilosia bergenstammi* was caught in 1972, 1973 and 1974 but not in later years, whereas *C.vernalis* was first caught in 1978, then again in 1980, after which it was taken every year, and in 1981 and 1982, commonly. The creation of a state of permanent succession is thus a source of further enrichment of the fauna.

STRUCTURAL HETEROGENEITY

A typical garden is a complex mosaic of tall and low vegetation, of open spaces and shade, forming an intricately structured and extremely patchy environment. Contrived plant diversity and permanent succession are thus combined with extreme structural heterogeneity. Within a small area there are elements of meadow, woodland and other habitats, so closely juxtaposed that a garden can be considered as a system of ecotones (transition zones between habitats). There are edges everywhere, between herbaceous borders and lawn, shrubbery and path, vegetable patch and compost heap, and so on. Where two habitats meet, say where the trees of woodland give way to the open grass of a meadow, the edge effect leads to an increase in abundance and variety of organisms. Transition zones accommodate some species characteristic of each community, others that belong to both, and also characteristic ecotone species. Gardens are like vastly extended woodland edge, which may explain, in part, the richness of their insect fauna.

The occurrence of hoverflies in my garden may be used to illustrate the concept of a garden as a system of ecotones. As described in Chapter 7, 30 of the species recorded in the period 1972–86 are usually associated with woodland or woodland edge, 11 with more open places, and 50, including such abundant species as *Platycheirus cyaneus*, range over both types of habitat. It is not so easy to distinguish ecotonal species, because adults of woodland species must regularly move to the woodland edge where there are more flowers, in order to feed. In the abundance of their flowers, gardens resemble wood-land edge, and in utilizing this resource, many garden hoverflies are ecotonal species. Hoverflies as a group are particularly suited to exploiting the patchi-ness of the garden habitat because adult and larval requirements are so different, and because larval life styles are so diverse. Structural hetero-geneity together with high plant diversity probably also account for the high diversity of Ichneumonidae (533 species), in that a garden encompasses a multitude of potential ecological niches, so that the diverse egg-laying requirements of females are met. It has often been shown that animal diver-sity is positively correlated with foliage height diversity (e.g. MacArthur & MacArthur 1961 for birds in deciduous forests; Murdoch *et al.* 1972 for Homoptera in old fields). Foliage height diversity is one of the aspects of structural heterogeneity in a garden, and forms another element of the garden habitat that leads to diversity of the fauna.

The extreme patchiness of the garden habitat is illustrated by patchiness in the occurrence of insects. This environmental heterogeneity is not always apparent to the observer, although the behaviour of insects confirms its reality. Some sprawling plants of spotted dead-nettle *Lamium maculatum* in my garden support caterpillars of the angle shades moth *Phlogophora meti-culosa* in all seasons, year after year, whereas others nearby are never eaten. Four species of hibernating hoverfly larvae (*Platycheirus cyaneus, Melanostoma mellinum, M.scalare* and *Syrphus ribesii*) were found one winter in a patch of *Saxifraga* sp., while another patch less than 3 m away seemed not to shelter any. The size and composition of the hoverfly catches of two Malaise traps operated in different parts of the garden in 1978 were quite different (see Chapter 7). Clumps of rosebay willowherb *Epilobium angustifolium* in direct summer sunlight have been found to support complex foodwebs including aphids on the leaves, adults and larvae of the 2-spot ladybird *Adalia bipunctata* and larvae of the hoverfly *Syrphus ribesii* eating the aphids, diplazontine ichneumonids parasitizing the hoverfly larvae, and vespid wasps, bumblebees (*Bombus* spp.) and hoverflies, such as *S. ribesii*, feeding on aphid honeydew. Full development of this food web seems to depend upon the plant being in a sunny situation.

In his study of the organization of a plant–arthropod association in simple and diverse habitats, Root (1973) found that the number of herbivore species was greater in the diverse habitats, and suggested that the complexity of a community arises from the juxtaposition of a large number of plant-based component communities. This is another way in which extreme patchiness, coupled with plant diversity, enriches the insect fauna of a garden. Insect diversity is high as a consequence of plant and structural diversity. At the same time, high plant diversity coupled with permanent succession make gardens extraordinarily bountiful yet markedly unstable environments. High animal diversity and narrow ecological niches are a feature of climax com-munities and were once generally assumed to be related to the stability of a community (Odum 1971). MacArthur (1955), for instance, thought that the number of possible energy pathways through a food web were a good indication of its stability. However, more recent investigation of community structure, particularly by model-building often using randomly assembled food webs, has tended to show that the stability of a community declines with the complexity of its interrelationships (Begon, Harper & Townsend 1986). What is clear is that the complex communities of stable environments, such as tropical rainforest, are particularly susceptible to man-made perturbation from outside, whereas the simpler communities of more variable environments in temperate regions are more robust and resist perturbation. The nature of the relationship between the stability and complexity of a community seems to vary in different environments and situations, but the consensus now is that there is a tendency for complex communities to be less stable than simpler ones.

The garden habitat accords with more recent analyses of the relationship between stability and complexity, in that it supports a complex yet unstable community. It is, however, a man-made and hence artificial situation, where both instability and complexity are inherent to the management of the habitat, permanent succession and high plant and structural diversity being intentionally induced by the gardener. The richness and variability of the animal component of the community derives from the way in which the plants are managed.

It is no accident that insects dominate the garden fauna, but rather a consequence of the attributes of insects and the structurally complex nature of the habitat. Southwood (1978) argued that small size, metamorphosis, seasonal activity, short generation times, and rapid breeding, all characteristics of insects, contribute to habitat exploitation in differing ways by sympatric species and therefore lead to greater diversity. Hutchinson & MacArthur (1959) suggested more than 30 years ago that small, diversified elements of an environmental mosaic permit a greater species diversity of small than of large animals. The garden habitat, unstable though it is, can thus support an astonishing diversity of insects. It is unusual in that it supports complex communities in an inherently unstable habitat.

THE SUBURBAN LAWN

The lawn is so different from the rest of a garden, and is usually so sharply delimited from it both physically and functionally, that it is perhaps best regarded as a separate ecosystem. Maintenance of a lawn by mowing is akin to grazing, the lawn-mower being the grazer, and the attributes of grasses that make for a good lawn are adaptations to being grazed. Most good lawn grasses are tufted perennials, which spread by means of underground stems that produce a high proportion of leafy shoots. Grass leaves grow from basal meristems, and so quickly make good loss from cropping. Cutting the leaves prevents production of flowering stems, and stimulates both thickening of the tuft and development of buds into leafy shoots. The buds are at or below ground level, where they are protected by the bases of old leaf sheaths and the soil from grazing (mowing) and trampling. Repeated mowing thus improves grass quality as well as promoting establishment of thick, dense turf. To understand grass and its reaction to mowing, it is necessary to acknowledge the interdependence, indeed co-evolution, of grass and large grazing ungulates. Grazers keep grasslands at an early successional stage by preventing the establishment of other sorts of plants, a successional stage so well-defined that it is sometimes called a grazing climax. A lawn-mower similarly maintains a lawn as a grazing climax. Some plants, however, can co-exist with lawn-mowers, because of low, compact growth and short flowering stems. My garden lawn, while being a compact, resilient turf, is not a pure stand of grass, but is interspersed with such plants as buttonweed *Cotula coronopifolia*, yarrow *Achillea millefolium*, bird's-foot-trefoil *Lotus corniculatus*, field wood-rush *Luzula campestris*, white clover *Trifolium repens*, hop trefoil *T.campestre*, dandelion *Taraxacum officinale*, greater plantain *Plantago major*, selfheal *Prunella vulgaris* and daisy *Bellis perennis*.

Conventional practice is to remove not only weeds, dead grass and fallen leaves from a lawn, but also to collect and remove the grass-cuttings. Although I remove fallen leaves, and smothering weeds, such as dandelions, from my lawn, most so-called weeds are tolerated and cut grass is left *in situ*. Lawns are usually regarded as needing fertilizers, but this is to make good the losses caused by removal of grass-cuttings. Leaving the cuttings to decompose on the lawn is probably better for the lawn and obviates the need for fertilizers. From a conventionally managed 110-square-metre lawn in California, 63 kg dry weight of living and dead vegetation was removed by raking and mowing in the course of a year. This contained the equivalent of 3300 g of nitrogen, 960 g phosphorus and 1850 g potassium. Fertilizers and manure equivalent to 1720 g nitrogen, 430 g phosphorus and 860 g potassium had to be applied to maintain a healthy lawn. Maintenance of the lawn also entailed the expenditure of considerable quantities of energy in the form of manpower for mowing etc., petrol for driving the mower, electricity to pump water for irrigation, energy used in manufacturing fertilizers and other chemicals, and the energy content of seeds sown. Keeping the lawn 'reasonably attractive' required 1865 Calories per square metre per year, more than twice the 715 Cal/sq.m/yr used to grow an edible grass, maize (Falk 1976).

Many animals utilize a lawn as a source of food in one form or another, and so transfer and recycle energy and nutrients, although the Californian study identified man as by far the dominant herbivore and scavenger in an intensively managed lawn community (Falk 1976). Homoptera, moth caterpillars and other animals feed from grass-blades, and predatory ants, spiders and beetles run across the surface of the ground. Most of the activity, however, is within the litter layer and in the soil beneath, where leatherjackets (crane fly larvae) and moth caterpillars, such as those of the large yellow underwing *Noctua pronuba*, feed on grass roots. Earthworms, nematode worms, various sorts of fly larvae, Collembola, bacteria and Fungi contribute to the breakdown of dead leaves or roots in the litter layer and the soil. Birds, such as starlings and song thrushes, do particularly well for food on lawns, where they mainly extract lepidopteran larvae, leatherjackets and earthworms. For

Photograph 14.4. Not only is the lawn the most heavily used part of a garden, it also involves most expenditure of energy in its up-keep.

birds, a lawn represents concentrated food resources, in a site where they can feed freely right up to the edge but out in the open able to see potential danger. Falk (1976) found that food utilization per unit of area of lawn by suburban birds, at 46 Cal/sq.m/yr, greatly exceeded food utilization by birds in natural grassland (1.01–2.33 Cal/sq.m/yr). The flocking habit of such birds as starlings probably maximizes food utilization from a lawn, because they respond to the downward swoop of one to a good food source by following, and the cumulative alertness of a flock means that on average each individual can spend more time feeding, rather than looking around.

FOOD SUPPLY

As is to be expected in a habitat where primary production (production of plant material) is high, green leaves are utilized as food by a wide variety of abundant insects and other animals in the Leicester garden. The most important chewers of leaves are moth caterpillars and sawfly larvae, but they are also consumed by other herbivorous insects, such as butterfly larvae and flea beetles, and the abundant slugs and snails rasp away considerable quantities of green plant tissue. Heteroptera and, particularly, Homoptera suck nutrients from leaves and green stems, and presumably have an appreciable impact on the growth of plants. A variety of insects, including moth caterpillars, aphids, leatherjackets and wireworms (elaterid beetle larvae) feed in different ways on plant roots, as do abundant, though minute, Symphyla. Butterflies, moths, bees, wasps, and such flies as Syrphidae, take nectar from flowers, and bees, *Meligethes* beetles and many hoverflies utilize pollen. The actual substance of flowers, as opposed to products, is consumed by moth caterpillars, some beetles and earwigs. Fruit is eaten whole by birds, and

353

nibbled or sucked by moth caterpillars, adult moths and butterflies, some wasps, and flies of various sorts. Seeds are eaten by birds and also nibbled by beetles, moth caterpillars (e.g. the 'grubs' in pea pods) and other insects. Once it is dead, vegetation becomes much more uniform in quality, and animals that feed on it (primary decomposers) are less specific in what they eat. Slugs, earthworms, woodlice, millipedes, Symphyla, larvae of some hoverflies and other flies, staphylinid beetles, Collembola and a range of other small animals eat decomposing plant material.

Diverse and abundant predators attack the herbivores, and each other. As in all habitats, most of the species are predatory or parasitic, whereas most of the individuals are herbivorous. Adult and larval ladybirds, lacewing larvae and many hoverfly larvae eat aphids; vespid wasps feed to their larvae any soft-bodied insect they can catch; solitary wasps catch a range of insects and spiders as food for their larvae; carabid and staphylinid beetles and centipedes patrol the soil surface and climb plants after prey; predatory bugs hunt over flowers and vegetation; carabid larvae and centipedes hunt within the soil; ubiquitous ants attack anything they can overwhelm; spiders set snares all over the garden, or hunt over the soil and vegetation; long-legged harvestmen hunt on the ground and high on plants; birds search the vegetation, the ground and the air for prey; and so on. Nothing is safe from predation. Sometimes it is devious rather than direct, as in the case of ichneumonids, whose larvae develop in or on other insects and spiders, slowly killing them.

Many individual insects and other animals do, however, escape predation, but there is no accumulation of dead animals for, as in all habitats, scavengers abound. These are known as secondary or higher order decomposers, depending on whether they eat dead herbivores or dead predators. Many animals are omnivorous and, although usually predatory, feed on dead animal material when it is available. Calliphorid fly larvae abound in vertebrate corpses, which are often buried by scavenging silphid beetles, and vespid wasps chew meat for their larvae from damaged corpses. Woodlice, harvestmen, ants, calliphorid and sarcophagid flies, and a multitude of minute insects, mites and bacteria feed on dead animal material, and such scavengers as *Eristalis* hoverfly larvae ingest minutely fragmented animal matter in wet micro-habitats, such as rot-holes in trees or the bottom of the pond. Nothing is wasted but, whether alive or dead, is consumed by some organism. The ultimate consumption of dead and fragmented animal and plant material is rarely obvious, because it is achieved by microscopic bacteria and Fungi in the soil, but is evident from the eventual disappearance of all dead animals.

Much of the decomposition and consumption of dead plant material in a garden is concentrated by the gardener in the compost heap, creating a nutrient-rich compost that can be utilized on the garden. Unmethodical, rather casual, compost heaps, such as that in my garden, tend to be the focus of much visible activity. Wasps and various sorts of flies buzz around it, and hoverflies, such as *Syritta pipiens*, and psychodid flies visit to lay eggs. The top, drier layers of the heap abound with woodlice, and the moist lower regions are a writhing mass of brandling worms. These are the readily visible decomposers but, on a smaller scale, the compost heap is alive with nematode worms, mites and collembolans, and at a microscopic level, it teems with bacteria and Fungi.

I have identified many food chains in my garden, and many more, and the way in which they interlock with others to form food webs, can be inferred. Rearing moth caterpillars, sawfly larvae and predatory hoverfly larvae found in the garden has led to recognition of plant–moth and plant–sawfly food links and plant–aphid–hoverfly food chains. Sixty-eight species of moth caterpillars were found on 115 species of plants, giving a total of 298 plant–moth links identified to species (Table 6.2, p. 136). In many instances, however, reared moth caterpillars of 11 species ended up producing not moths but ichneumonids (see Table 8.5, p. 220), and in a number of these cases, the moth caterpillar could be identified to species, giving five accurate plant–moth–ichneumonid food chains. In most other cases, the caterpillar could be identified to family, but no further; nevertheless, this still enabled recognition of a food chain to a fairly detailed level. More than 40 per cent of the species of ichneumonids captured in the Malaise trap in 1972–73 and more than 29 per cent of the individuals parasitize Lepidoptera (Table 8.3, p. 216), so the plant–moth–ichneumonid food chain must be acknowledged as one of the important nutrient and energy pathways in the garden. Ten species of sawflies have been reared from larvae eating nine different species of plants (Chapter 11), establishing ten plant–sawfly feeding links; two species of ich-neumonid have been reared from sawfly larvae on birch (Table 8.5), giving two plant–sawfly–ichneumonid food chains. A total of 11.9 per cent of the ichneumonid species trapped in the garden 1972–73, and 6.4 per cent of the individuals, parasitize tenthredinid sawflies. Of the nine species of hoverflies bred from predatory larvae found in the garden (see Chapter 7), five were feeding on six known species of aphids, giving a total of nine accurate plant–aphid–hoverfly food chains. Eight species of ichneumonids were reared from hoverfly larvae giving seven known plant–aphid–hoverfly–ichneumonid food chains, in five of which all the elements were identified to species (Table 8.5). The plant–aphid–hoverfly–ichneumonid food chain is evidently another important nutrient and energy pathway in the garden, 9 per cent of the species of ichneumonid trapped in 1972–73 and more than 27 per cent of the individuals being parasites of aphid predators (Table 8.3).

Ichneumonids are by no means the only consumers of hoverflies with predatory larvae. For instance, the wasp *Vespula vulgaris* has been seen to catch and eat *Syrphus ribesii*, and one was also caught and eaten by the spider *Enoplognatha ovata*, which had a scaffolding web on a cabbage leaf. This latter gives another accurate food chain: cabbage – *Brevicoryne brassicae* (aphid) – *Syrphus ribesii* – *Enoplognatha ovata*. Other elements of the food web including *Syrphus ribesii* can be inferred because *Ectemnius* spp. (solitary wasps) catch

and store hoverflies for their larvae to eat, and swifts catch hoverflies, including *S.ribesii* in the air, both to eat and to feed to their young (Parmenter &
Owen 1954). Some hoverfly larvae are almost certainly eaten by marauding
general predators, such as carabid and staphylinid beetles, and they, of
course, are not the only predators of aphids. Ladybirds were abundant in the
garden, particularly in some years (see Chapter 10), and these and their larvae
must make considerable inroads on the garden's aphids. Ladybirds, too, are
eaten by swifts, and may fall foul of other predators, as when newly hatched
7-spot larvae were seen to fall into a hammock web of the spider *Tegenaria
domestica*. Other spiders, such as *Tibellus oblongus* seen hunting among aphids
and ladybird larvae on the thistle *Cirsium arvense*, undoubtedly also eat the
larvae.

Ichneumonids consume many other sorts of garden insects (Table 8.3), and
they also parasitize spiders and spider eggs, when they are probably operating as fourth or higher-order consumers. Indeed, it is evident by observation
that both the major food chains described, i.e. plant–moth–ichneumonid and
plant–aphid–hoverfly–ichneumonid are only parts of complex food webs. For
instance, caterpillars of the dot moth *Melanchra persicariae* and cabbage moth
Mamestra brassicae have each been found feeding on 38 species of plants, and
such aphids as *Aphis fabae* are also markedly polyphagous. Spittlebugs,
whitefly larvae, and many beetles and bugs, as well as sawfly larvae, also
consume plants. Caterpillars are eaten by blue tits, blackbirds and other
insectivorous birds as well as being collected for their larvae by social and
solitary wasps, and many probably fall prey to predatory beetles. Scavenging
Volucella hoverfly larvae eat the excrement of wasp larvae and sometimes the
larvae themselves. On the other hand, such insects as hoverflies, wasps,
bumblebees and ants lap aphid honeydew, and when winged reproductive
ants rise into the air, they are eaten by house sparrows, starlings, house
martins, swifts, black-headed gulls and other birds. Lacewing larvae, as well
as predatory hoverfly larvae and ladybirds, eat aphids, which are often
parasitized by minute chalcid wasps. As has been shown, many other
animals consume hoverflies, and, conversely, many predators are catholic
feeders, the main constraint being size of prey relative to their own size and
ability to deal with prey.

So complex are the interconnections of garden food webs, that it is impossible to recognize trophic levels, other than to identify the instantaneous
trophic level of a feeding consumer. For instance, when a swift eats flying
aphids it is operating as a secondary consumer, but when it eats ladybirds it is
a tertiary consumer. A further complication is provided by the important,
although usually less evident, decomposer food chains. These cannot be
separated from consumer food webs but form part of an intricate feeding
nexus because, for instance, song thrushes and blackbirds consume earthworms, although the latter are decomposers. Furthermore, after death, even
top predators become incorporated into decomposer food chains, and such
animals as harvestmen, eat living prey and corpses. It is evident that in the
bountiful garden situation, food supply is good for a multitude of animals

with different feeding habits. This is a major factor in promoting the richness of the garden fauna and the complexity of the feeding relationships.

A garden such as mine evidently accommodates many complicated food webs which may, indeed, interlock, forming one vast, intricate food web. The garden community is not stable, but is constantly changing in composition as species disappear and new ones arrive. It remains, however, a large and varied community, because contrived plant diversity, permanent succession and structural heterogeneity diversify feeding opportunities, generate many different food chains and contribute to a multitude of ecological niches. This bountiful, intricately structured, yet unstable habitat is created and maintained by the activities of gardening *per se*. It might be asked why, when there are such abundant resources for herbivores in my garden, they do not regularly get out of hand as pests. Part of the answer is that the very complexity of feeding relationships in the garden means that herbivores are in large degree controlled by predators of one sort or another. Also, as explained in Chapter 3, mixed planting reduces damage from so-called pests and, together with maintenance of good ground cover, encourages predators. Finally, there is the question of what is a pest in a garden. For most people, it is whatever interferes adversely with their enjoyment of what they grow. However, pollen beetles can be shaken from cut flowers, the odd caterpillar or slug can be picked off a cabbage, and anyway, like most gardeners, I grow more than I can cut or eat. There are no pests, because everything in my garden is a source of interest and enjoyment.

15

England's most important nature reserve?

The rich fauna of my garden begs comparison with the fauna not only of other gardens but also of other sorts of areas including nature reserves. For many of the groups investigated, the garden fauna includes an astonishingly large proportion of the species recorded in the British Isles, and it is worthwhile assessing what all these animals are doing in the garden. The diversity and abundance of the garden fauna is a consequence of the activities of gardening, and so it is highly probable that other gardens, perhaps gardens in general, are as rich. That being the case, gardens would seem to be of considerable significance for conservation.

GARDEN FAUNA IN RELATION TO THAT OF THE BRITISH ISLES

The fauna of the British Isles has been extensively investigated over many years, and is not only well-known but also well-documented. It is a rare occurrence for new species to be added to the British list, other than in three sorts of circumstance. Firstly, when a group is taxonomically difficult, individuals are relatively small, and few people have specialized in its study; secondly, consequent on new arrivals in the country, either naturally or as a result of man's activities; and thirdly, when taxonomists split what had hitherto been regarded as one species into two. Overall, however, the British fauna is so well known, that it provides a firm background against which to view the garden fauna, and with which the latter can be compared. The total numbers of animals and plants in different taxonomic groups recorded in the study garden in the 15 years 1972–86 are shown in Table 15.1. This list cannot be considered as exhaustive because, for many practical reasons, only selected groups have been identified; the soil fauna, for instance, is totally unknown. The total of 2204 species is a minimum figure anyway because, for instance, unidentified Aleocharinae (Staphylinidae) and *Meligethes* (Nitidulidae) among the beetles, probably include more than one species. The 2204 species comprise 422 plants, 1602 insects, 121 other invertebrates, and 59 vertebrates.

A number of animal groups are known to be represented in the Leicester garden by an astonishingly large proportion of British species (Table 15.2),

Table 15.1 *Numbers of species of plants and animals recorded in the Leicester garden,*
1972–86

Flowering plants (including grasses)	384
Flowerless plants	38
Flatworms (Turbellaria)	3
Snails and slugs (Gastropoda)	17
Earthworms (Oligochaeta)	5
Leeches (Hirudinea)	1
Woodlice (Isopoda)	8
Centipedes (Chilopoda)	7
Millipedes (Diplopoda)	5
False-scorpions (Pseudoscorpiones)	1
Harvest-spiders (Opiliones)	10
Spiders (Araneae)	64
Dragonflies (Odonata)	4
Grasshoppers and crickets (Orthoptera)	3
Earwigs (Dermaptera)	1
Psocids (Psocoptera)	14
Bugs (Hemiptera)	94
Lacewings and allies (Neuroptera)	18
Butterflies (Lepidoptera)	21
Macro-moths (Lepidoptera)	263
Micro-moths (Lepidoptera)	80
Caddisflies (Trichoptera)	4
Hoverflies (Syrphidae–Diptera)	91
Other flies (Diptera)	49
Sawflies (Symphyta)	58
Ichneumons (Ichneumonidae–Hymenoptera)	533
Other parasitic wasps (Serphidae and Heloridae–Hymenoptera)	24
Ants (Formicidae–Hymenoptera)	2
Wasps (Hymenoptera)	41
Bees (Hymenoptera)	51
Beetles (Coleoptera)	251+
Amphibians	3
Birds	49
Mammals	7
GRAND TOTAL	2204+

particularly bearing in mind that Leicestershire is relatively poor faunistically.
These groups tend to be flying insects, which is only to be expected, partly
because of their powers of dispersal and partly because trapping methods
used in the garden are particularly suited to catch them. In addition, many of
the well-represented groups are flower visitors, which are not only attracted
to gardens, but fare particularly well there. There are, however, exceptions to
these broad generalizations, harvest-spiders, for instance, being represented
by nearly half of British species, although representation of spiders, at about a
tenth of the British list, is probably more typical of non-insect invertebrates.

Table 15.2 *Numbers of species in Leicester garden and in British Isles of groups that have been well-listed in the garden*

	Species in garden	Species in British Isles	Garden as % British Isles
Native flowering plants	166	c.1500	11.1
Centipedes	7	46	15.2
Harvest-spiders	10	23	43.5
Grasshoppers and crickets	3	28	10.7
Earwigs	1	4	25.0
Lacewings and allies	18	55	32.7
Butterflies	21	62	33.9
Macro-moths	263	881	29.9
Hoverflies	91	256	35.6
Sawflies	58	481	12.1
Ichneumons	533	2028	26.3
Ants	2	36	5.6
Wasps	41	297	13.8
Bees	51	252	20.2
Ground beetles (Carabidae)	28	342	8.2
Ladybirds (Coccinellidae–part)	9	24	37.5

The groups known to be well-represented in the Leicester garden are among those that have been intensively investigated. I can say with confidence that they are well-represented because they have been well-listed. Species listed for other groups cannot be used in the same way; they have not been well-listed (studied in detail) and so representation in the garden of the British fauna cannot be assessed. For instance, many individual earwigs have been collected and identified, but all are the same species and I am confident that there is only one species of earwig in the garden; no effort has been made, however, to collect false-scorpions, and the garden list of one species means only that one was noticed and identified, not that this is the only species in the garden. Moreover, many groups, such as mites (Acari) and the majority of families of flies (Diptera), have not been investigated at all.

Among insects, the large sawfly family Tenthredinidae is represented by 14.1 per cent, more than a seventh, of the British species, and Coccinellidae (including ladybirds), Ichneumonidae, and the moth family Geometridae by 26.2 per cent, 26.3 per cent and 29.7 per cent, respectively, more than a quarter. Garden lacewings and their allies make up 32.7 per cent, almost a third of the British list, and garden butterflies, moths of the large family Noctuidae, and hoverflies (Syrphidae) 33.9 per cent, 39.2 per cent and 35.6 per cent, respectively, a third or more. Solitary bees represent 16.3 per cent, rather less than a sixth of British species, but social bees (Apidae) 56.0 per cent, more than half. Similarly with wasps, 12.1 per cent, or less than an eighth of British solitary species have been recorded in the garden, but all but one, i.e. 6/7, of the social species (Vespidae) have been. Evidently, butterflies, moths, hoverflies, and social bees and wasps are particularly well represented

Photograph 15.1. High plant diversity is one of the features that enriches a garden and leads to the presence in the Leicester garden of a large proportion of the British fauna of many animal groups. Among the plants here are marguerites Chrysanthemum frutescens, *runner beans* Phaseolus coccineus, *sweet peas* Lathyrus odoratus, *pot marigolds* Calendula officinalis, *tomatoes* Lycopersicon esculentum, *cabbages* Brassica oleracea, *radishes* Raphanus sativus, *cranesbill* Geranium cinereum, *African marigolds* Tagetes erecta *and dwarf convolvulus* Convolvulus tricolor.

in the garden and general observation would suggest that this is true of gardens in general.

Animal groups that have been well listed in the garden are represented there by between 5.6 per cent and 43.5 per cent of species known for the British Isles, native plants by 11.1 per cent (Table 15.2). Extrapolating these proportions for animals to the total terrestrial fauna of the British Isles, it is possible to estimate within upper and lower limits the total number of species of animals in the garden. According to the Biological Records Centre, Institute of Terrestrial Ecology, there are about 30 000 species of terrestrial animals in the British Isles, and on this basis there are between 1680 and 13 050 species in my garden. I know, despite the selective nature of my investigation, that there are more than the lower figure and so can anticipate vastly increasing the species list by looking at more groups. Insect groups are represented in the garden by up to 37.5 per cent of species known for the British Isles, and as the total British list of insects is about 22 500, there may be as many as 8450 species of insects in the garden.

The record of the flora and fauna of the 156.8 ha Monks Wood Nature

Reserve in Huntingdonshire, produced ten years after the opening of Monks Wood Experimental Station (Steele & Welch 1973), lists by group species recorded there. Numbers of species of woodlice, lacewings and micro-moths were more or less the same as were recorded in the garden, numbers of Ichneumonidae and of solitary and social bees and wasps were much smaller than in the garden, and numbers of species of harvest-spiders, spiders, tenthredinid sawflies, coccinellid and other beetles, hoverflies, Geometridae, Noctuidae and other macro-moths were greater, often twice as great. Many of these larger species lists at Monks Wood are to be expected in view of the much larger area involved, and a group such as Coleoptera is bound to be better represented where there is woodland. The smaller species lists of Ichneumonidae, and bees and wasps are more interesting. In the case of Ichneumonidae, the discrepancy clearly illustrates the unpopularity of the group and the difficulties inherent to its study. With bees and wasps, however, the explanation is almost certainly that a garden is a far more favourable habitat for them, particularly in view of the abundance of flowers there.

Relatively rare, or unexpected, species have turned up in the garden in most groups, such as the silver-washed fritillary among butterflies (Chapter 5), the pyralid moth *Sceliodes laisalis* (Chapter 6), and the hoverfly *Ferdinandea ruficornis* (Chapter 7), but it is in little-investigated groups, such as small parasitic wasps, that the real rarities have been recorded. Five species of Serphidae and 15 species of Ichneumonidae trapped or reared in my garden are believed to be additions to the British list, and four further ichneumonids are hitherto undescribed species (Chapter 8). This clearly illustrates the advantages, in terms of discoveries, of looking at a relatively little-known group.

The size of the species list for the garden and the inclusion in it of unexpected species demonstrate how much is to be gained by looking in detail, over a long period, at a small area of land. Gilbert White, the doyen of British naturalists, realized this as long ago as 1768, when he wrote to Thomas Pennant 'all nature is so full that that district produces the greatest variety which is the most examined' (White 1947). The only published long-term studies comparable to that of the garden are of much larger areas, such as the investigation by Francis Evans and collaborators of the natural history of an abandoned field in Michigan, USA (Evans 1975), and the ongoing study of Wytham Wood, near Oxford (Elton 1966). They have certainly borne out Gilbert White's dictum, as indeed did Frank Lutz, when he investigated the insect fauna of his suburban New York garden (Lutz 1941). The majority of nature reserves are far less well known than my garden, usually for lack of specialists with time to collect and identify species; evaluations of areas threatened by development or proposed as nature reserves have to be conducted on far less information. Complete faunal lists for nature reserves or equivalent areas are hard to come by, even if they exist, and, indeed, Sites of Special Scientific Interest have largely been designated on the basis of their floras.

The fauna of Buckingham Palace garden was investigated, as far as possible, in the 1960s (Bradley & Mere 1966, McClintock *et al.* 1964). Birds were
well-represented, with 60 species, but there were only two species of
amphibians and two of mammals. Fifty-seven species of spider were
recorded, one harvest-spider, one false-scorpion, 27 terrestrial mites, three
myriapods, two woodlice, four slugs but no snails, and 12 earthworms.
Among insects, Lepidoptera were intensively investigated, yielding 367 species, but the lists for other insect groups are far less complete, with one
Thysanura, 36 Collembola, two Orthoptera, one Dictyoptera, one Dermaptera, two Psocoptera, two Ephemeroptera, one Odonata, 11 Thysanoptera,
127 Hemiptera (including 59 aphids), three Neuroptera, 13 Trichoptera, 87
Coleoptera, 42 Hymenoptera (including 16 ichneumonids), and 95 Diptera
(including only seven syrphids). The large numbers of Collembola,
Thysanoptera, Hemiptera and mites, compared with numbers for the Leicester garden, show that specialists in these groups participated. On the other
hand, numbers of ichneumonids, in particular, on my garden list are high
because a specialist in the group has looked at trap samples. This illustrates
the bias introduced into species lists by the interests and expertise of those
involved in the investigation.

Often an area has been particularly well surveyed, using several collecting
methods, for one particular group, as at Malham Tarn Estate Nature Reserve
in North Yorkshire, which is one of the best-surveyed sites in the country for
Diptera. The survey was initiated in 1954–58, and by the end of 1983, the
Diptera list included more than 1000 species and was believed still to be
incomplete (Disney *et al.* 1982, Disney 1986). This ongoing survey has
involved a range of collecting methods, the expertise of many specialists, and
a lot of time and effort. The much smaller number of Diptera species recorded
in my garden (140) is partly a consequence of a much smaller area, but also
results from the incompleteness of the garden survey. One family of Diptera
(Syrphidae) has been looked at in detail in the garden, others have been
partially investigated, but many are totally unknown. That the difference
stems from a difference in effort is borne out by the sizes of the species lists of
Syrphidae (a group on which a lot of effort has been expended in the garden)
for the two sites – 91 for the garden, 79 for the far greater area of Malham Tarn
Estate Nature Reserve.

WHAT ARE SO MANY ANIMAL SPECIES DOING IN THE GARDEN?

A brief survey of the garden fauna and its activities would perhaps give the
impression that relatively few of the animals are residents and that most are
temporary visitors for food. This might be true of flying insects and
vertebrates, but it cannot be so for terrestrial invertebrates such as snails,
slugs, earthworms, woodlice, centipedes, millipedes, harvestmen and
spiders. These are evidently feeding and breeding in the garden, and are
undoubted residents. As for the vertebrates, all three amphibians have now
bred in the pond (a toad in 1987) and evidently feed in the garden, ten species

Photograph 15.2. *Trees provide deep shade in a garden and enhance its structural heterogeneity.*

of birds have definitely bred and another three possibly so, and among mammals, house mice have bred in an out-house. All these must be classed as garden residents. Another 18 species of birds, and probably four of mammals, regularly feed in or over the garden, although they had not bred there up to 1986.

The situation is, however, somewhat more difficult to assess for flying insects, especially as so many records came from Malaise trap captures rather than observation. The picture is clearer for herbivores than it is for carnivores, because they are, on the whole, more sedentary and easier to observe. Four species of butterflies definitely bred in the garden, and the meadow brown would also have done so had the lawn not been cut; by definition, their caterpillars fed in the garden, as did the adults of these and the other 16 recorded species (Chapter 5). At least four of the garden butterflies must thus be classed as residents. The feeding situation is harder to assess with moths, for some may have been drawn in to the mercury-vapour light without ever settling at flowers. However, at least 20 per cent of moths recorded in the garden have bred there and can thus be classed as residents. Like butterflies and moths, adult sawflies are mobile, and may be caught in the Malaise trap on temporary visits to the garden, but at least ten species are known to have bred (Chapter 11), so nearly 17 per cent of the recorded sawflies are residents. A rather larger percentage, 30 per cent, of Homoptera have been recorded feeding and breeding on garden plants (Chapter 11), and this too is undoubtedly a minimum figure as many small Auchenorrhyncha are high on trees out of view. The general impression is thus that at least 20 per cent, and probably more than 30 per cent, of the insects recorded in the garden, that are herbivorous at some stage of their lives, are residents.

A large proportion, if not all, carnivorous adult insects trapped in the garden probably feed there, predators tending to take food as and when it is available. Such insects as hoverflies with predatory larvae are flower-feeders as adults, when they certainly feed in the garden. Residence, as with herbivores, can be recognized on the basis of breeding, but there are relatively fewer records of breeding carnivores, largely because their presence and activities are more difficult to monitor. Probably the majority of the 52 species of Heteroptera trapped (not all carnivores, however) and of the 28 species of carabid beetles, bred in the garden, because they are not particularly prone to flight. Of the 57 species of hoverflies with predatory larvae recorded in the garden, nine (16 per cent) have been found breeding, and about 27 per cent of hoverflies in general are assumed to do so (Chapter 7). Twenty species of Ichneumonidae, which as parasitoids are specialized predators, have been bred from parasitized larvae of moths, sawflies and hoverflies found in the garden, and ichneumonid larvae have been found on spiders (Chapter 8). This is only 4 per cent of the number of ichneumonid species recorded in the garden, but the proportion that breed is certainly larger, because I have reared in captivity only a few types of insects. The overall figure for carnivorous insects breeding in the garden is probably at least 20 per cent and possibly higher. That it is erroneous to assume that insects trapped rarely in

Photograph 15.3. Are suburban gardens England's most important nature reserve?

the garden and rare in county or country are chance visitors, is demonstrated by the ichneumonid *Hyposoter tricolor*. When two individuals were caught in the Malaise trap 1972–74 it was a new record for Britain, but subsequently two were bred from geometrid larvae feeding in the garden (Chapter 8).

The answer to the question 'What are so many animal species doing in the garden?' would thus appear to be that a large proportion, if not the majority, are feeding, and from 20 per cent to all, depending on taxonomic group and mobility, are breeding and hence resident. The many species of flying insect caught only once in the Malaise trap (about 25 per cent in total, ranging from 12.1 per cent in hoverflies to 32.7 per cent in Heteroptera) are regarded as vagrants, chance visitors that happen to get trapped in the course of their wanderings.

GARDENING FOR CONSERVATION

I manage my garden in such a way as to enhance it as a habitat. Every effort is made to grow plants whose flowers are attractive to insects or fruits attractive to birds. Furthermore, flowering is prolonged, where possible, by dead-heading. High plant diversity is maintained, and cultivation of plants of different heights and growth form contributes to structural heterogeneity. Pruning and clearing is kept to a minimum, and delayed until winter whenever possible. Good ground cover is maintained, both by cultivated plants and, failing that, by weeds. Above all, and of paramount importance in a wildlife garden, no

Photograph 15.4. The planting of French marigolds Tagetes patula *among the cabbages is one of the features that enrich a garden.*

pesticides or other noxious chemicals are used. The result is a lush garden, alive with animals, although one that still furnishes produce including flowers, herbs, salads, vegetables and fruit, because mixed planting keeps so-called pests down to manageable proportions.

As has been stressed earlier, the study garden with its rich wildlife is not a wilderness, but is a neat, productive suburban garden. There is no reason to suppose that it differs fundamentally from other gardens, and every reason to suppose that gardens in general are good habitats, rich in animal life. This is particularly so with the prevalence since about 1975 of so-called wildlife gardening. A recent book on creating a wildlife garden (Gibbons & Gibbons 1988) describes such gardens as 'oases of diverse, semi-natural, and controlled environments, where nature can find some sanctuary from pesticides and habitat loss.'

Wildlife gardens are seen as one of the answers to a dilemma that increasingly perplexes Western man – how, in this modern civilization, to preserve the plants, animals and habitats that continue to be destroyed by the process of civilization? Transport, housing, amenities for all, and generation of the wealth that pays for emancipation from poverty, disease and ignorance, have eroded the natural world and threaten its very existence. Towns have spread to the detriment of the countryside, and modern agriculture has made many wild plants and animals rare. The same processes of civilization have allowed for the development of new feelings, sensibilities and values about the world that is being destroyed. Nineteenth and twentieth century legislation on conservation has its roots in an increasing wish among the many, particularly middle class, amateur natural historians of the eighteenth century, to protect wild plants and animals that for centuries had been despised and systematically destroyed. An attempt to resolve the dilemma of how to retain elements of the past while living in the present, has led to a vigorous conservation movement and the establishment of nature reserves and other conservation areas, although these have been described as 'fantasies which enshrine the values by which society as a whole cannot afford to live' (Thomas 1983). Nevertheless, nature reserves of one sort or another, that conserve animals and plants, species and communities, and habitats, satisfy intellectual curiosity and give aesthetic pleasure, as well as providing interest and recreation.

The most cited criteria for evaluation of a wildlife conservation area are diversity of habitats and/or species, naturalness, rarity of habitats and/or species, and area (Usher 1986). Gardens would appear to meet the criterion of diversity of species, some of which may be rare, and, collectively, the 485 000 ha of gardens in Britain meet the criterion of area. Gardens are by definition unnatural, and the medley of micro-habitats that they contain are created. On diversity of species alone, however, a garden such as mine would rate highly for conservation. Describing the total garden area in terms of satisfying the area criterion is no doubt misleading, because the garden area is fragmented into millions of individual gardens, each under different management. This, however, may be no bad thing. The value given to a large area for conservation purposes stems from island biogeography theory, which states that larger areas hold more species (MacArthur & Wilson 1967). Nature reserves have been likened to habitat islands in an inhospitable sea of environment, and so island biogeography theory is thought to apply to them. However, a number of small reserves may have more species than a single large one of equivalent area (Higgs & Usher, 1980), and this almost certainly arises from habitat differences between reserves (Gilpin & Diamond 1980). Indeed, it is almost certainly increased habitat diversity that makes a larger island richer in species than a small one. It seems that it is quality and diversity of habitat that make for a good nature reserve, because it can serve as a refuge for more species, and that the island biogeography theory should not be applied uncritically to nature reserves (Simberloff 1986).

The vast number of small gardens in England, no two exactly alike, are, and with the recent popularity of wildlife gardening are increasingly becoming, a complex of small nature reserves, whose collective species list must be far larger than would exist on the land under one management policy, or perhaps even naturally. The 1700 reserves held in 1987 by the 46 county naturalists' trusts affiliated to the Royal Society for Nature Conservation collectively accommodated 17 000 ha of woodland and 7600 ha of meadows and pastures (Perring 1987). This is a major holding in terms of conservation, and valuable because the reserves are natural areas, but the area is small compared with total garden area. The reserves held by the county naturalists' trusts are in many cases last-ditch attempts to conserve a fragment of a habitat that is under threat locally or nationally. They are a bastion against the destruction of communities and habitats by development. But the garden habitat is in no sense threatened. On the contrary, the encroachment of towns and cities on the countryside means the creation of more gardens, and with the movement towards gardening for wildlife, and widespread anxieties about use of pesticides, gardens are becoming more and more hospitable as habitats and refuges. It is perhaps even more relevant now than it was in 1975 (Owen & Owen 1975) to ask 'Are suburban gardens England's most important nature reserve?'

Scientific papers, articles and books based wholly or partly on ecological research in the Leicester garden

Archer, M.E. (1985). Population dynamics of the social wasps *Vespula vulgaris* and *Vespula germanica* in England. *Journal of Animal Ecology* **54**: 473–85.

Archer, M.E. (1990). The solitary aculeate wasps and bees (Hymenoptera: Aculeata) of an English suburban garden. *Entomologist's Gazette* **41**: 129–142.

Gilbert, F.S. & Owen, J. (1990). Size, shape, competition, and community structure in hoverflies (Diptera: Syrphidae). *Journal of Animal Ecology* **59**: 21–39.

Owen, D.F. (1975a). Estimating the abundance and diversity of butterflies. *Biological Conservation* **8**: 173–83.

Owen, D.F. (1975b). The efficiency of blue tits *Parus caeruleus* preying on larvae of *Phytomyza ilicis. Ibis* **117**: 515–16.

Owen, D.F. (1976a). The suburban nature reserve. *Country Life*, 20 May.

Owen, D.F. (1976b). Ladybird, ladybird, fly away home. *New Scientist* **71**: 686–7.

Owen, D.F. (1976c). The conservation of butterflies in garden habitats. *Environmental Conservation* **3**: 285–90.

Owen, D.F. (1977a). Hoverflies in the garden. *Country Life*, 17 March.

Owen, D.F. (1977b). Buddleia is beautiful. *Country Life*, 1 September.

Owen, D.F. (1977c). Painted ladies in the suburbs. *New Scientist* **75**: 9–11.

Owen, D.F. (1978a). Insect diversity in an English suburban garden. In *Perspectives in Urban Entomology*, ed. G.W. Frankie & C.S. Koehler, pp. 13–29. Academic Press.

Owen, D.F. (1978b). The effect of a consumer, *Phytomyza ilicis*, on seasonal leaf-fall in the holly, *Ilex aquifolium. Oikos* **31**: 268–71.

Owen, D.F. (1978c). Striped friend or fearsome foe? *Country Life*, 18 May.

Owen, D.F. (1978d). Abundance and diversity of bumblebees and cuckoo bees in a suburban garden. *Entomologist's Record* **90**: 242–4.

Owen, D.F. (1979a). Hawk moths at home. *Country Life*, 17 May.

Owen, D.F. (1979b). The holly tree's leaf miner. *Country Life*, 19 July.

Owen, D.F. (1980a). Feeding on foreigners. *Country Life*, 20 March.

Owen, D.F. (1980b). Fooled by trickery. *Country Life*, 26 June.

Owen, D.F. (1981). Dining off one's host. *Country Life*, 19 March.

Owen, D.F. (1982/83). Fluctuations in abundance of Coccinellidae. *Entomologist's Record* **94**: 225–8, **95**: 29–31.

Owen, D.F. (1983a). A hole in a tent or how to explore insect abundance and diversity. *Contributions of the American Entomological Institute* **20**: 33–47.

Owen, D.F. (1983b). Lepidoptera feeding on the leaves and flowers of *Buddleia davidii. Entomologist's Record* **95**: 20.

Owen, D.F. (1986a). *Sceliodes laisalis* (Walker) (Pyralidae) in Leicestershire. *Entomologist's Record* **98**: 203.

Owen, D.F. (1986b). Do native plants support a richer lepidopteran fauna than alien plants? *Environmental Conservation* **13**: 359–62.

Owen, D.F. (1986c). 'It is found . . . on all sorts of low plants'. *Entomologist's Record* **98**: 179–80.

Owen, D.F. (1987a). Insect species richness on the Rosaceae: are the primary data reliable? *Entomologist's Gazette* **38**: 209–13.

Owen, D.F. (1987b). *Otiorhynchus porcatus* (Herbst.) (Col.: Curculionidae) in Leicestershire. *Entomologist's Record* **99**: 96.

Owen, D.F. (1988). Native and alien plants in the diet of *Philaenus spumarius* (L.) (Homoptera: Cercopidae). *Entomologist's Gazette* **39**: 327–8.

Owen, D.F. (1989a). How to monitor annual fluctuations in numbers of the immigrant moth, *Autographa gamma* (L.) (Lepidoptera: Noctuidae). *Entomologist's Gazette* **40**: 197–8.

Owen, D.F. (1989b). *Heliothis armigera* (Hubn.) (Lep.: Noctuidae) in Hampshire and Leicestershire. *Entomologist's Record* **100**: 66.

Owen, D.F. & Owen, J. (1974). Species diversity in temperate and tropical Ichneumonidae. *Nature* **249**: 583–4.

Owen, D.F. & Whiteway, W.R. (1980). *Buddleia davidii* in Britain: history and development of an associated fauna. *Biological Conservation* **17**: 149–55.

Owen, J. (1979). Hoverflies (Diptera: Syrphidae) of Leicestershire: an annotated checklist. *Transactions of the Leicester Literary and Philosophical Society* **73**: 13–31.

Owen, J. (1981a). Trophic variety and abundance of hoverflies (Diptera: Syrphidae) in an English suburban garden. *Holarctic Ecology* **4**: 221–8.

Owen, J. (1981b). Hoverflies in gardens. *Journal of the Royal Horticultural Society (The Garden)* **106**: 191–5.

Owen, J. (1982). The energy budget of a lawn. *Journal of the Royal Horticultural Society (The Garden)* **197**: 357–9.

Owen, J. (1983a). *Garden Life*. Chatto & Windus.

Owen, J. (1983b). Effects of contrived plant diversity and permanent succession on insects in English suburban gardens. In *Urban Entomology: Interdisciplinary Perspectives*, ed. G.W. Frankie & C.S. Koehler, pp. 395–422. Praeger.

Owen, J. (1983c). A garden and its ichneumonids: a microcosm of the countryside. *Countryside* **25**: 125–7.

Owen, J. (1983d). In my garden. *Natural World* **7**: 41.

Owen, J. (1983e). The most neglected wildlife habitat. *New Scientist* **97**: 9–11.

Owen, J. (1984). Marigolds among the cabbages. *Journal of the Royal Horticultural Society (The Garden)* **109**: 13–17.

Owen, J. (1988a). *Sericomyia silentis* (Harris) (Diptera: Syrphidae): the 44,834th hoverfly. *Entomologist's Record* **100**: 43–4.

Owen, J. (1988b). Wonder on the wing. *Organic Gardening* **1**(2): 46–8.

Owen, J. (1988c). Suspended animation. *Organic Gardening* **1**(3): 45–7.

Owen, J. (1988d). Who's who in Grub Street? *Organic Gardening* **1**(5): 37–40.

Owen, J. (1989a). Strictly for the birds. *Organic Gardening* **1**(6): 36–8.

Owen, J. (1989b). A bumble of joy. *Organic Gardening* **2**(1): 46–9.

Owen, J. (1989c). The fruits of their labours. *Organic Gardening* **2**(2): 25–8.

Owen, J. (1989d). Spitting image. *Organic Gardening* **2**(3): 27–30.

Owen, J. & Gilbert, F.S. (1989). On the abundance of hoverflies (Syrphidae). *Oikos* **55**: 183–93.

Owen, J. & Owen, D.F. (1975). Suburban gardens: England's most important nature reserve? *Environmental Conservation* **2**: 53–9.

Owen, J., Townes, H. & Townes, M. (1981). Species diversity in Ichneumonidae and Serphidae (Hymenoptera) in an English suburban garden. *Biological Journal of the Linnean Society* **16**: 315–36.

Other references

Alford, D.V. (1973). *Bumblebee distribution maps scheme: guide to the British species.*
   Reprinted from *Entomologist's Gazette* **21, 22** and **23**. Bee Research Association.
Allan, P.B.M. (1949). *Larval Foodplants.* Watkins & Doncaster.
Arvill, R. (1973). *Man and Environment.* Penguin.
Askew, R.R. (1971). *Parasitic Insects.* Heinemann Educational Books.
Baines, C. (1984). *Chris Baines' Wildlife Garden Notebook.* The Oxford Illustrated
   Press.
Baines, C. (1985). *How to make a Wildlife Garden.* Elm Tree Books.
Baker, R.R. & Sadovy, Y. (1978). The distance and nature of the light-trap response
   in moths. *Nature* **276**: 818–21.
Banks, C.J. (1959). Experiments with suction traps to assess the abundance of
   Syrphidae (Diptera), with special reference to aphidophagous species. *Entomologia
   Experimentalis Applicata* **2**: 110–24.
Barnard, P.C. (1978). A check-list of the British Neuroptera with taxonomic notes.
   Entomologist's Gazette **29**: 165–76.
Barnard, P.C. (1985). *Coniopteryx borealis* Tjeder in southern England (Neuroptera:
   Coniopterygidae). *Entomologist's Gazette* **36**: 2.
Bean, W.J. (1976). *Trees and Shrubs Hardy in the British Isles.* John Murray.
Begon, M., Harper, J.L. & Townsend, C.R. (1986). *Ecology.* Blackwell Scientific
   Publications.
Bell, G.A.C. (1970). The distribution of amphibians in Leicestershire. *Transactions of
   the Leicester Literary and Philosophical Society* **64**: 122–43.
Bellamy, D. (1984). *The Queen's Hidden Garden: Buckingham Palace's Treasury of Wild
   Plants.* David & Charles.
Betts, C. (1986). Symphyta. In *The Hymenopterist's Handbook,* ed. C. Betts, pp. 21–3.
   Amateur Entomologist 7.
Bhatia, M.L. (1939). Biology, morphology and anatomy of aphidophagous Syrphid
   larvae. *Parasitology* **31**: 78–129.
Birch, M.C. (1970). Structure and function of the pheromone-producing brush-
   organs in males of *Phlogophora meticulosa* (L.) (Lepidoptera: Noctuidae).
   Transactions of the Royal Entomological Society of London **122**: 277–92.
Bouskell, F. (1907). Insects. In *The Victoria History of the Counties of England:
   Leicestershire,* Vol. I, ed. W. Page, pp. 64–94. Archibald Constable, London.
Bradley, J.D. & Mere, R.M. (1966). Natural history of the garden of Buckingham
   Palace. Further records and observations 1964–65. Lepidoptera. *Proceedings of the
   South London Entomological and Natural History Society 1966*: 15–17.

Brattsten, L.B., Wilkinson, C.F. & Eisner, T. (1977). Herbivore–plant interactions: mixed function oxidases and secondary plant substances. *Science* **196**: 1349–52.

Bretherton, R.F. (1982). Lepidoptera immigration to the British Isles, 1969 to 1977. *Proceedings and Transactions of the British Entomological and Natural History Society* **15**: 98–110, **16**: 1–23.

Bristowe, W.S. (1971). *The World of Spiders*. Collins.

Bryan, P.W. (ed.) (1933). *A Scientific Survey of Leicester and District*. Prepared for the Leicester meeting 1933 of the British Association for the Advancement of Science.

Burton, J.A. (1974). *The Naturalist in London*. David & Charles.

Chinery, M. (1976). *A Field Guide to the Insects of Britain and Northern Europe*. Collins.

Chinery, M. (1977). *The Natural History of the Garden*. Collins.

Chinery, M. (1986). *The Living Garden*. Dorling Kindersley.

Churcher, P.B. & Lawton, J.H. (1987). Predation by domestic cats in an English village. *Journal of Zoology, London* **212**: 439–55.

Clark, J.T. (1979). Psocoptera activity as monitored by a Malaise trap. *Entomologist's Monthly Magazine* **114**: 94–5.

Clements, H.A.B. & Evans, I.M. (1973). Leicestershire bugs. *Transactions of Leicester Literary and Philosophical Society* **68**: 50–68.

Cloudsley-Thompson, J.L. (1968). *Spiders, Scorpions, Centipedes and Mites*. Pergamon Press.

Coe, R.L. (1953). Syrphidae. *Handbooks for the Identification of British Insects*, X(1). Royal Entomological Society of London.

Colebourn, P. & Gibbons, B. (1987). *Britain's Natural Heritage: Reading Our Countryside's Past*. Blandford.

Colyer, C.N. & Hammond, C.O. (1968). *Flies of the British Isles*. Frederick Warne & Co. Ltd.

Conn, D.L.T. (1972a). The genetics of the bee-like patterns of *Merodon equestris*. *Heredity* **28**: 379–86.

Conn, D.L.T. (1972b). The genetics of mimetic colour polymorphism in the large narcissus bulb fly, *Merodon equestris* Fab. (Diptera: Syrphidae). *Philosophical Transactions of the Royal Society of London* **264**: 353–402.

Cook, L.M., Mani, G.S. & Varley, M.E. (1986). Postindustrial melanism in the peppered moth. *Science* **231**: 611–13.

Coombs, C.F.B., Isaacson, A.J., Murton, R.K., Thearle, R.J.P. & Westwood, N.J. (1981). Collared doves (*Streptopelia decaocto*) in urban habitats. *Journal of Applied Ecology* **18**: 41–62.

Corbet, S.A. & Backhouse, M. (1975). Aphid-hunting wasps: a field study of *Passaloecus*. *Transactions of the Royal Entomological Society of London* **127**: 11–30.

Crowson, R.A. (1981). *The Biology of the Coleoptera*. Academic Press.

Danks, H.V. (1971). Biology of some stem-nesting aculeate Hymenoptera. *Transactions of the Royal Entomological Society of London* **122**: 323–99.

Davis, B.N.K. (1978). Urbanization and the diversity of insects. *Symposia of the Royal Entomological Society of London* **9**: 126–38.

Dempster, J.P. (1969). Some effects of weed control on the numbers of the small cabbage white (*Pieris rapae* L.) on Brussels sprouts. *Journal of Applied Ecology* **6**: 339–45.

Dempster, J.P. & Coaker, T.H. (1974). Diversification of crop ecosystems as a means of controlling pests. In *Biology in Pest and Disease Control*, ed. D. Price-Jones & M.E. Solomon, pp. 106–14. Blackwell.

Dickman, C.R. (1987). Habitat fragmentation and vertebrate species richness in an urban environment. *Journal of Applied Ecology* **24**: 337–51.

Dickman, C.R. & Doncaster, C.P. (1987). The ecology of small mammals in urban habitats. I. Populations in a patchy environment. *Journal of Animal Ecology* **56**: 629–40.

Disney, R.H.L. (1986). Assessments using invertebrates: posing the problem. In *Wildlife Conservation Evaluation*, ed. M.B. Usher, pp. 271–93. Chapman & Hall.

Disney, R.H.L., Erzinclioğlu, Y.Z., Henshaw, D.J. de C., Howse, D., Unwin, D.M., Withers, P. & Woods, A. (1982). Collecting methods and the adequacy of attempted fauna surveys, with reference to the Diptera. *Field Studies* **5**: 607–21.

Dixon, A.F.G. (1985). *Aphid Ecology*. Blackie.

Dohmen, G.P., McNeill, S. & Bell, J.N.B. (1984). Air pollution increases *Aphis fabae* pest potential. *Nature* **307**: 52–3.

Dony, J.G., Jury, S.L. & Perring, F.H. (1986). *English Names of Wild Flowers*. Botanical Society of the British Isles.

Eason, E.H. (1964). *Centipedes of the British Isles*. Frederick Warne & Co. Ltd.

Edwards, R. (1980). *Social Wasps: Their Biology and Control*. Rentokil Ltd.

Ellis, C.D.B. (1948). *History in Leicester 55 BC – AD 1900*. City of Leicester Publicity Department.

Elton, C.S. (1966). *The Pattern of Animal Communities*. Methuen.

Emery, M. (1986). *Promoting Nature in Cities and Towns*. Croom Helm.

Emmet, A.M. (ed.) (1988). *A Fieldguide to the Smaller British Lepidoptera*. British Entomological and Natural History Society.

Evans, F.C. (1975). The natural history of a Michigan field. In *Prairie: A Multiple View*, ed. M.K. Wali, pp. 27–51. University of North Dakota Press.

Evans, I.M. & Block, W.C. (1972). Fauna. In *Leicester and Its Region*, ed. N. Pye, pp. 160–80. Leicester University Press, for the British Association for the Advancement of Science.

Falk, J.H. (1976). Energetics of a suburban lawn ecosystem. *Ecology* **57**: 141–50.

Fisher, J. (1982). *The Origins of Garden Plants*. Constable.

Fisher, R.A., Corbet, A.S. & Williams, C.B. (1943). The relation between the number of species and the number of individuals in a random sample of an animal population. *Journal of Animal Ecology* **12**: 42–58.

Fitter, R.S.R. (1945). *London's Natural History*. Collins.

Fitter, R., Fitter, A. & Blamey, M. (1974). *The Wild Flowers of Britain and Northern Europe*. Collins.

Fitton, M.G., Graham, W.W.R. de V., Bouček, Z.R.J., Ferguson, N.D.M., Huddleston, T., Quinlan, J. & Richards, O.W. (1978). Hymenoptera. *Handbooks for the Identification of British Insects*, **11**(4). Royal Entomological Society of London.

Fitton, M.G. & Rotheray, G.E. (1982). A key to the European genera of diplazontine ichneumon-flies, with notes on the British fauna. *Systematic Entomology* **7**: 311–20.

Foelix, R.F. (1982). *Biology of Spiders*. Harvard University Press.

Forsythe, T.G. (1987). *Common Ground Beetles*. Naturalists' Handbooks no. 8. Richmond Publishing Co. Ltd.

Frank, K.D. (1988). Impact of outdoor lighting on moths: an assessment. *Journal of the Lepidopterists' Society* **42**: 63–93.

Frankie, G.W. & Ehler, L.E. (1978). Ecology of insects in urban environments. *Annual Review of Entomology* **23**: 367–87.

Free, J.B. & Butler, C.G. (1959). *Bumblebees*. Collins.

Gault, S.M. (1976). *The Dictionary of Shrubs in Colour*. Ebury Press and Michael Joseph.

Gemmell, R.P. (1977). *Colonization of Industrial Wasteland*. Edward Arnold.

Gibbons, B. & Gibbons, L. (1988). *Creating a Wildlife Garden*. Hamlyn.

Gilbert, F.S. (1981). Foraging ecology of hoverflies: morphology of the mouthparts in relation to feeding on nectar and pollen in some common urban species. *Ecological Entomology* **6**: 245–62.

Gilbert, F.S. (1986). *Hoverflies*. Naturalists' Handbooks no. 5. Cambridge University Press.

Gilbert, O. (1983). The wildlife of Britain's wasteland. *New Scientist* **97**: 824–9.

Gill, D. & Bonnet, P. (1973). *Nature in the Urban Landscape: A Study of City Ecosystems*. York Press, Baltimore.

Gilpin, M.E. & Diamond, J.M. (1980). Subdivision of nature reserves and the maintenance of species diversity. *Nature* **285**: 567–8.

Goodwin, D. (1978). *Birds of Man's World*. British Museum (Natural History)/Cornell University Press.

Greenslade, P.J.M. (1964). Pitfall trapping as a method for studying populations of Carabidae (Coleoptera). *Journal of Animal Ecology* **33**: 301–10.

Harding, P.T. & Sutton, S.L. (1985), *Woodlice in Britain and Ireland: Distribution and Habitat*. Institute of Terrestrial Ecology, Monks Wood Experimental Station, Huntingdon.

Harris, S. & Rayner, J.M.V. (1986a). Urban fox (*Vulpes vulpes*) population estimates and habitat requirements in several British cities. *Journal of Animal Ecology* **55**: 575–91.

Harris, S. & Rayner, J.M.V. (1986b). A discriminant analysis of the current distribution of urban foxes (*Vulpes vulpes*) in Britain. *Journal of Animal Ecology* **55**: 605–11.

Harrison, J. & Grant, P. (1976). *The Thames Transformed: London's River and its Waterfowl*. Andre Deutsch.

Hawksworth, D.L. & McManus, P.M. (1989). Lichen recolonization in London under conditions of rapidly falling sulphur dioxide levels, and the concept of zone skipping. *Botanical Journal of the Linnean Society* **100**: 99–109.

Hay, R. & Synge, P.M. (1971). *The Dictionary of Garden Plants*. Ebury Press and Michael Joseph.

Heath, J., Pollard, E. & Thomas, J. (1984). *Atlas of Butterflies in Britain and Ireland*. Viking.

Hessayon, D.G. & Hessayon, J.P. (1973). *The Garden Book of Europe*. Hamish Hamilton.

Hickling, R. (1978). *Birds of Leicestershire and Rutland*. Leicestershire and Rutland Ornithological Society.

Higgs, A.J. & Usher, M.B. (1980). Should nature reserves be large or small? *Nature* **285**: 568–9.

Holloway, J.D. & Hebert, P.D.N. (1979). Ecological and taxonomic trends in macrolepidopteran host plant selection. *Biological Journal of the Linnean Society* **11**: 229–51.

Hopkins, B. (1965). *Forest and Savanna: An Introduction to Tropical Plant Ecology with Special Reference to West Africa*. Heinemann.

Horwood, A.R. (1933). The flora of Leicestershire. In *A Scientific Survey of Leicester and District*, ed. P.W. Bryan, pp. 25–33. Prepared for the Leicester meeting 1933 of the British Association for the Advancement of Science.

Hoskins, W.G. (1957). *Leicestershire: An Illustrated History of the Landscape*. Hodder & Stoughton.

Hoskins, W.G. (1973). *The Making of the English Landscape*. Penguin.

Hubbard, C.E. (1974). *Grasses*. Penguin.

Hutchinson, G.E. & MacArthur, R.H. (1959). A theoretical ecological model of size distributions among species of animals. *American Naturalist* **93**: 117–25.

Huxley, A. (1978). *An Illustrated History of Gardening*. Paddington Press Ltd.

Imms, A.D. (1964). *A General Textbook of Entomology*. Methuen & Co. Ltd.

Jenkins, A.C. (1982). *Wildlife in the City*. Webb & Bower.

Johnsen, I. & Søchting, U. (1973). Influence of air pollution on the epiphytic lichen vegetation and bark properties of deciduous trees in the Copenhagen area. *Oikos* **24**: 344–51.

Jones, D. (1983). *The Country Life Guide to Spiders of Britain and Northern Europe*. Country Life Books.

Jones, D. et al. (1981). *Garden Wildlife: The Living World of Your Garden*. Ebury Press.

Joy, N. H. (1932). A Practical Handbook of British Beetles. H.F. & G. Witherby.

Keble Martin, W. (1967). *The Concise British Flora in Colour*. Ebury Press and Michael Joseph.

Kennedy, C.E.J. & Southwood, T.R.E. (1984). The number of species of insects associated with British trees: a re-analysis. *Journal of Animal Ecology* **53**: 455–78.

Kerney, M.P. & Cameron, R.A.D. (1979). *A Field Guide to the Land Snails of Britain and North-west Europe*. Collins.

Kettlewell, B. (1973). *The Evolution of Melanism*. Clarendon Press.

Kibby, G. (1979). *Mushrooms and Toadstools*. Oxford University Press.

Kieran, J. (1959). *A Natural History of New York City*. Houghton Mifflin.

Killington, F.J. (1936). *A Monograph of the British Neuroptera*. Ray Society, London.

Kloet, G.S. & Hincks, W.D. (1964). A checklist of British insects. *Handbooks for the Identification of British Insects*, XI(1). Royal Entomological Society of London.

Kloet, G.S. & Hincks, W.D. (1972). Lepidoptera. *Handbooks for the Identification of British Insects*, XI(2). Royal Entomological Society of London.

Kloet, G.S. & Hincks, W.D. (1975). Diptera and Siphonaptera. *Handbooks for the Identification of British Insects*, XI(5). Royal Entomological Society of London.

Krieger, R.J., Feeny, P.P. & Wilkinson, C.F. (1971). Detoxification enzymes in the guts of caterpillars: an evolutionary answer to plant defences? *Science* **172**: 579–81.

Lange, M. & Hora, F.B. (1981). *Mushrooms and Toadstools*. Collins.

Laurence, B.R. (1950). Syrphidae (Diptera) in Bedfordshire. *Entomologist's Monthly Magazine* **86**: 351–3.

Leather, S.R. (1986). Insect species richness of the British Rosaceae: the importance of host range, plant architecture, age of establishment, taxonomic isolation and species area relationships. *Journal of Animal Ecology* **55**: 841–60.

Leclercq, J. & Remacle, A. (1974). Recherches sur les Hyménoptères de la ville de Liège. In *Agriculture et Environnement*, pp. 600–4. Faculte des Sciences Agronomiques de l'Etat Centre de Recherches Agronomiques Gembloux (Belgique); September 2–6.

Leicestershire Regional Planning Report (1932). Prepared by Regional Town Planning Joint Advisory Committee and Allen & Potter, London, Town Planning Consultants.

Le Quesne, W.J. & Payne, K.R. (1981). Cicadellidae (Typhlocybinae) with a checklist of the British Auchenorrhyncha (Hemiptera, Homoptera). *Handbooks for the Identification of British Insects*, II(2c). Royal Entomological Society of London.

Lever, C. (1977). *The Naturalized Animals of the British Isles*. Hutchinson.

Liddle, P. (1982a). *Leicestershire Archaeology – The Present State of Knowledge. Vol. 1. To the End of the Roman Period*. Leicestershire Museums Service. Archaeological Reports Series No. 4.

Liddle, P. (1982b). *Leicestershire Archaeology – The Present State of Knowledge. Vol. 2. Anglo-Saxon and Medieval Periods*. Leicestershire Museums Service. Archaeological Reports Series No. 5.

Lindroth, C.H. (1974). Coleoptera: Carabidae. *Handbooks for the Identification of British Insects*, IV(2). Royal Entomological Society of London.

Litsinger, J.A. & Moody, K. (1976). Integrated pest management in multiple cropping systems. In *Multiple Cropping*, ed. R.I. Papendick, P.A. Sanchez & G.B. Triplet, Chapter 15. Special publication no. 27, American Society of Agronomy, Madison, Wisconsin.

Locket, G.H. & Millidge, A.F. (1951). *British Spiders. Vols. I & II*. Ray Society, London.

Locket, G.H., Millidge, A.F. & Merrett, P. (1974). *British Spiders. Vol. III*. Ray Society, London.

Lomax, P. (1987). *Leicester Habitat Survey*. City Wildlife Project, Leicester.

Long, J.L. (1981). *Introduced Birds of the World*. David & Charles.

Lowe, E.E., Mayes, W.E., Wagstaffe, R. & Taylor, S.O. (1933). The zoology of Leicestershire. In *A Scientific Survey of Leicester and District*, ed. P.W. Bryan, pp. 33–40. Prepared for the Leicester meeting 1933 of the British Association for the Advancement of Science.

LROS (1974–1987). *The Birds of Leicestershire and Rutland*. Annual reports, 1972–1986, of the Leicestershire and Rutland Ornithological Society.

Luff, M.L. (1968). Formalin as an attractant. *Entomologist's Monthly Magazine* **104**: 115–16.

Luff, M.L. (1975). Some features influencing the efficiency of pitfall traps. *Oecologia* **19**: 345–57.

Lussenhop, J. (1973). The soil arthropod community of a Chicago expressway margin. *Ecology* **54**: 1124–37.

Lutz, F.E. (1941). *A Lot of Insects: Entomology in a Suburban Garden*. Putnam.

Mabey, R. (1981). *The Common Ground*. Arrow.

MacArthur, R. (1955). Fluctuations of animal populations, and a measure of community stability. *Ecology* **36**: 533–6.

MacArthur, R.H. (1965). Patterns of species diversity. *Biological Reviews* **40**: 510–33.

MacArthur, R.H. & MacArthur, J.W. (1961). On bird species diversity. *Ecology* **42**: 594–8.

MacArthur, R.H. & Wilson, E.O. (1967). *The Theory of Island Biogeography*. Princeton University Press.

Magurran, A.E. (1988). *Ecological Diversity and its Measurement*. Croom Helm.

Majerus, M. & Kearns, P. (1989). *Ladybirds*. Naturalists' Handbook no. 10. Richmond Publishing Co. Ltd.

Martin, J.D. & Bird, R. (1958). Humberstone. In *A History of the County of Leicester*. Victoria County History, Vol. IV, ed. R.A. McKinley, Oxford University Press.

Mathias, J.H. (1975). A survey of amphibians in Leicestershire gardens. *Transactions of Leicester Literary and Philosophical Society* **69**: 28–41.

May, R.M. (1987). How many species are there? *Nature* **324**: 514.

McClintock, D. *et al.* (1964). Natural history of the garden of Buckingham Palace. *Proceedings of the South London Entomological and Natural History Society 1963*.

Miess, M. (1979). The climate of cities. In *Nature in Cities*, ed. I.C. Laurie, pp. 91–114. John Wiley.

Mitchell, A. (1974). *A Field Guide to the Trees of Britain and Northern Europe*. Collins.

Morrison, B. (1974). Observations on ground beetles (Col. Carabidae) in a Perthshire garden. *Proceedings of the British Entomological and Natural History Society* **6**: 97–103.

Mourier, H. & Winding, O. (1975). *Wildlife in House and Home*. Collins.

Murdoch, W.W., Evans, F.C. & Peterson, C.H. (1972). Diversity and pattern in plants and insects. *Ecology* **53**: 819–29.

Murton, R.K. (1971). *Man and Birds*. Collins.

Nelson, B.C. (1978). Ecology of medically important arthropods in urban environments. In *Perspectives in Urban Entomology*, ed. G.W. Frankie & C.S. Koehler, pp. 87–124. Academic Press.

New, T.R. (1971). An introduction to the natural history of the British Psocoptera. *Entomologist* **104**: 59–97.

New, T.R. (1974). Psocoptera. *Handbooks for the Identification of British Insects*, I(7). Royal Entomological Society of London.

Nicholson, B.E., Wallis, M., Anderson, E.B., Balfour, A.P., Fish, M. & Finnis, V. (1972). *The Oxford Book of Garden Flowers*. Oxford University Press.

Nixon, G.E.J. (1938). A preliminary revision of the British Proctotrupinae (Hym., Proctotrupoidea). *Transactions of the Royal Entomological Society* **87**: 431–66.

Odum, E.P. (1971). *Fundamentals of Ecology*. Saunders.

Olkowski, W., Olkowski, H., Kaplan, A.I. & van den Bosch, R. (1978). The potential for biological control in urban areas: shade tree insect pests. In *Perspectives in Urban Entomology*, ed. G.W. Frankie & C.S. Koehler, pp. 311–48. Academic Press.

Owen, D.F. (1975c). Lessons from a caterpillar plague in London's Berkeley Square. *Environmental Conservation* **2**: 171–7.

Owen, D. (1978e). *Towns and Gardens*. Hodder & Stoughton.

Owen, D.F. (1980c). *What is Ecology?* Oxford University Press.

Owen, D.F. (1984). The geographical distribution of Buddleia-feeding in *Cucullia verbasci* (L.) (Noctuidae) in the British Isles. *Entomologist's Record* **96**: 49–51.

Owen, J. (1985). Leicester Urban Fox Survey. *Bulletin of the British Ecological Society* **16**: 214–16.

Pappa, B. (1976). Zierpflanzenschädlinge in und um Hamburg. *Entomologische Mitteilungen aus dem Zoologischen Museum Hamburg* **5**: 25–47.

Parmenter, L. & Owen, D.F. (1954). The swift, *Apus apus* L., as a predator of flies. *Journal of the Society for British Entomology* **5**: 27–33.

Perring, F. (1987). A last word. *Natural World* **19**: 5.

Perrins, C. (1974). *Birds*. Collins.

Phillips, R. (1981). *Mushrooms and Other Fungi of Great Britain and Europe*. Pan Original.

Phythian-Adams, C. (1986). *The Norman Conquest of Leicestershire and Rutland: A Regional Introduction to Domesday Book*. Leicestershire Museums Publication no. 73. Leicester.

Pollard, E. (1971). Hedges VI. Habitat diversity and crop pests: a study of *Brevicoryne brassicae* and its syrphid predators. *Journal of Applied Ecology* **8**: 751–80.

Pollard, E., Hall, M.L. & Bibby, T.J. (1986). *Monitoring the Abundance of Butterflies*. Research and survey in nature conservation series, no. 2. Nature Conservancy Council.

Pope, R.D. (1977). Coleoptera and Strepsiptera. *Handbooks for the Identification of British Insects*, XI(3). Royal Entomological Society of London.

Port, G.R. & Thompson, J.R. (1980). Outbreaks of insect herbivores on plants along motorways in the United Kingdom. *Journal of Applied Ecology* **17**: 649–56.

Primavesi, A.L. & Evans, P.A. (eds.) (1988). *Flora of Leicestershire*. Leicestershire Museums Service, Leicester.

Pye, N. (1972a). The regional setting. In *Leicester and its Region*, ed. N. Pye, pp. 1–15. Leicester University Press for the British Association for the Advancement of Science.

Pye, N. (1972b). Weather and climate. In *Leicester and Its Region*, ed. N. Pye, pp. 84–113. Leicester University Press for the British Association for the Advancement of Science.

Quicke, D. (1986). Host location, selection and suitability in the parasitic Hymenoptera. In *The Hymenopterist's Handbook*, ed. C. Betts, pp. 30–7. Amateur Entomologist 7.

Ratcliffe, D. & Conolly, A.P. (1972). Flora and vegetation. In *Leicester and its Region*, ed. N. Pye, pp. 133–59. Leicester University Press for the British Association for the Advancement of Science.

Rathcke, B.J. & Price, P.W. (1976). Anomalous diversity of tropical ichneumonid parasitoids: a predation hypothesis. *American Naturalist* **110**: 889–93.

Richards, O.W. (1964). Insects other than Lepidoptera. In *Natural History of the Garden of Buckingham Palace*, ed. McClintock, D., *Proceedings of the South London Entomological and Natural History Society* 1963: 75–98.

Risch, S. (1980). The population dynamics of several herbivorous beetles in a tropical agroecosystem: the effect of intercropping corn, beans and squash in Costa Rica. *Journal of Applied Ecology* **17**: 593–612.

Root, R.B. (1973). Organization of a plant–arthropod association in simple and diverse habitats: the fauna of collards (*Brassica oleracea*). *Ecological Monographs* **43**: 95–124.

Rotheray, G.E. (1981). Host searching and oviposition behaviour in some parasitoids of aphidophagous Syrphidae. *Ecological Entomology* **6**: 79–87.

Routledge, R.D. (1979). Diversity indices: which ones are admissible? *Journal of Theoretical Biology* **76**: 503–15.

Salisbury, E.J. (1935). *The Living Garden*. G. Bell & Sons Ltd.

Sankey, J.H.P. (1988). *Provisional Atlas of the Harvest-spiders (Arachnida: Opiliones) of the British Isles*. Institute of Terrestrial Ecology, Natural Environment Research Council.

Satchell, G.H. (1947). The ecology of the British species of *Psychoda*. *Annals of Applied Biology* **34**: 611–21.

Schneider, F. (1969). Bionomics and physiology of aphidophagous Syrphidae. *Annual Review of Entomology* **14**: 103–24.

Scott, E.I. (1939). An account of the developmental stages of some aphidophagous Syrphidae and their parasites. *Annals of Applied Biology* **26**: 509–32.

Shaw, M. & Askew, R. (1976). Parasites of Lepidoptera. In *Butterflies and Moths of Great Britain and Northern Ireland*, ed. J. Heath, Blackwell Scientific Publications.

Simberloff, D. (1986). Design of nature reserves. In *Wildlife Conservation Evaluation*, ed. M.B. Usher, pp. 315–37. Chapman & Hall.

Simmons, J. (1974). *Leicester: Past and Present*. Eyre Methuen.

Simms, E. (1975). *Birds of Town and Suburb*. Collins.

Skinner, B. (1984). *Colour Identification Guide to Moths of the British Isles*. Viking.

Smith, J.G. (1976). Influence of crop background on natural enemies of aphids on Brussels sprouts. *Annals of Applied Biology* **83**: 15–29.

Smith, K.G.V. (1969). Diptera: Conopidae. *Handbooks for the Identification of British Insects*, X(3a). Royal Entomological Society of London.

Snow, D.W. (1958). The breeding of the blackbird *Turdus merula* at Oxford. *Ibis* **100**: 1–30.

Southwood, T.R.E. (1961). The number of species of insect associated with various trees. *Journal of Animal Ecology* **30**: 1–8.

Southwood, T.R.E. (1978). The components of diversity. *Symposia of the Royal Entomological Society of London* **9**: 19–40.

Southwood, T.R.E. & Leston, D. (1959). *Land and Water Bugs of the British Isles*. Frederick Warne & Co. Ltd.

Speight, M.C.D. (1978). The genus *Paragus* (Dipt.: Syrphidae) in the British Isles, including a key to known and possible British Isles species. *Entomologist's Record and Journal of Variation* **90**: 100–7.

Steele, R.C. & Welch, R.C. (1973). *Monks Wood: A Nature Reserve Record*. The Nature Conservancy: Natural Environment Research Council.

Stellemann, P. (1978). The possible role of insect visits in pollination of reputedly anemophilous plants, exemplified by *Plantago lanceolata*, and syrphid flies. *Linnean Society Symposium* **6**: 41–6.

Stokoe, W.J. & Stovin, G.H.T. (1948). *The Caterpillars of British Moths*. Frederick Warne & Co. Ltd.

Stork, N.E. (1988). Insect diversity: facts, fiction and speculation. *Biological Journal of the Linnean Society* **35**: 321–37.

Stubbs, A. & Chandler, P. (eds.) (1978). *A Dipterist's Handbook*. Amateur Entomologist 15.

Stubbs, A.E. & Falk, S.J. (1983). *British Hoverflies: An Illustrated Identification Guide*. British Entomological and Natural History Society.

Sutton, S. (1972). *Woodlice*. Ginn & Company Ltd.

Tahvanainen, J.O. & Root, R.B. (1972). The influence of vegetational diversity on the population ecology of a specialized herbivore, *Phyllotreta cruciferae* (Coleoptera: Chrysomelidae). *Oecologia* **10**: 321–46.

Taylor, L.R. (1978). Bates, Williams, Hutchinson – a variety of diversities. *Symposium of the Royal Entomological Society of London* **9**: 1–18.

Taylor, L.R. & Carter, C.I. (1961). The analysis of numbers and distribution in an aerial population of Macrolepidoptera. *Transactions of the Royal Entomological Society of London* **113**: 369–86.

Theunissen, J. & den Ouden, H. (1980). Effects of intercropping with *Spergula arvensis* on pests of Brussels sprouts. *Entomologia Experimentalis Applicata* **27**: 260–8.

Thomas, K. (1983). *Man and the Natural World: Changing Attitudes in England 1500–1800*. Allen Lane.

Thompson, F.C., Vockeroth, J.R. & Speight, M.C.D. (1982). The Linnean species of flower flies (Diptera: Syrphidae). *Memoirs of the Entomological Society of Washington* **10**: 150–65.

Thorns, D. (1973). *Suburbia*. Paladin.

Todd, V. (1948). The habits and ecology of the British harvestmen (Arachnida, Opiliones), with special reference to those of the Oxford district. *Journal of Animal Ecology* **18**: 209–29.

Torp, E. (1984). [The Danish hoverflies (Diptera, Syrphidae).] *Danmarks Dyreliv*, Vol. 1. Fauna Bøger, København (in Danish).

Townes, H. (1969). The genera of Ichneumonidae, Part 1. Ephialtinae to Agriotypinae. *Memoirs of the American Entomological Institute* **11**.

Townes, H. (1972a). A light-weight Malaise trap. *Entomological News* **83**: 239–47.

Townes, H. (1972b). Ichneumonidae as biological control agents. *Proceedings of the Tall Timbers Conference on Ecological Animal Control by Habitat Management*, no. 3: 235–48.

Townes, H. (1977). A revision of the Heloridae (Hymenoptera). *Contributions of the American Entomological Institute* **15**: 1–12.

Townes, H. & Townes, M. (1981). A revision of the Serphidae (Hymenoptera). *Memoirs of the American Entomological Institute* **32**.

Tutin, T.G. (1973). Weeds of a Leicester garden. *Watsonia* **9**: 263–7.

US Council on Environmental Quality and Department of State (1980). *The Global 2000 Report to the President: Entering the Twenty-first Century*. Technical Report, Vol. 2. US Government Printing Office.

Usher, M.B. (1983). Species diversity: a comment on a paper by W.B. Yapp. *Field Studies* **5**: 825–32.

Usher, M.B. (1986). Wildlife conservation evaluation: attributes, criteria and values. In *Wildlife Conservation Evaluation*, ed. M.B. Usher, pp. 3–44. Chapman & Hall.

Verlinden, L. & Decleer, K. (1987). *The Hoverflies (Diptera, Syrphidae) of Belgium and their Faunistics: Frequency, Distribution, Phenology*. Institut Royal des Sciences Naturelles de Belgique. Documents de travail no. 39.

Vinson, S.B. (1976). Host selection by insect parasitoids. *Annual Review of Entomology* **21**: 109–34.

Vinson, S.B. & Iwantsch, G.F. (1980). Host regulation by insect parasitoids. *Quarterly Review of Biology* **55**: 143–65.

Walpole, M. (1972). Nature conservation. In *Leicester and Its Region*, ed. N. Pye, pp. 181–92. Leicester University Press for the British Association for the Advancement of Science.

Weaver, C.R. & King, D.R. (1954). *Meadow Spittlebug Philaenus leucophthalmus (L.)*. Research Bulletin 741. Ohio Agricultural Experiment Station.

White, G. (1947). Letter XX to Thomas Pennant. *The Natural History of Selbourne*, ed. J. Fisher. The Cresset Press, London.

White, W. (1846). *History, Gazetteer and Directory of Leicestershire, and the Small County of Rutland*. Robt. Leader, Independent Office, Sheffield.

Wiegert, R.G. & Owen, D.F. (1971). Trophic structure, available resources and population density in terrestrial vs. aquatic ecosystems. *Journal of Theoretical Biology* **30**: 69–81.

Williams, G. (1963). Seasonal and diurnal activity of harvestmen (Phalangida) and spiders (Araneida) in contrasted habitats. *Journal of Animal Ecology* **31**: 23–42.

Williams, P.H. (1982). The distribution and decline of British bumblebees (*Bombus*). *Journal of Apicultural Research* **21**: 236–45.

Wilson, R. (1979). *The Back Garden Wildlife Sanctuary Book*. Astragal.

Witherby, H.F., Jourdain, F.C.R., Ticehurst, N.F. & Tucker, B.W. (1946). *The Handbook of British Birds*, H.F. and G. Witherby Ltd.

Wright, A. (1988). Hoverflies in a city environment: experiences in Coventry. *Dipterists' Digest* **1**: 37–40.

Yeo, P.F. & Corbet, S.A. (1983). *Solitary Wasps*. Naturalists' Handbooks no. 3. Cambridge University Press.

Index

Tilia 114, *T. cordata* – see lime, small-leaved
Tineidae 126
Tingidae 287–8
Tiphia 227
Tipula paludosa 297–8
Tipulidae 10, 297–8
tits 11, 12, tit, blue 13, 31, 38, 334, 356, coal 336, 337, great 13, 31, 38, 337, long-tailed 337, marsh 337
toad 5, 14, 32, 38, 302, 331–3
toadflax, common 31, pale 31, small 31
tobacco 53, flowering 85, 261, 344, 345 – see *Nicotiana alata*
tomato 53, 128, 345, 361 – see *Lycopersicon esculentum*
toothwort 28
Toronto 11
Tortricidae 126–7, 136
Trachilepus rathkei 27
Tragopogon pratensis – see goat's-beard
traps 54–8 – see light trap, mercury vapour, Malaise trap, pitfall trap
treecreeper 35, 38, 39, 337
tree-of-heaven 9
trefoil, hop 80, 352
Tretoserphus 224
Trichia striolata 312
Trichiosoma tibiale 305
Trichoceridae – see gnats, winter
Trichoniscidae 315, 316
Trichoniscus pusillus 316, *T. pygmaeus* 316
Trichoptera 359, 363 – see caddisflies
Tricladida 310
Tridax procumbens 9
Trifolium 52, *T. campestre* – see trefoil, hop, *T. repens* – see clover, white
Triglyphus 156
Triturus cristatus – see newt, crested, *T. helveticus* – see newt, palmate, *T. vulgaris* – see newt, smooth
Troglodytes troglodytes – see wren
Trogulidae 322
Tromatobia 216
Tropaeolaceae 71
Tropaeolum majus – see nasturtium
Tropidia 156
Tryphoninae 200, 202, 216, 217, 220
Tryphonini 216
Trypoxylon 229, 233
tulip 17
Tulipa 44
Turbellaria 310, 359
Turdidae – see thrushes
Turdus ericetorum – see thrush, song, *T. merula* – see blackbird, *T. migratorius* – see robin, American, *T. musicus* – see redwing, *T. pilaris* – see fieldfare, *T. viscivorus* – see thrush, mistle
Tussilago 128, *T. farfara* – see colt's-foot
tutsan 79
twayblade, common 28

Tymmophorus graculus 220
Typhaeus typhoeus – see minotaur
Tyto alba – see owl, barn

Ulex europaeus – see gorse
Ulmaceae 71
Ulmus 46, 114, *U. glabra* 122 – see elm wych, *U. procera* – see elm
Umbelliferae 71, 78, 120, 139, 141, 142, 145, 303
Umbilicus rupestris – see navelwort
United Nations 1
Uppingham 285
Ursus maritimus – see bear, polar
Urtica 128, *Urtica dioica* 119, 128, 135, 139, 140, 141 – see nettle
Urticaceae 78, 139, 140, 145

Vaccinium 113, 128, *V. vitis-idaea* – see cowberry
valerian 332
Valerianaceae 71
Vallonia costata 312
Vallonidae 312
Vanellus vanellus – see lapwing
Vanessa atalanta – see butterfly, red admiral
venation 148, 195, 221, 293
Venice 6
verges 25, 34, 36, 337, 341
Vespa crabro 235
Vespidae 226, 231, 350, 354, 360
Vespula 151, 229, 244, 302, *V. austriaca* 229, 231, 235, *V. germanica* 231, 235, 240–2, 247–8, *V. rufa* 229, 231, 235, 300, *V. vulgaris* 229, 231, 235, 240–2, 247–8, 355
Viburnum lantana – see wayfaring tree, *V. opulus* – see guelder-rose
Vicia 127, *V. faba* 139 – see bean, broad, *V. hirsuta* – see tare, hairy
Vinca major – see periwinkle, greater
Viola – see violets
Violaceae 71, 78
violets 91, sweet 79
viper's bugloss 49, 72, 345
Viperus berus – see adder
viruses 287
Vitaceae 71
Vitrea contracta 312
vole, bank 341, field 29, 341
Volucella 156, 356, *V. bombylans* 156, 175, 184–5, *V. pellucens* 149, 175, 184–5
Voria ruralis 301
Vulpes vulpes – see fox
vultures 7–8, black 7

wagtail, grey 337–8, pied 2, 13, 35, 337–8, yellow 35
wallflower 85
walnut 347